Laser-Generated Functional Nanoparticle Bioconjugates

Annette Barchanski

Laser-Generated Functional Nanoparticle Bioconjugates

Design for Application in Biomedical Science and Reproductive Biology

 Springer Spektrum

Annette Barchanski
Laser Zentrum Hannover e.V.
Nanotechnology Department
Hannover, Germany

Dissertation, University of Duisburg-Essen, Faculty of Chemistry, 2015

Date of thesis defense: 22.10.2015

1st Assesor: Prof. habil. Dr.-Ing. Stephan Barcikowski, University of Duisburg-Essen

2nd Assesor: Prof. Dr. med. vet. Detlef Rath, Friedrich-Loeffler-Institut, Institut für Nutztiergenetik, Biotechnologie, Neustadt

Chairman: Prof. Dr. rer. nat. Sebastian Schlücker, University of Duisburg-Essen.

OnlinePlus material to this book can be available on
http://www.springer-gabler.de/978-3-658-13515-7

ISBN 978-3-658-13514-0   ISBN 978-3-658-13515-7 (eBook)
DOI 10.1007/978-3-658-13515-7

Library of Congress Control Number: 2016936435

Springer Spektrum
© Springer Fachmedien Wiesbaden 2016

Printed on acid-free paper

This Springer Spektrum imprint is published by Springer Nature
The registered company is Springer Fachmedien Wiesbaden GmbH

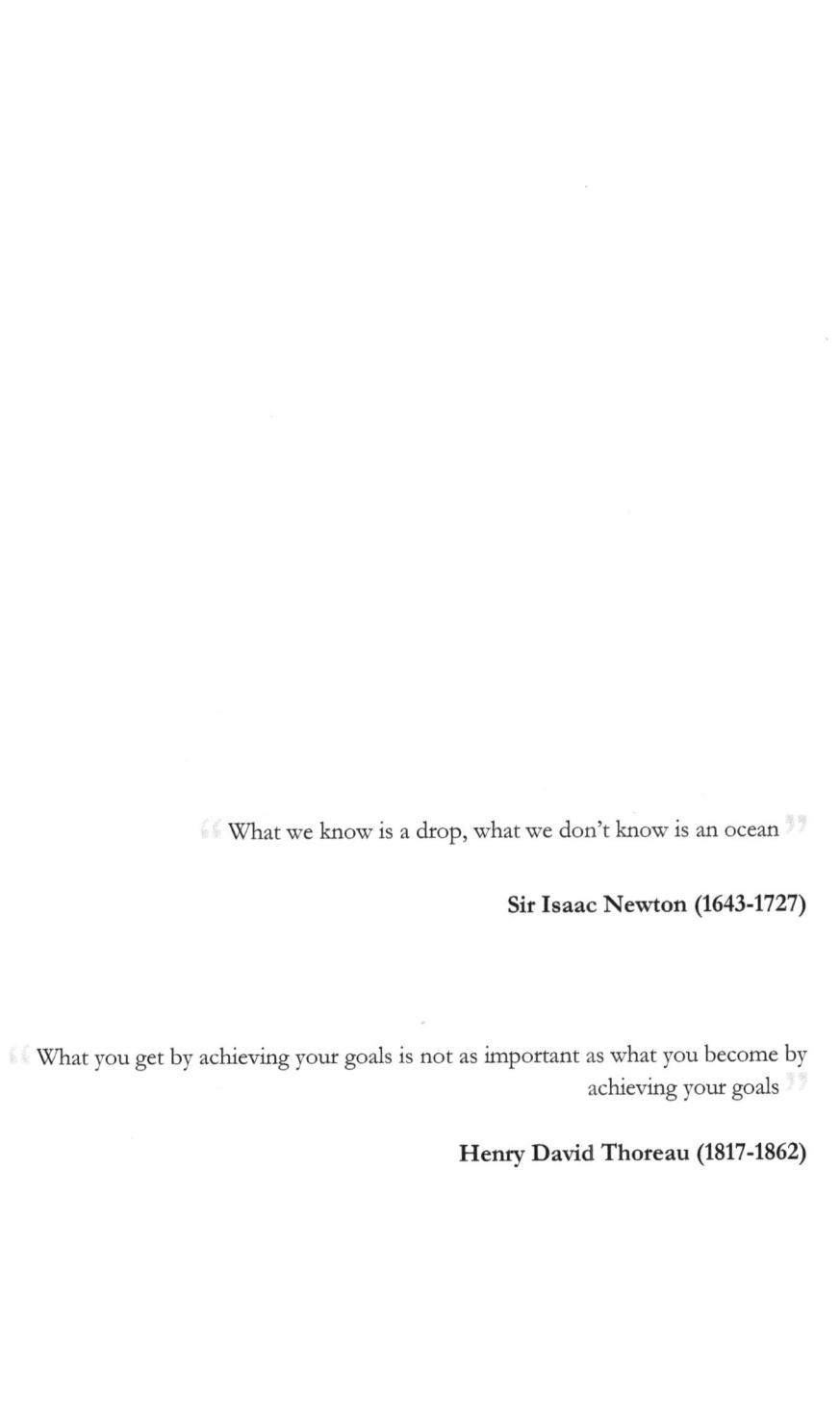

" What we know is a drop, what we don't know is an ocean "

Sir Isaac Newton (1643-1727)

" What you get by achieving your goals is not as important as what you become by achieving your goals "

Henry David Thoreau (1817-1862)

Preface

This work was established during my practice as researcher in the workgroup 'Nanomaterials' at the Laser Zentrum Hannover e.V. (LZH) where I focused on various national and international cooperations and two main research projects:

(I) 'Functionalized nanoparticles for the sex-specific selection of bovine spermatozoa' (Masterrind, NBank), aiming for the development of a biocompatible and specific gold nanomarker for the bovine Y-chromosome in cooperation with the Friedrich-Loeffler-Institut (FLI) Mariensee (duration: 02/10–02/11).

(II) The DFG Excellence Cluster REBIRTH (from **RE**generating **BI**ology to **R**econstructive **TH**erapy) within the Research Unit (RU) 7.3. 'Nanoparticles', where researchers from the LZH, the Hannover Medical School (MHH), the Leibniz University of Hannover (LUH), the University of Veterinary Medicine Hannover (TiHo), the FLI and other institutions worked on the establishment of novel regenerative technologies and tools. Therein, the RU 7.3 used the technique of pulsed laser ablation in liquids for the fabrication of ultrapure mono- and multi-material nanobioconjugates. These constructs were applied e.g. as vectors for directed ion- drug- and gene-delivery purposes with stimulus-induced release or as medical nanomarkers for advanced immunolabeling, high-resolution bio-imaging and sensitive nanosensory applications (duration: since 02/10).

The process of *in situ* bioconjugation during pulsed laser ablation in liquids was established at LZH in 2007 and various gold-DNA and gold-protein conjugates have been successfully fabricated since that time. However, there remains a deficiency of knowledge regarding the optimal conjugation conditions and the controlled functionalization of gold nanoparticles with two or more different bio-ligands. With my combined educational background as an engineer and a biomedical scientist, I concentrated on this deficiency and tried to understand the interplay and implications of the various process parameters. I also worked to establish guidelines for the customized bioconjugation. This thesis presents a condensed summary of my findings and offers general instruction for the individual configuration of novel and functional gold nanoparticle bioconjugates for biological applications.
All research results were collected from my own lab experiments and from 17 national and international student internships as well as two Bachelor theses, which I mentored. Parts of the results and contributions to the work of collaborative research partners were published in international, peer-reviewed research journals, which are summarized as a *list of own publications* at the end of this thesis and labeled in Roman numerals at the beginning of the results chapters.

Abstract

The question of how to design gold nanoparticles for biomedical research has become crucial since the application of nano-scaled tools increased significantly over the last decade. From a biologist's point of view, properties such as size, shape, charge, biocompatibility and functionalization of nanoparticles must be carefully considered in order to achieve specific cellular responses in combination with a controllable stimulus. Since it is known that all of these properties may influence each other in performance, a comprehensive portrait of the results from numerous *in vitro* and *in vivo* studies is required for a better understanding of each attributes' impact. Thus, in this work, the structure-function relationship of gold nanoparticle conjugates derived from a laser-based synthesis method will be discussed. Both, the limits and perspectives of tunable conjugate functions will be presented, providing a general outline for researchers to configure functionalized gold nanoparticles with a specifically optimized design for biomedical requests, e.g. in biomedical and regenerative science, reproductive biology and biotechnology.

Keywords: Nanobiohybrids, biological functionality, variability, biomedicine

Kurzzusammenfassung

Wie sollten Gold Nanopartikel für die biomedizinische Forschung gestaltet sein? Diese Frage hat besonders in dem letzten Jahrzehnt an Bedeutung gewonnen, seit nanoskalige Werkzeuge verstärkt zum Einsatz kommen. Aus der Sicht eines Biologen müssen Eigenschaften wie die Größe, Form, Ladung, Biokompatibilität sowie die Funktionalisierung von Nanopartikeln genau angepasst werden um in Kombination mit einem kontrollierten Stimulus spezifische zelluläre Antworten zu initiieren. Seitdem herausgefunden wurde, dass sich all diese Eigenschaften gegenseitig in ihrer Auswirkung beeinflussen können, ist ein umfangreiches Portrait der Ergebnisse zahlreicher *in vitro* und *in vivo* Studien nötig um den Beitrag jedes einzelnen Attributes zu verstehen. Aus diesem Grund wird in der vorliegenden Arbeit das Struktur-Funktions-Verhältnis von lasergenerierten Goldnanopartikel-Konjugaten kritisch diskutiert. Sowohl die Grenzen, als auch die Möglichkeiten der einstellbaren Konjugat-Funktionen werden aufgezeigt, um Wissenschaftlern eine allgemeine Richtlinie zu schaffen für die Zusammenstellung von funktionalisierten Goldnanopartikeln mit spezifischem, optimierten Design für biomedizinische Fragestellungen z. B. in der Biomedizin und regenerativen Medizin, der Reproduktionsbiologie sowie der Biotechnologie.

Schlagwörter: Bio-Nano-Hybride, biologische Funktionalität, Variabilität, Biomedizin

All figures in this book have been converted into grayscale for the print version. The colored originals are provided as additional material on the product page for this book on www.springer.com.

Contents

1. Introduction and Objectives of the Research

1.1. Nano-revolution of biomedical science and reproductive biology

The development of detection and treatment strategies for medical disorders and diseases[1;[1]] are the major aims of **biomedical research**. Clinical approaches are mainly concentrated on diagnostic blood testing; imaging via X-ray, computer tomography (CT) scans and magnetic resonance imaging (MRI); pharmaceutical full-body medication; or surgery. However, the main challenge in research is to focus on the genetic origin and molecular development of the medical disorder/disease and to establish novel methods for the analysis and manipulation at the cellular and even sub-cellular levels. In this context, the **site-specific targeting**,[2;3] **sensing**,[4] **malignant cell destruction**[5] and **drug/gene delivery**[6;7] are topics that are strongly focused on the current research.

Likewise, in the scientific area of **reproductive biology**, the sub-cellular level is targeted, because the situation and activity of chromosomes and genes, as well as the molecular processes inside of gametes, need to be studied and manipulated. For instance, genetic transformation is a crucial topic of investigation, because the delivery of foreign genes into oocytes by sperm-mediated gene transfer (SMGT) may generate transgenic animals (genetically modified organisms, GMOs).[8] Moreover, the specific **genetic labelling**, analysis and **sorting of spermatozoa**[9;10] could aid with making a **pre-fertilization diagnosis** during artificial insemination[11;10] or with the **selection** of beneficial traits in livestock.[12]

To address these sub-cellular dimensions, **nano-scaled tools** are required for accurate targeting, imaging, and for therapeutic issues.[13] In this context, a broad variety of nanodevices and nanotechnologies have revolutionized the research in the last decades, covering, for example, nanomaterials for tissue engineering,[14-16] biocompatible nanostructured surfaces for implants,[17-19] biochips,[20] carbon nanotubes as suitable scaffold materials for cellular proliferation and bone formation,[21] nanowires for sensing applications[22;23] and nanorods/**nanoparticles** for delivery purposes.[24;25]

Within the fields of biomedical science and reproductive biology, the number of potential applications for colloidal particles is growing rapidly,[26-28;12] due to their **biomole-**

---

[1] Besides communicable diseases like lower respiratory diseases (e.g. cold, influenza or pneumonia), diarrhoeal diseases (e.g. ebola or cholera), malaria or acquired immunodeficiency syndrome (AIDS), the major non-communicable diseases according to the World Health Organization include cancer, diabetes, cardiovascular diseases (e.g. ischemic heart disease or stroke), mental diseases (e.g. Alzheimer disease or Parkinson disease) and asthma).[1]
WHO, *World Health Statistics 2014 - A Wealth of Information on Global Public Health*. **2014**, Geneva.

cule-related size and their unique electronic, optical, magnetic or catalytic properties when compared with the corresponding bulk material.[29]. As proposed recently by a BBC research study, the global market for nanotechnology will effectively grow from an estimated $11.7 billion in 2009 to nearly $26.7 billion in 2015 and to $48.9 billion in 2017, while approximately 80 % of the shares will be captured by nanoparticle-based components and devices.[30]

Nanoparticles may be categorized by their material into organic (e.g. polymer-based) and inorganic species.[31] Among the inorganic materials such as iron(oxide), semiconductor quantum dots, silicon and silver, gold nanoparticles (AuNPs) have triggered a significant, emerging interest for biomedical and reproduction-relevant purposes. This is due to their outstanding optical characteristics,[32;33] a good biocompatibility[34] and the ease of surface functionalization with thiolated, bioactive ligands, yielding stable AuNP bioconjugates with e.g. a targeting or delivering function.[35;36]

The AuNP synthesis by chemical reduction[37;38] with an optional, subsequent (bio)functionalization step is an established and frequently adopted fabrication method for AuNP bioconjugates. However, a novel, physically-based technique called pulsed laser ablation in liquids (PLAL) with *in situ* bioconjugation[39] has become an appealing alternative in the past decades, due to a simplified single-step process and a wider scale of technical capabilities.[40;41] In fact, the number of published papers on the PLAL topic has increased by a factor of 15 between 1998 and 2008[42] and continues to rise.

However, there is often an imbalance of understanding between the technical/physical laser fabrication fundamentals, the chemical (surface) properties of the synthesized nanoparticles that influence their bioconjugation behavior and the optimal process and solvent conditions that are required to preserve the biological activity of attached ligands for further bio-application.

Therefore, this area of study is overdue for a critical and broad-interdisciplinary evaluation that covers the advantages and disadvantages of laser-generated AuNPs and AuNP bioconjugates. The evaluation should also address their controllable, biofunctional design and fabrication feasibility as well as their contribution to research on biomedical science and the reproductive biology sector. This evaluation shall be provided in this thesis.

1.2.   Objectives and outline of the thesis

If chemical synthesis is avoided, it is difficult to fabricate gold nanoparticle bioconjugates with pre-set specifications regarding particle size and a defined number of covalently attached, functional ligands in a biocompatible solvent.
Therefore, the main objective of this thesis is to provide a guideline for the laser-based fabrication of distinct gold nanoparticle bioconjugates considering the process limitations as well as customers' needs for specific biological applications.

As a basis for this discussion, a detailed literature review on gold nanoparticle's characteristics, their biological interactions and applications as well as the fabrication process of PLAL will be given in **Chapter 2**. Thereafter, the adopted experimental techniques and procedures will be presented in **Chapter 3**.

Chapter 4 will cover the presentation of experimental results; organized as follows:
A crucial drawback of the PLAL process is generally that there is a mismatch between an efficient production yield and the maintenance of optimal conditions for the fabrication of functional nanobioconjugates. Thus, an approach for yield enhancement and two methods to increase the nanoparticle concentration by post-processing will be presented and discussed in **Chapter 4.1**. This will facilitate a check on the competitiveness of PLAL method for AuNP bioconjugate fabrication.
Because most PLAL and bioconjugation process parameters influence each other, the comprehension through an understanding of their roles and interactions is crucial knowledge for the manufacturer. Thus, a detailed overview and critical remarks regarding PLAL-fabrication settings and the specific structure-function relationship of customized AuNP bioconjugates will be provided in **Chapter 4.2**. For that intent, four consideration areas have been subdivided, which cover: (**I**) the modulation of particles' intrinsic parameters; (**II**) the manipulation of the conjugation process and a discussion of basic issues such as choice of bond type, ligand amount and surrounding medium; (**III**) the ligand characteristics as their length, dimension or amphiphilic nature and the adoption of diverse ligands for bivalent (and multivalent) functionalization; (**IV**) the biological function of fabricated AuNP bioconjugates.
The material gold will be the focus of this thesis. However, the transferability of the laser-based bioconjugation technique to other materials is also an important factor for the broadband-compatibility of a method and should not be ignored. Thus, the *in situ* bioconjugation of PLAL-fabricated silicon and magnetic, iron-based nanoparticles and the significance of the findings for the future research will be presented in **Chapter 4.3**

Chapter 5 will conclude the thesis by presenting and analyzing the most significant aspects that were discussed and by offering suggestions for further study in this area.

2. Fundamental Background

2.1. Gold and gold nanoparticles

An overview

The element gold (Latin, *aurum*, Au) was discovered in approximately the 6th century B.C. in free elemental form as nuggets or grains in rocks and has been used for coinage and jewelry since that time.[43]
Up to the 17th century, early chemists (alchemists) such as Paracelsus believed that all materials were a mixture of mercury, sulfur and salts. They thought that by altering the proportions of these base substances and by using the mystic *Philosopher's stone* they could transform them into a noble metal such as gold.[44] However, proof of this hypothesis has never been shown.

Most of the Earth's gold is found at its core and is extracted by screening river sand or with the reduction of rocks. According to the World Gold Council, less than 174,000 tons of gold have been mined in human history.[45]
Currently, gold is widely used in medicine and electronics due to its high malleability, ductility, resistance to corrosion, inertness and conductivity.[46;47]

Gold is a group 11 chemical transition element with the atomic number 79. In bulk, it is solid, dense, extremely soft and ductile under standard conditions. Gold is one of the least reactive chemical elements and it is resistant to oxide formation. It can only be dissolved in aqua regia or in alkaline solutions of cyanide and mercury. Gold appears nearly solely in its elementary form instead of in a compound. Further characteristics will be summarized in **Table SI 1**.[48-50]

Gold features a face-centered cubic crystal structure (fcc, ABCABC). In this type of packing, the atoms of the second layer B are seated in the depressions between the atoms of the first layer A, while the third layer C is placed in octahedral voids[51] (**Figure 2.1a**). Thus, each metal atom is ultimately surrounded by 12 equidistant neighbors and 4 potential adsorption positions on the (111) surface can be defined. It can be situated on top of an atom in the A layer (atop), on top of an atom in the B layer (hcp), on top of an atom in the C layer (fcc) and between two atoms in the A layer (bridge)[51] (**Figure 2.1a**). Adsorption that falls between these positions is also possible, however it is unlikely. The adsorption energies for the four positions was calculated for the methanethiol-gold(111) system and an energy minimum was determined for the fcc position, characterizing it as optimal.[52] The distance between two Au atoms is 288.4 pm and each atom has six shells of electrons yielding the electron configuration: [Xe] $4f^{14}5d^{10}6s^1$ (**Figure 2.1b**).

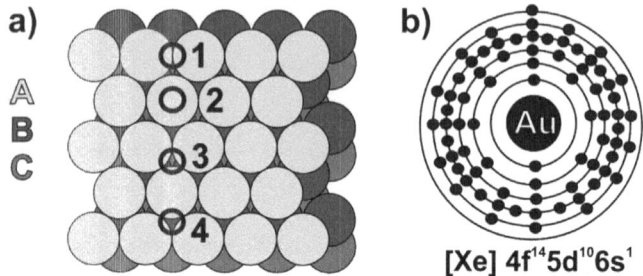

$$[Xe]\ 4f^{14}5d^{10}6s^1$$

Figure 2.1.Fundamental configuration of gold. a) The face-centered cubic crystal structure (fcc) of gold with three atom layers (**A** = yellow, **B** = red, **C** = blue) and four potential adsorption positions (**1** = bridge, **2** = atop, **3** = fcc and **4** = hcp). **b)** The electron configuration [Xe] $4f^{14}5d^{10}6s^1$ of a gold atom (Au) with electrons (dots) positioned on six atom shells.

The dominant oxidation states for organogold compounds are:

- **+I** featuring the coordination number 2, a linear molecular geometry, diamagnetic properties and 14 electron species [d^{10} ions] and
- **+III** with coordination number 4, a square-planar molecular geometry, diamagnetic properties, toxic behavior and 16 electron species [d^8 ions].[53]

Although gold is inert in bulk form, it features complex ligand chemistry on the nanometer-scale. For instance, an oxidative addition of the S-S bond to a gold surface is enabled by the mechanism as summarized in **eq 2.1**.

$$RS\text{-}SR + 2\ Au_{(s)} \rightarrow 2\ RS\text{-}Au_{(s)}$$

eq 2.1

The binding energy of the RS-SR bond is ~ 65 kcal mol^{-1}.[54]

Whereas, the mechanism for an oxidative addition of the S-H bond to the gold surface, followed by a reductive elimination of the hydrogen is shown in **eq 2.2**.

$$R\text{-}SH + 2\ Au_{(s)} \rightarrow 2\ RS\text{-}Au_{(s)} + H_2$$

eq 2.2

The binding energy of RS-H bond is ~ 86 kcal mol^{-1}.[54]

The released hydrogen could either adsorb on the gold surface or desorb as molecular hydrogen (solubility 1.6 mg L^{-1} in water).[55]

The time it takes to form *full* monolayers has been reported to vary from seconds[56] to minutes[57], up to several hours[58] and sometimes days[59] and both, thiols and disulfides were found to adsorb at the same rate, limited mainly by mass transport.

Even though colloidal gold has been used for centuries e.g. to manufacture the famous Lycurgus Cup (5th century) and later as *Purple of Cassius* in stained church glass (17th century), the first systematic study of its synthesis and characterization was performed in 1857 by Michael Faraday. Faraday discovered that the optical properties of gold submicrometer-sized particles were different from those of the bulk metal[60] and he related the variety of fluid colorations to the particle sizes.[2;[60]

However, details on gold nanoparticles' (AuNPs) optical characteristics were declared later by researchers such as Richard Adolf Zsigmondy or Gustav Mie and will be summarized in **Chapter 2.2.**

2.2.   Optical aspects of gold nanoparticles and their imaging

Shedding light on cellular processes

A flexible and adjustable size as well as tunable shapes and conformations may be provided for gold nanoparticles, depending on synthesis strategies, the surrounding medium and particle density. These physicochemical differences result in distinct changes in the optical properties, rendering the particles differentiable with microscopy techniques[61;62] such as transmission electron microscopy and by the colloid coloration[63;64] (**Figure 2.2a**).

The unique optical properties of gold nanoparticles are mainly caused by collective oscillations (*localized surface plasmons*, LSP) of conduction band electrons upon excitation with an alternating, electric field of incident electromagnetic radiation.[65;61;66] The field induces a polarization/displacement of the surface electrons relative to the nuclei which develop a restoring force due to the Coulomb attraction between the electrons and the nuclei (**Figure 2.2b**).[32;67]

The electromagnetic radiation is scattered by particle sizes that are similar to or larger than the wavelength of incident light. This scattering is predominantly explained with the Mie solution to Maxwell's equations.[68] In this context, the relationship of extinction, absorption and scattering is given in **eq 2.3**, for the distinct wavelength, particle radius and shape, particle refractive index and refractive index of the medium.

---

[2] '[...] known phenomena seemed to indicate that a mere variation in the size of [gold] particles gave rise to a variety of resultant colors.'

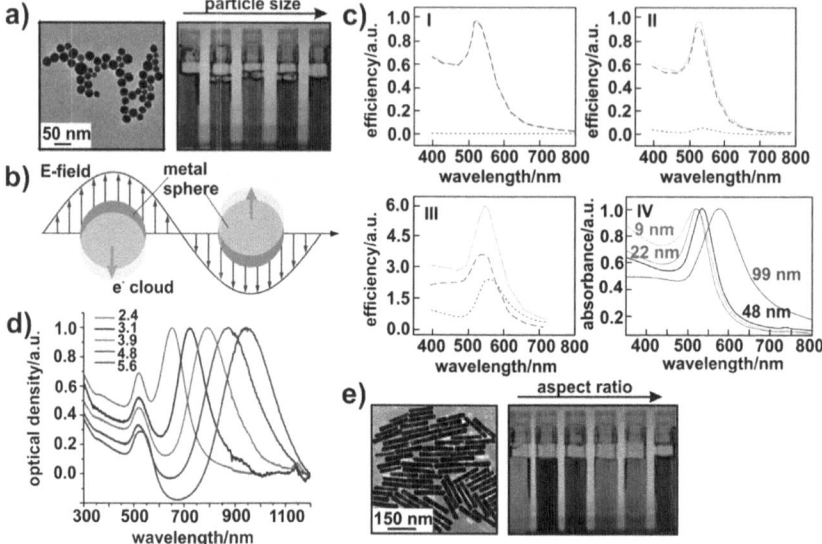

Figure 2.2. Size-dependent and shape-dependent optical properties of AuNPs. a) Left panel: Transmission electron micrograph of gold nanoparticles. Reprinted with permission from Besner et al., copyright 2007 by Springer Science+Business Media.[69] **Right panel:** Photography of gold nanoparticle colloids with increasing particle size. Adapted with permission from Mody et al., copyright 2010 by the Journal of Pharmacy and Bioallied Sciences.[63;64] **b)** Schematic representation of movements in the electron cloud in an electric field that are responsible for the surface plasmon resonance effect of AuNPs. Reprinted with permission from Kelly et al., copyright 2003 by the American Chemical Society.[32] **c)** The extinction of AuNPs (green solid line) is exclusively represented by the absorption cross section (red dashed line) of particles smaller than 20 nm (**I**). With increasing size, a growing component of scattering (blue dotted line) appears as shown for 40 nm (**II**) and 80 nm (**III**) particles. Adapted with permission from Jain et al., copyright 2006 by the American Chemical Society.[70] In addition, the LSPR band broadens and shifts to longer wavelengths with nanoparticle size increase (**IV**) from 9 to 99 nm. Adapted with permission from Link et al., copyright 1999 by the American Chemical Society.[71] **d)** The LSPR band of elongated AuNPs (gold nanorods) splits into two bands and is presented for different aspect ratios from 2.4 to 5.6. Adapted with permission from Huang et al., copyright 2006 by the American Chemical Society.[72] **e) Left panel:** Transmission electron micrograph of gold nanorods. Reprinted with permission from Vigderman et al., copyright 2013 by the American Chemical Society.[73] **Right panel:** Photography of gold nanorod colloids with increasing aspect ratio. Adapted with permission from Mody et al., copyright 2010 by the Journal of Pharmacy and Bioallied Sciences.[63;64]

$$c_{ext} = c_{abs} + c_{sca}$$

eq 2.3

C_{ext} = extinction cross section, C_{abs} = absorption cross section, C_{sca} = scattering cross section.

Mie scattering is not strongly wavelength-dependent, but the particle size dependence may be illustrated in **Figure 2.2c**. While for small particles of 20 nm in diameter the extinction is exclusively represented by the absorption cross section of the particles (**I**), the Mie scattering contribution starts with particles of 40 nm sizes (**II**) and significantly adds to the extinction of 80 nm AuNPs (**III**).

Conversely, there is another scattering type called *Rayleigh scattering*, which explains the elastic scattering for nanoparticles of much smaller size than the wavelength of incident light. The Rayleigh scattering intensity is proportionate to the 6th power of the particle's diameter and is strongly related to wavelengths in the UV region (230–320 nm).

The LSP can be excited in the UV-vis spectral range and the excitation of LSP resonances (LSPR) leads to an enormous increase in the absorption and scattering cross sections,[74] which are accessible to various imaging strategies as discussed later in this thesis.

Due to the LSPR, AuNPs have an extinction maximum in the green spectral region around 520 nm and the colloid features an intense red coloration. Modifications in the size or shape of AuNPs result in a shift and/or broadening of the LSPR band and in a change of the colloidal color (**Figure 2.2a**, **Figure 2.2d–e**). Therefore, increased scattering and a further red-shifted absorption maximum are mainly related to e.g. increasing particle size, deformation or agglomeration.[71]

Furthermore, for an elongated particle (nanorod) the surface plasmon band splits into two bands (**Figure 2.2d**).[72] The band that absorbs at short wavelengths characterizes the oscillation of the electrons perpendicular to the long rod axis and is referred to as transverse plasmon absorption. The other band that absorbs at higher wavelengths in the near-infrared (NIR) region is defined as longitudinal surface plasmon absorption and is caused by the oscillation of free electrons along the long rod axis.

In addition, the LSPR wavelength is very sensitive to changes in the dielectric properties of the surrounding medium as shown in the framework of the *Drude model* (**eq 2.4–eq 2.6**).[70;75]

$$\lambda_{SPR\,max}^{\ 2} = \lambda_p^{\ 2}(\epsilon_\infty + 2\epsilon_m)$$

eq 2.4

$$\lambda_p^{\ 2} = \frac{2\pi\,c^2}{w_p^{\ 2}}$$

eq 2.5

$$w_p^{\ 2} = \frac{N\,e^2}{m_e\,\epsilon_0}$$

eq 2.6

$\lambda_{SPR\,max}$ = wavelength of the surface plasmon resonance peak of gold nanoparticles, λ_p = bulk plasmon resonance wavelength of gold, ϵ_∞ = high-frequency dielectric constant of gold due to interband and core transitions, ϵ_m = dielectric constant of the surrounding medium, c = speed of light in vacuum, w_p = bulk plasma frequency, N = density of free electrons in the nanoparticle, m_e = effective mass of an electron, ϵ_0 = permittivity in vacuum.

In this context, media with high refractive indices couple stronger with the surface plasmon electrons and the required energy to excite the collective oscillation is reduced. Thus, the LSPR absorbance shifts to lower energy, which correlates with longer wavelengths. By this means, molecules that increase the refractive index near the nanoparticle surface by adsorption will also induce an LSPR shift, enabling sensing applications.[76]

The interband absorption of gold is found at 380 nm and correlates with the atomic concentration of colloidal gold according to results from Muto et al.[77] However, the scattering of aggregates, agglomerates and the couplings of primary particles and nanoparticles with a large diameter is found in the NIR region (**Figure 2.2c IV**, high offset for 90 nm particles at 800 nm wavelength).

Based on optical characteristics of AuNPs, such as a high quantum yield, good signal to noise ratio and the disability of bleaching[71;78;61;79] with the described tunable spectroscopic characteristics, a ground-breaking study was conducted to replace conventional fluorescent dyes by the AuNPs, especially in relation to longer time-scale observations in living specimen.[65;80;61] Various imaging strategies were developed and adopted for this intent. The classical standard for the visual characterization and quantification of gold nanoparticles in dispersion and in their spatial distribution in fixed samples is the transmission electron microscopy (TEM).[81] The very high contrast of gold renders these particles easily distinguishable within cellular structures. The method can only be applied on very thin preparation sections and detailed information about size distributions of the particles and their arrangements within intracellular structures can be validated.[82-85] For example, Chithrani et al. imaged and counted single gold nanoparticle spheres of different sizes from 14 to 100 nm in high resolution within endosomal vesicles (**Figure 2.3**).[82]

In addition, Murphy et al. presented differently shaped gold nanostructures with increasing diameter and aspect ratio, that were easily distinguishable from each other on TEM micrographs.[84] However, the detection of lead or osmium artefacts from the fixation protocol cannot be excluded and the investigation of *in vivo* processes is not feasible with this method. Thus, electron microscopic approaches such as TEM or STEM (scanning transmission electron microscopy) provide indispensable basic knowledge about nanoparticle composition and intracellular distribution, but are limited in their application on living biosystems.

Nanoparticles are per definition less than 100 nm in diameter, which is only half the size necessary to meet the refraction limit for light microscopic detection of single particles. However, due to the high quantum yield in absorption and scattering of AuNPs it was demonstrated repeatedly that single particles could be visualized by optical microscopy from 5 nm onwards using the absorption cross section for differential interference contrast (DIC) microscopy[86-88] and using the scattering for 40 nm onwards for optical coherence tomography (OCT)[89;90] and reflection based (dark field) microscopy.[76;91;79;92]

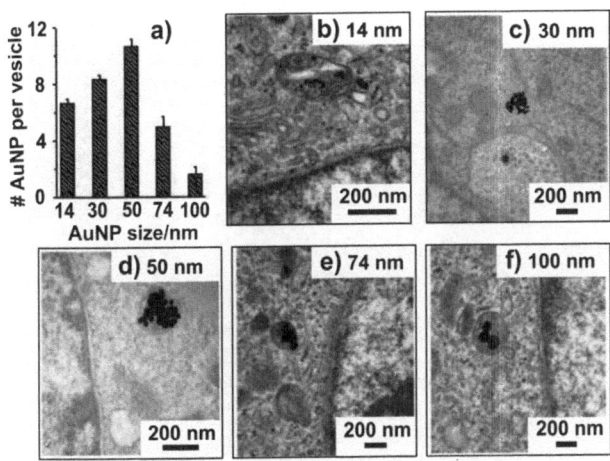

Figure 2.3. Example of AuNP imaging by TEM. Transmission electron microscopy provides information on the number, size and arrangement of particles in organic structures and allows for the quantification of nanoparticles within distinct cell compartments. **a)** Number of counted AuNPs per vesicle as function of particle size. **b)–f)** Transmission electron micrographs illustrating endocytosed AuNPs inside of intracellular vesicles for AuNP diameters of **b)** 14 nm, **c)** 30 nm, **d)** 50 nm, **e)** 74 nm, **f)** 100 nm. Adapted with permission from Chithrani et al., copyright 2006 by the American Chemical Society.[82]

In this context, the differentiation of single or aggregated AuNPs was presented using the LSPR scattering-based confocal laser scanning microscopy (CLSM).[92;93] Klein et al. successfully displayed dispersed single gold nanoparticles from diameters of 60 nm onwards (**Figure 2.4a–b**), while particles of 15 nm diameter could be imaged after aggregation (**Figure 2.4c**).[93]
Since intracellular particle aggregation is common, all size classes were imaged after co-incubation with bovine immortalized endothelial cells (**Figure 2.4d–f**). Thus, a differential imaging of aggregated versus dispersed gold nanoparticles by specific light scattering of aggregates and single particles in different wavelength regions is feasible for imaging throughout the visible spectrum.

As explained earlier, the spectral properties of imaged nanoparticles are indicative of the current status of the nanoparticle dispersion and could provide information on size, shape and functionalization of gold nanoparticle constructs *in vitro* and even *in vivo*.
In this context, the LSPR scattering-based, vibrational, surface-enhanced Raman spectroscopy (SERS) may be used to gain information about the chemical composition of a sample in the close vicinity of AuNPs. For instance, Sezgin et al. applied SERS for prob-

ing the cellular environment of AuNPs after their introduction into living cells and differentiated their accumulation areas based on molecular level differences.[94]

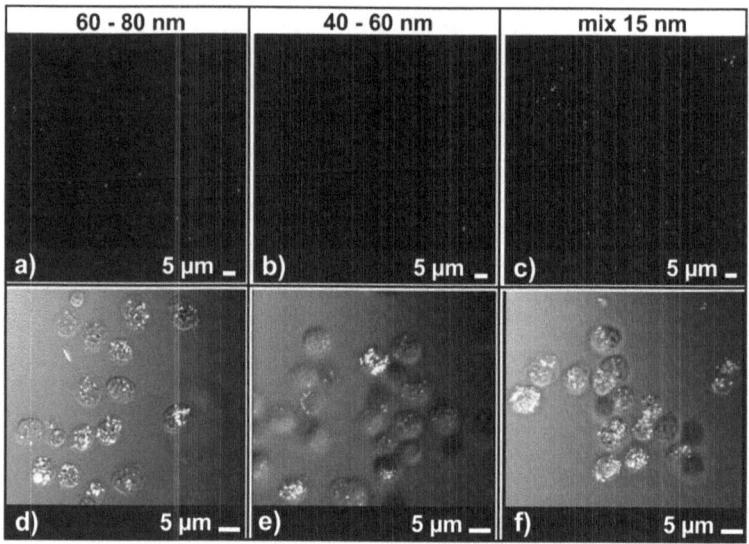

Figure 2.4. Example of AuNP imaging by CLSM. a)–c) Confocal imaging of the LSPR scattering (green dots) from differently sized AuNPs at equal number concentrations in dispersion. **d)–f)** The LSPR scattering from differently sized AuNPs after 48 h co-incubation with bovine endothelial cells. Only dispersions of 60–80 nm-sized particles allowed visualization of single particles in dispersion. However, the aggregation and containment of particles within cells enhanced the scattering cross section to visualize also the smaller particles. Adapted with permission from Klein et al., copyright 2010 by the Society of Photo Optical Instrumentation Engineers.[93]

Addressing *in vivo* imaging, a very unique property of a noble metal such as gold is the photothermal effect[95] that allows the independent localization[96] and treatment of gold nanoparticles after excitation with light independently from other, non-metal particles in the same field. Werner et al. and Lasne et al. demonstrated the tracking of 5 nm gold beads in living cells using this technique.[97;98] Both reports emphasized the extremely low background noise that also exists in scattering environments such as cells and tissues. Furthermore, the photoacoustic detection of AuNPs by shock wave generation upon pulsed particle heating was also reported.[99;100]

It can be summarized, that gold nanoparticles feature unique optical characteristics that are mainly due to the localized surface plasmons. The resulting flexibility of the spectral properties that correspond to the particle's size, shape and surface functionalization enables the clear identification of the specific state of AuNPs and highlights them as suitable material for biomarker research.

Various imaging techniques were adopted in order to detect AuNPs within biological samples and living specimens. A size-selective imaging of small gold nanoparticles after cellular penetration is applicable using light and electron microscopic methods. Thereby, it is feasible to visualize particle aggregation and their quantitative accumulation selectively by employing scattering-based approaches such as confocal laser scanning microscopy. Other methods can also be used, such as optical coherence tomography or the photo-thermal and photoacoustic detection that enable even the *in vivo* visualization of AuNPs.

2.3.   Toxicological aspects of gold nanoparticles

The dose makes the poison

The toxicity of AuNPs is a very complex topic because the level of toxicity is highly de-pendent on the size, concentration, shape and surface chemistry of the particles as well as on the experimental design involving animal or cell culture models and the methods used to characterize particle localization and distribution.[101] Standardized methods for toxico-logical assays with nanoparticles are not widely applied, although the Organization for Economic Co-operation and Development (OECD) has recommended general guidelines in 2012.[102] Thus, it is not surprising that results are very different for each case and vary significantly. However, by reviewing a variety of toxicological studies, an estimation of the toxicity mechanisms and thresholds can be determined.

In contrast to the bulk material, nanoscaled particles feature a higher reactivity and toxici-ty, which is mainly derived from their high mass-specific surface area per mass ratio with surface-specific dose-response.[103;104] The mechanisms of cellular damage caused by na-noparticles are explained by the interactions at the nano-bio interface,[105-108] resulting mainly in the affection of DNA[109;110] and the production of reactive oxygen spe-cies[104] (ROS). This can cause inflammation or even malignant transformation of somatic cells. For instance, Liu et al. described an interaction of 1.4 nm sized AuNPs with B-form DNA while causing its transition to A-form DNA.[109] However, ROS formation is gener-ally found for nanoparticles featuring a band gap structure, e.g. semiconductor metal na-noparticles.[111]

The parameters that affect nano-bio interactions are the size, shape, charge and function-alization of AuNPs. This variety makes it difficult to find a satisfactory comparison of toxicity studies. However, a selective overview of recent publications and their AuNP parameter dependence is shown in **Table SI 12**, while for deeper insight, the readers are referred to other review articles.[104;84;112;34;101;113-115]

Most of the available literature tends to cover the *in vitro* toxicity of AuNPs (**Table SI 12**) due to the experimental freedom that allows for a much broader dosing and testing range compared to *in vivo* studies, whose numbers are usually kept low for ethical reasons. The toxicity discovered in those *in vitro* studies ranged from negligible and regardless of the used particle type[116-118] to intermediate[119-123] and even severe.[124-127] However, various cell lines were applied for the studies which may respond completely different on the same nanoparticle exposure. It was further demonstrated, that primary cells differ in sensitivity towards nanoparticle-derived toxicity than the corresponding cancer and immortalized cell lines.[128;129]

A size-dependent cytotoxicity of AuNPs was found by Pan et al., when they analyzed the size ranges from 0.8 to 15 nm.[124] They found that AuNPs with sizes from 1 to 2 nm raised significant toxic effects in four cell types, while even a minor decrease in size (from 1.8 to 1.4 nm) may increase the toxic effect by 4-6 factors.[124] 15 nm-sized particles provided comparably no cytotoxic effect. Further, Pan et al. identified that 1.2 nm-sized AuNPs mainly induced cell death by apoptosis while 1.4 nm-sized particles were responsible for cell death by necrosis (**Figure 2.5a**).[124] However, Pan et al. applied a high particle number dose and surface area, which is critical according to a study on airborne nanoparticles from Oberdorster et al., who demonstrated that the main parameter for adverse effects in biological systems is the particle surface area.[104]

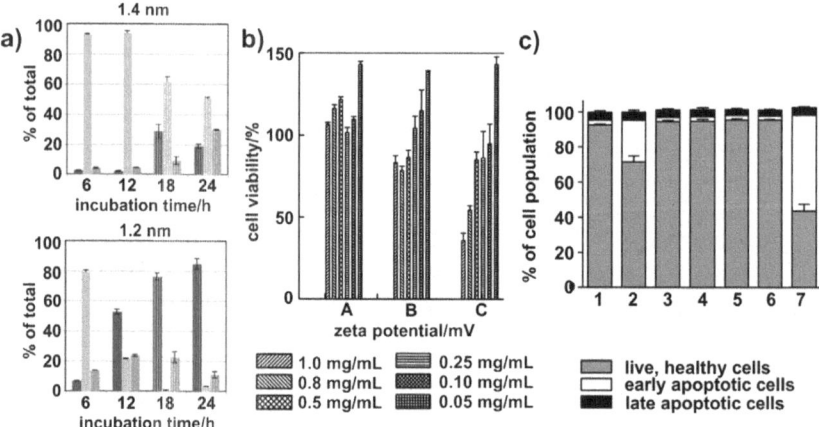

Figure 2.5. Examples on the *in vitro* toxicity of AuNPs and AuNP bioconjugates. a) Influence of particle size on the *in vitro* toxicity of AuNPs. Red bars = necrotic cells, yellow bars = live cells, green bars = apoptotic cells. Adapted with permission from Pan et al., copyright 2007 by John Wiley & Sons Inc.[124] b) Influence of particle charge on the *in vitro* toxicity of AuNPs. A = +20 mV, B = +30 mV, C = +40 mV. Adapted with permission from Ding et al., copyright 2010 by the American Chemical Society.[127] c) Influence of particle surface modification on the *in vitro* toxicity of AuNPs. 1 = untreated control, 2 = citrate, 3 = BSA, 4 = ssDNA, 5 = dsRNA, 6 = Doxorubicin. Adapted with permission from Massich et al., copyright 2010 by the American Chemical Society.[121]

The impact of AuNP charge was analyzed at the particle-liquid-interface by Ding et al., who discovered a distinct correlation between cytotoxicity and an increase in positive surface charge[127] (**Figure 2.5b**). This result was further confirmed in a study by Goodman et al., in which they compared 2 nm-sized cationic and anionic AuNPs.[130] While the cationic AuNPs provoked pronounced cell lysis, the anionic AuNPs featured a low cytotoxicity which was most likely due to the weak electrostatic interactions with the cell membrane.[130] In addition, Bartneck et al. found in a more detailed study that carboxyl groups on the particle surface induced the expression of mRNAs which encode pro-inflammatory proteins, while amino groups on the particle surface induced mRNAs which encode anti-inflammatory proteins.[131]

During chemical synthesis, AuNPs are often stabilized by sodium citrate molecules, which change the surface conditions of the nanoparticles. Such citrate-stabilized AuNPs were found to induce significant changes in the gene expression profile of HeLa cells[121] (**Figure 2.5c**). However, when the stabilizing agents were disclaimed, Salmaso et al. did not observe any toxic effect up to a concentration of 0.74 nM gold,[117] while Taylor et al. noticed an effect, but only at five times higher AuNP concentrations of ~ 5 nM for ultrapure, laser-generated AuNPs.[122] Thus, the cytotoxic impact of nanoparticle surface ligands should always be considered and systematically studied in comparison to ligand-free reference nanoparticles (e.g. AuNPs fabricated by pulsed laser ablation in pure *MilliQ*).

Acknowledging Paracelsus' doctrine,[3],[132] an AuNP threshold concentration/dose for the initiation of toxic effects must exist. Unfortunately the AuNP concentration in all studies (**Table SI 12**) varied widely and different sizes, shapes and charges were adopted. Thus to date, no comparability can be given for a universal threshold determination.
Khlebstov and Dykman proposed a general limiting dose of ~ 10^{12} particles per mL for AuNPs within the size-range from 3 to 100 nm.[101] However, mass dose, number dose and surface dose have to be differentiated in this context (**Figure 2.6a**).
As illustrated on **Figure 2.6a 1** the same particle mass concentrations can be obtained by using many small or fewer large particles with completely differing surface area.[115] Thus, studying the toxicological effect of nanoparticles with different sizes at equal mass doses it is impossible to clarify which variable (particle size, particle number, particle concentration or particle surface area) is the determining factor (**Figure 2.6b**).[115] Moreover, Taylor et al. recently declared that 'drastic effects caused by very high *unrealistic* exposure values may be over-interpreted, while subtle effects due to low-dose *realistic* exposures may be overlooked'.[115]

---

[3] 'Dosis facit venenum' = it is the dose that makes the poison.

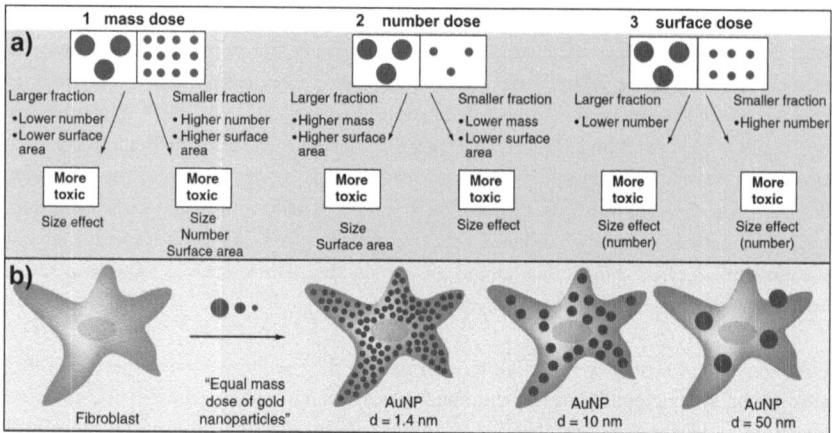

Figure 2.6. Schematic illustration of nanoparticle dose expression. a) Comparison of two nanoparticle size classes (larger fraction = left box, smaller fraction = right box) that are equivalent to the same nanoparticle mass dose (**1**), nanoparticle number dose (**2**) and nanoparticle surface dose (**3**), showing that most information is accessible from the surface area of the dose. **b)** The degree of surface coverage with differently sized nanoparticles of the same nanoparticle mass dose is presented on the example of a fibroblast. Reprinted with permission from Taylor et al., copyright 2014 by Taylor et al., licensee Beilstein-Institute.[115]

It is important to consider further, that the applied particle dose in the cell culture medium is rarely identical with the delivered dose that the cells come into contact with and the cellular dose that is internalized by the cells.[115] In the style of the toxicological testing of airborne particles, Oberdorster et al. recommended to express the applied dosage as particle surface area concentration (e.g. cm^2 of nanoparticles per mL).[104] However, as the toxicological effect is highly dependent on the cell number, an expression of dose per cell density, organ mass or organ surface area (e.g. cm^2 nanoparticles per cell number or cm^2 nanoparticles per g of biomass) is highly recommended by the OECD.[102]

The ability of AuNPs to effectively cross the blood-testis-barrier after intravenous injection was recently documented by Balasubramanian et al.[133] Therefore, concerning biological reproduction, the toxic effects of gold nanoparticles on gametes should be considered, because this might result in impaired fertility and/or congenital defects of the offspring. Unfortunately, only a few studies have focused on this topic thus far (**Table SI 12**).[134] For instance Wiwanitkit et al. discussed the morphological defects and motility decrease of human spermatozoa after treatment with citrate-stabilized AuNPs.[135] However, no information on the adopted particle dose was provided. Tiedemann et al. determined no toxic effects on boar spermatozoa after incubation with 10 μg mL^{-1} AuNPs,[136] while Taylor et al. and Moretti et al. reported membrane-attachment and a dose-dependent decrease of bovine/human spermatozoa motility after incubation with

50–500 µM AuNPs[137;138] (**Figure 2.7a**). However, Tiedemann et al. adopted serum protein-stabilized, non-aggregated particles, while aggregated AuNPs were applied in the other studies. Gold aggregates are subject of fast sedimentation in cell culture and feature a different cellular interaction and uptake behavior than non-aggregated nanoparticles (see **Chapter 2.4.**).

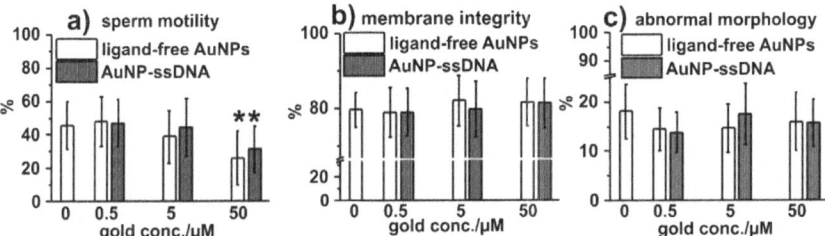

Figure 2.7. Examples on the reprotoxicity of AuNPs and AuNP bioconjugates. The effects of ligand-free AuNPs (white bars) and AuNP-ssDNA bioconjugates (colored bars, 0.5–50 µM concentration) on bovine spermatozoa characteristics. **a)** Effect on sperm motility. * ANOVA, $p \leq 0.5$. **b)** Effect on membrane integrity. **c)** Effect on sperm morphology. Results are presented in comparison to an untreated negative control (0 µM). Based on data from Taylor et al., copyright 2014 by Taylor & Francis.[137]

The concentration range from 50–500 µM AuNP significantly exceeds the number of nanoparticles that are required for scientific or medical applications. Furthermore, Taylor et al. found that there was no effect on membrane integrity and spermatozoa morphology after AuNP incubation (**Figure 2.7b–c**).[137] However, Zakhidov et al. found that very small AuNPs with diameter of 2.5 nm disrupted nuclear chromatin decondensation in mouse spermatozoa.[139]

The penetration of AuNPs into ovaries or follicles has not been studied to date, but Tiedemann et al. found no toxic effects on oocyte maturation after treating the cumulus-oocyte complex with AuNPs up to a concentration of 30 µg mL^{-1}.[136]

When looking at the developmental toxicity and fetal impairment that can occur with AuNPs, two separate studies with rodent models analyzed and confirmed their transfer across the placental membrane (**Table SI 13**).[140;141] Interestingly, two other studies could not find any particle transfer.[142;143] In addition, in an *ex vivo* model by Myllynen et al. no placental AuNP transfer was detected, which illustrates the difficulty to determine whether or not this transfer actually occurred.[144]

Further developmental toxicity of gold nanoparticles was analyzed in zebrafish,[145;146] chicken[147;148] and murine[149] embryos. Although the presence of AuNPs inside the embryos was proven, [145;146;149] no toxic effects were determined in the studies. However, a recent publication on zebrafish detected an embryotoxic effect of gold clusters after applying a number dose of 10^{14} NP per embryo.[150] Furthermore, the toxicity depended on particle size and ligand chemistry and AuNPs with covalently bound ligands were determined to be less toxic than AuNPs with electrostatically bound ligands.[150]

Examining the *in vivo* studies of AuNP toxicity on developed (adult) animals (**Table SI 12**), a size-dependent effect on BALB/C mice was determined by Chen et al. In their study they investigated AuNPs with diameters from 3 to 100 nm and found a significant lethality for mice treated with AuNPs with diameters between 8 and 37 nm, while other sizes did not induce any effect.[151] Other studies described various effects of AuNPs, covering expressed changes of the inner organs[96] and abnormal up and down regulation of genes[133] in rats, induced inflammation and apoptosis in mice[152] and induced oxidative stress in mytilus edulis.[153]

Conversely, other studies did not discover any toxic effects of 1.9 to 100 nm-sized AuNPs in mice and pigs although a dose-dependent accumulation in various organs was determined.[154-158]

In 1997, a clinical study was performed by Abraham et al. on 10 rheumatoid arthritis (RA)-affected humans, using 20 nm-sized Aurasol® AuNPs at a daily oral administration of 30 mg, over a period of up to 5 months.[159] Interestingly, they not only found the administration to be non-toxic, but determined that various RA-relevant factors were significantly suppressed when the subjects reported an improvement in joint pain, swelling and mobility.[159] However, the production of Aurasol® has now been discontinued without any information about the reasons.

In general, the *in vivo* studies indicate that low doses of AuNPs ($< 400 \ \mu g \ kg^{-1}$) do not appear to cause appreciable toxicity, [158;160] although at higher concentrations, severe sickness, shortened survival time and liver inflammation were observed.[151;152]

In summary, the toxicological aspects of gold nanoparticles remain up for debate because of the variety of parameters that influence the particles' toxic behavior and the incomparability among toxicological studies. Especially the expression of *particle dose* is a crucial aspect for the evaluation and comparability of toxicological studies. Up to now there are no general regulations or standardized methods for toxicology assays with colloidal nanoparticles. For instance the same particle mass concentration dose may be obtained by using many small or fewer large particles with completely differing surface area. However, the indication of a *nanosurface per bio* dose additionally to the common mass concentration dose in the format of 'surface area dose of nanoparticles referenced to the cell number or organ mass' may overcome this issue in the future.[115]

2.4.   Gold nanoparticle-membrane interactions

How to cross barriers and where to go

Controlled nutrient uptake and the disposal of contaminants are essential for cells to sustain metabolism. To provide differentiation and maintain controllability of this survival process, evolution has naturally developed several uptake and transport mechanisms.[161-164] Ions and small molecules usually migrate along concentration gradients and may enter cells via an unspecific diffusion process. Conversely, macromolecules are mainly internalized energy-dependent by endocytosis after a specific interaction occurs with their texture and membrane-associated surface receptors.

In this context, it is obvious that the cellular uptake and the uptake mechanism of gold nanoparticles is not only affected by their size and shape but also by their steric (e.g. nature/amount of conjugated ligand) and electrochemical (e.g. surface charge) properties. They may also be dependent on the studied cell line because each cell line exhibits different phenotypes and receptor expression levels (**Figure 2.8a**).

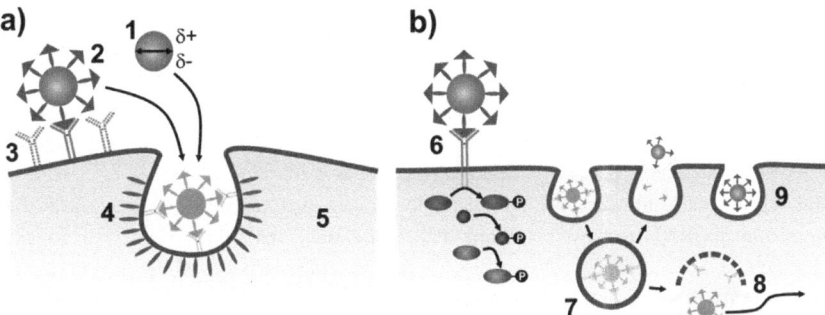

Figure 2.8. Cellular uptake and intracellular fate of AuNPs. a) Factors that can influence the interactions between nanoparticles and cells at the nano-bio interface (**1** = size, shape, charge; **2** = ligand density; **3** = receptor expression level; **4** = internalization mechanism; **5** = cell properties such as phenotype and location). **b)** Potential interaction of nanoparticle bioconjugates with cells (**6** = antibody-coated NPs bind specifically to membrane receptors and induce a signaling cascade without entering the cell; **7** = endocytosis/exocytosis of nanoparticle bioconjugates without leaving the vesicle; **8** = endocytosed nanoparticle bioconjugates escape from the vesicle and interact with organelles such as the nucleus, mitochondria or actin filaments; **9** = unspecific internalization of nanoparticle bioconjugates into cells without membrane receptor interaction). Based on a figure from Albanese et al., copyright 2012 by Annual Reviews.[112]

In addition to the possibilities of unspecific diffusion, NP-receptor interaction and specific, receptor-mediated uptake, the intracellular fate of nanoparticles is also of great importance for biomedical application, especially if distinct organelles need to be targeted e.g. the nucleus for gene silencing issues or lysosomes for treatment of the lysosomal storage disease. Thus, specific ligands may be required to induce intracellular signaling cascades, to enable the escape of endocytotic vesicles and to reach the area of interest (**Figure 2.8b**).

A number of studies are currently being done on the influence of the intrinsic properties or gold nanoparticles and their functionalization for cellular penetration. A selection of these studies has been summarized in **Table SI 13**. Readers are also encouraged to examine other review articles to gain a deeper insight into this subject.[112;34;165;166;101;167]

Several studies have demonstrated that the cellular effects of AuNPs depend on their size, shape and surface charge (**Table SI 13**) [82;168-171].

With regard to size, the cell membrane wrapping time of nanoparticles was investigated by Gao et al.[172] In their hypothesis, the cellular uptake was considered to be a result of competition between the thermodynamic driving force for wrapping (amount of free energy to drive nanoparticles inside cells) and the receptor diffusion kinetics (kinetics of recruitment of receptors to the binding site). These two factors determine how rapidly and how many nanoparticles are taken up by the cell. Gao et al. proposed that the docking of a nanoparticle with a size that is smaller than 50 nm would not produce enough free energy to be completely wrapped and that the fastest wrapping time would occur for nanoparticles that were 55 nm in diameter.[172] Thus, to facilitate the efficient cellular internalization of small nanoparticles, they must be clustered. However, for nanoparticles that are larger than 60 nm, the receptor diffusion kinetics and thereby the wrapping time is slower, which leads to a fewer number of particles being internalized.[172] According to this suggestion, Chithrani et al. investigated the uptake of spherical gold nanoparticles that ranged in diameter from 10 to 100 nm. They determined that 50 nm primary nanoparticles were able to enter the cells with high efficiency (**Figure 2.9a**), while the 14 nm species required approximately 6 nanoparticles to cluster together before uptake occurred.[168]

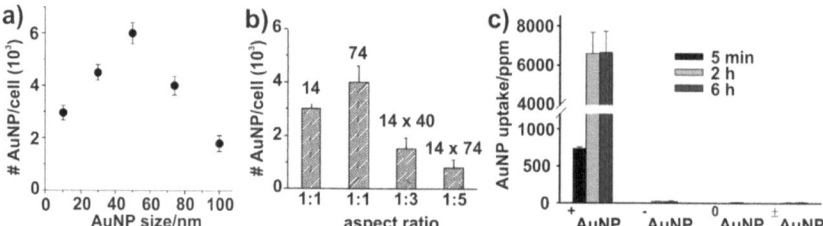

Figure 2.9. Examples on the cellular uptake of AuNPs. a) Number of internalized AuNPs per cell as function of AuNP size. Reprinted with permission from Chithrani et al., copyright 2007 by the American Chemical Society.[168] **b)** Number of internalized AuNPs per cell as function of AuNP size and shape. Aspect ratio 1:1 = spheres, aspect ratio 1:3 & 1:5 = rods. Reprinted with permission from Chithrani et al., copyright 2006 by the American Chemical Society.[82] **c)** Cell uptake of AuNPs as function of particle surface charge. Reprinted with permission from Arvizo et al., copyright 2010 by the American Chemical Society.[171]

In line with Gao's suggestion, they confirmed that particles smaller and larger than 50 nm were internalized to a lesser extent (**Figure 2.9a**).[168] These results were further confirmed by other studies that analyzed size-dependent cell internalization of AuNPs.[169;173;174]

However, concerning the intracellular fate and nuclear internalization, predominantly small AuNPs with diameters < 10 nm were detected inside the cell core, which indicates a size limitation for translocation through the tight nuclear pores.[175-179]

Various studies concerning the influence of AuNP shape on cellular uptake were also performed and are summarized in **Table SI 13**.[168;170;85] Thereby, the uptake of spherical 14 nm and 74 nm gold nanoparticles was determined by Chithrani et al. to be approximately 3 orders of magnitude higher when compared to rod-shaped gold particles with aspect ratios of 1:3 and 1:5[82] (**Figure 2.9b**). They considered this effect to be potentially curvature-dependent, because the contact area of rod-shaped particles is larger than for spherical NPs when the longitudinal axis of the rods interacts with the cell membrane receptors.[82] On the contrary, Bartneck et al. found that there was an up to 230 times more efficient uptake of gold nanorods into macrophages than of gold spheres with the same diameter.[131] However, considering the function of macrophages as a non-specific immune defense to engulf and ingest pathogens, Bartneck et al. speculated that the morphological similarity of nanorods to protein capsules of virus particles may support their increased uptake.[131]

With regard to surface charge, several studies concluded that cells in serum-free media internalize positively charged gold nanoparticles with higher efficiency than negatively charged or uncharged particles (**Table SI 13, Figure 2.9c**).[180;171;181-183] This is most likely due to the high affinity of positively charged species to a negatively charged cellular membrane.[184] However, Arvizo et al. found this effect to be based on the depolarization of plasma membrane potential.[171] Since baseline membrane potential was determined to range between -75 and -55 mV, only positively charged gold nanoparticles were determined to induce this depolarization.[171] The depolarization effect might cause the loss of rigidity and initiate morphological changes of the cell.[185] Thus, to maintain the original charge distribution, Cho et al. suggested that the plasma membrane must remove the attached gold nanoparticles e.g. by delivering AuNPs in intracellular vesicles (endocytosis),[185;180] or by other mechanisms that deliver AuNPs directly into the cytosol.[117;122] Since positively charged nanoparticles are attached more affine to negatively charged membranes than negatively or neutral charged species, cationic particles will consequently be internalized more easily and efficiently by the cells than anionic, zwitterionic or neutral NPs. In addition, Ding et al. correlated the zeta potential of AuNPs with their transmembrane efficiency and found that those with higher potential were internalized more quickly with enabled nucleus targeting than NPs with lower zeta potential.[127] However, if surface potential is too high, the particles may destabilize the cell membrane and induce cell damage and cytotoxicity.

A mechanism for the AuNP uptake of negatively charged species was discovered by using serum-containing cell culture media.[82;186] Chithrani et al. found that serum proteins were binding preferential to negatively charged gold nanoparticles and assisting them to enter the cells. By this means, gold nanoparticles that are functionalized with both positively charged and negatively charged ligands may be inserted into cells, if the culture medium is carefully considered.

Moreover, the surface charge of nanoparticles has also been examined in order to potentially determine the intracellular fate of particles. In this regard, Panyam et al. reported that endosomal escape was preferentially observed for cationic nanoparticles compared to anionic ones.[187]

In a similar manner, the surface modification of specific ligands has been reported to provide inefficiently internalized gold nanoparticles with the ability to specifically overcome obstacles such as cellular membranes or vesicles.[123;188-190;175] They were also shown to accumulate at the area of interest. For instance, Verma et al. reported on the highly efficient cellular uptake of gold nanoparticles which were covered with ordered arrangements of hydrophilic and hydrophobic functional groups (**Figure 2.10a**, inset).[189] This specific arrangement facilitated the transport of particles with 4-5 nm in diameter into the cytosol (**Figure 2.10a**) by enhancing the free energy for membrane wrapping, while other ligand distributions on the particle surface resulted in lower uptake efficiencies.

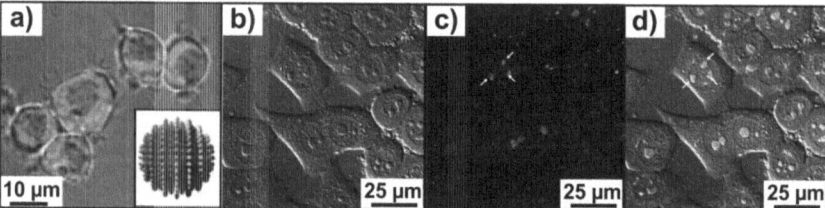

Figure 2.10. Examples on the cellular uptake of AuNP bioconjugates. a) The intracellular distribution (red coloration) of nanoparticles with ordered arrangements of hydrophilic and hydrophobic surface functional groups on confocal image with schematic illustration of ligand shell structure presented in the inset. Adapted with permission from Verma et al., copyright 2008 by the Nature Publishing Group.[189] **b)–d)** confocal images showing **(b)** the DIC, **(c)** the fluorescent image and **(d)** the merged DIC and fluorescence image of HeLa cells treated with AuNP@MPA-PEG-FITC conjugates and the intranuclear accumulation and clustering of nanoparticles (green dots). Reprinted with permission from Gu et al., copyright 2009 by Elsevier.[176]

Specific biological ligands have also been adopted as *Trojan horses* to support the cellular internalization of AuNPs. Besides viral vectors[191] and dendrimers,[192] also peptides that contain protein transduction domains (PTD) have been used in several studies to efficiently transport particles across cell membranes.[193-195] The internalization mechanism of these so-called *cell-penetrating peptides* (CPP) is strongly dependent on the cargo molecule as well as on the adopted peptide concentration.[188;196-198] Both the translocation of AuNPs directly into the cytosol[175;190;83] and their controlled endosomal uptake by endocytosis[196;197;199] have been discussed by CPP support. Tkachenko et al. and Berry et al. reported on the most common CPP termed *TAT* (transactivator of transcription), which was derived from the human immunodeficiency virus type-1 (HIV-1). This TAT was shown to efficiently deliver gold nanoparticles into cells and even the nucleus if the particle size was smaller than the nuclear pore size.[123;200] The nuclear transport was further compared with tiopronin-TAT functionalized gold nanoparticles and a tiopronin-functionalized species without TAT.[175] Numerous tiopronin-functionalized particles were detected in the cytosol, while the tiopronin-Tat conjugates were visualized inside the nucleus. In addition, the functionalization of small-sized gold nanoparticles (< 5 nm) with (poly)ethylene glycol (PEG)[176] or a nucleus translocating signal (NLS) was reported to be highly effective for particle accumulation in the cell core (**Table SI 13, Figure 2.10b–d**).[123;83;196;198] However interestingly, Krpetic et al. found that after 24 hours, both the AuNP-CPP and the AuNP-NLS bioconjugates were exocytosed from the cells again.[201]

When looking at the medical application of gold nanoparticles it is important to note that not only their distribution and fate inside a cell, but also their biodistribution in the organism and their clearance from the body must be considered. For this intent, several *in vivo* studies were performed to analyze AuNP accumulation in specific organs of mice, rats and pigs (**Table SI 13**). Most reports describe a significant accumulation of AuNPs in the liver and spleen of the animals, indicating that the nanoparticles most likely bind to plasma antibodies and are subsequently recognized by the phagocyte-rich reticulo-endothelial system (RES).[133;202;143;203;204] In one study, Sadauskas et al. actually described the differentiated particle accumulation in the immune Kupffer cells of the liver,[143] which were in line with observations from Fent et al.[154]

Further accumulation areas of AuNPs within the organism were found in the kidneys and the testis,[133;202] in the lungs,[133;154;156;204] in the heart and the thymus,[202] in the retina[157] and also in the neural tissue after crossing the blood-brain-barrier.[158;204] A size-dependence in the tissue penetration of AuNPs was thereby determined by Sonavane et al.[204] In one study, AuNPs with 15, 50, 100 and 200 nm in diameter were administered intravenously in mice. The 15 nm-sized particles were found to yield the highest amounts in all organs including the blood, liver, lung, spleen, kidney, brain, heart and stomach. In contrast, only a minute presence of 200 nm AuNPs was found in all organs 24 hours after injection.[204] Moreover, Semmler-Behnke et al. found a size-dependent crossing of the air-blood-barrier after intratracheal administration of AuNPs, with small particles < 2 nm being efficiently transferred, while larger-sized particles (18 nm) were trapped in the lung.[141]

Once they have penetrated into different tissues, NPs can have a long retention time. For example, in the respiratory tract, the mid-life for NPs was found to be approximately 700 days in humans.[104] Moreover, Terentyuk et al. found that smaller AuNPs (15 nm) circulated longer in the organism than for larger-sized ones (50 nm).[96] In this regard, Zhang et al. further determined that 20 nm-sized AuNPs have best blood pool activity and tumor uptake, while 40 and 80 nm-sized AuNPs were cleared readily from the body by uptake in the liver and the spleen.[160]

Because the particle concentration in the organs decreased over time, it appears likely that AuNPs are re-translocated into the bloodstream through lymphatic vessels.[133;204;96] However, a different study presented that AuNPs were efficiently released into the urine after 5 h through filtration in the renal glomeruli.[205] In a more focused study, Zhang et al. detected BSA-conjugated AuNPs that aggregated to 40–80 nm-sized clusters *in vivo* mainly accumulated in liver and spleen, while glutathione-conjugated AuNPs of 5–30 nm cluster size were highly efficiently cleared by the kidneys (**Figure 2.11**).[206]

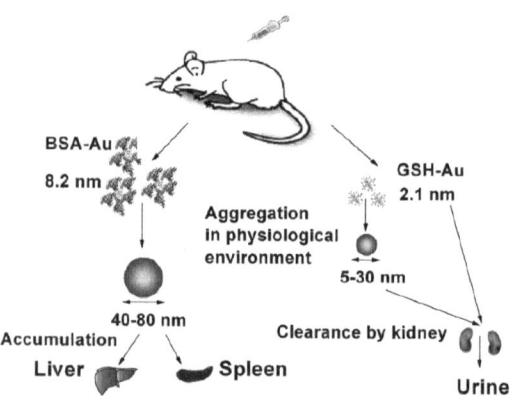

Figure 2.11. Size-dependent accumulation and clearance of AuNP bioconjugates from the organism. Aggregated BSA-AuNPs of 40-80 nm cluster size accumulated in liver and spleen, while GSH-AuNPs of 5–30 nm cluster size were removed from the body by renal clearance. Reprinted with permission from Zhang et al., copyright 2012 by Elsevier.[206]

Similar results were also determined by Zhou et al. who found that after 24 hours, more than 50 % of the 2 nm-sized glutathione-AuNPs that had been administered to mice, was in the urine. This was a 10 to 100 times better clearance than for comparable cysteine-AuNPs.[207] Thus, it can be stated that both the particle size and the attached ligand can influence the biodistribution and the clearance of AuNPs.

In summary, the intrinsic properties of gold nanoparticles, such as particle size, shape and charge as well as their functionalization with penetration agents have a strong influence on their cellular uptake behavior. The intracellular fate and internalization efficiency of particles may be tuned individually by the nanoparticle design, according to the desired uptake mechanism and biomedical indication. Regarding biodistribution, it is assumed that a size- and ligand-dependent biodistribution with tissue accumulation of AuNPs in the liver and spleen is favored, while small particle sizes (< 30 nm or < 10 nm) may be removed from the organism by renal clearance.

2.5. Biological application areas of gold nanoparticles

Golden age of modern diagnostics and therapy

Although gold nanoparticles and gold nanoparticle bioconjugates are monitored for a multitude of novel research concepts, three main application areas in biomedical and reproduction-related research may be defined thus far:
(**I**) Selective targeting [208;2] and sensing [209;4] of molecules or cells, e.g. for detection, imaging and sorting issues.
(**II**) Localized, photothermal cancer therapy by plasmonic heating of malignant tissue.[5]
(**III**) Delivery and switchable release of effector molecules to specific receptors/at the area of interest.[210;6]

Selective targeting and sensing
The targeting, sensing and imaging of molecules by gold nanoparticle bioconjugates relies mainly on their interaction with light. As presented in **Chapter 2.2**, gold colloids display an intense red color due to the LSPR, which may be utilized as a non-photobleaching alternative label to fluorophores. Therefore, particles must be functionalized with recognition moieties (e.g. antibodies) for the specific detection of target molecules (e.g. antigens) according to the lock-and-key principle (**Figure 2.12a**).[211;212] Upon specific targeting of gold-antibody bioconjugates to an antigen, a red dot or band depicts a positive binding result, which enables evaluation with the naked eye (immunolabeling/sensing) (**Figure 2.12b**).[213;214]

Currently, multiple sensing applications based on this principle have been established for gold nanoparticle bioconjugates, such as ultrafast detection assays for deoxyribonucleic acid (DNA) hybridization,[216;217;212] locked nucleic acid (LNA) triplex formation,[218] enzyme-linked immunosorbent assay (ELISA) for protein testing,[219-221] sugar sensing[222;20] and cell sensing.[223-225] Researches have also found applications that can be used in daily life around the world. Examples of this are ready-to-use test strips that function based on the immunoflow method for e.g. pregnancy[226] (*'Clearblue'* — SPD Swiss Precision Diagnostics GmbH, Switzerland), cancer (ScheBo Biotech AG, Germany), myocardial infarct[227] (Novamed Israel) (**Figure 2.12b**) or drug screening[228] (*'DrugCheck'* — Express Diagnostics Int'l Inc., USA).

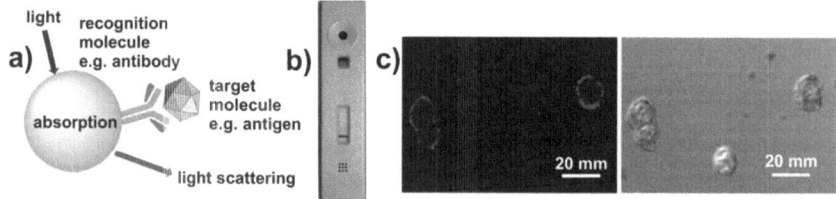

Figure 2.12. **Sensing of target molecules and imaging of cells with AuNP bioconjugates.** a) Scheme of sensing process. b) Commercial test stripe for the sensing of myocardial infarcts, based on immunoflow principle. © 2015 Novamed Israel. c) Specific detection of cancer cells by light scattering of AuNP bioconjugates (red coloration) which coupled to the cell membrane by cancer cell-specific antibodies. Reprinted with permission from Sokolov et al., copyright 2003 by SAGE Publications Ltd.[215]

The specific immunolabeling may further provide an accumulation of AuNPs in a distinct area of interest, e.g. on target cells via cell-specific membrane markers (cluster of differentiation (CD) molecules). With this accumulation, adequate contrast can be provided for the optical imaging of tissue and cells via the light scattering detection of AuNPs both *in vitro* and *in vivo*.[215;229-233] In this context, Sokolov et al. visualized cancer cells that were over-expressing the epidermal growth factor receptor (EGFR) by CLSM after incubation of cells with anti-EGFR antibody-coupled AuNPs (**Figure 2.12c**).[215]

In addition, the specifically targeted and AuNP-accumulated cells may be separated from a mixture by sorting methods such as flow cytometry. Either the light scattering property of AuNPs is used for contrast differentiation of the cells [234;235] or fluorophore-functionalized AuNP bioconjugates are applied for fluorescence-activated cell sorting (FACS).[236;237]

Photothermal cancer therapy

A common medical treatment for cancer is the local temperature increase of malignant tissue by hyperthermia therapy.[238] The restricted overheating in the range of 40 to 44 °C increases the blood flow and thereby enhances the efficiency of chemotherapy. Moreover, the heat is confined to the tumor because dissipation into the surrounding tissue cannot be established due to a simple and compact vascularization. This causes nutrient depletion, which leads to a reduced metabolism and reparability. That in turn, might induce a cellular dieback.[238] If very high temperatures > 50 °C are applied, an ablative destruction of tumor cells could be achieved; however this method involves the insertion of a cannula directly into the tumor.[239]

There are several approaches to cancer hyperthermia, such as microwave, radiowave or ultrasound treatment, the adoption of magnetic nanoparticles and the laser-induced thermotherapy. Among those approaches, the number of publications on hyperthermia therapy with magnetic nanoparticles has increased significantly in the past decades.[240-243] The method requires the oscillation of the particles in an alternating electric field with local heat development.[244] Thus, tumor cells that contain the magnetic nanoparticles are thermally destroyed.

However recently, biofunctionalized gold nanoparticles have become an area of interest and several workgroups have utilized them with heat therapy for cancer.[245-247;5;96]

This approach is based on the plasmon-coupled heat release into the environment using light irradiation of the gold nanoparticles and is also known as plasmonic photodynamic therapy (PPTT).[249] Due to heating zone development and the explosive evaporation of water around the particles, emerging bubbles cause the formation of irreversible pores in the tumor cell membrane (**Figure 2.13a**).

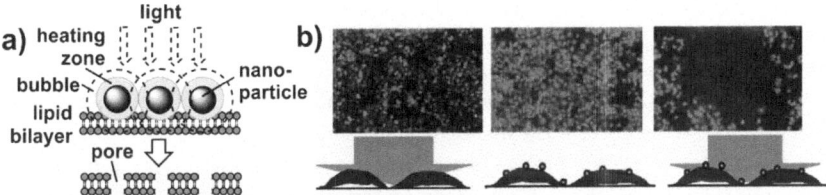

Figure 2.13. **Photothermal effect of AuNP bioconjugates on malignant cells.** a) Scheme of membrane pore formation principle upon light irradiation of membrane-attached gold nanoparticle bioconjugates. b) Photographies (upper row) and schematic illustration (lower row) of cancer cell destruction by photothermal treatment. Green coloration = cancer cells, yellow dots = AuNP bioconjugates. Reprinted with permission from Loo et al., copyright 2004 by SAGE Publishing Ltd.[248]

With this photothermal effect, malignant tissue is destroyed locally when AuNPs that are coated with tumor-specific marker bind to the cancer cells. This was shown in a successful study by Loo et al. (**Figure 2.13b**).[248] When attempting to achieve a high tissue penetration and less off-target absorption, near-infrared lasers are often applied.[249]

Moreover, the transient permeability of the cell membrane after laser heating enables the cellular uptake of extracellular molecules, which may support cancer treatment.[250;251]

Delivery and switchable release of effector molecules

To achieve optimal effectivity of pharmaceutics and to perform gene silencing using short-interfering RNA (siRNA), large numbers of the effective molecules must be delivered to a defined intracellular area of interest that contains specific CD-bearing cells such as the nucleus, while an overall *in vivo* distribution is hindered.

A variety of methods for the intracellular delivery of effector molecules has been developed for this intent, including the adoption of viral vectors and the transient pore formation that was mentioned previously.[250;252;253] However, most of these methods lack specificity and suffer from insufficient delivery efficiency and low throughput.

Regarding solid tumors, the passively-targeted accumulation of gold nanoparticles at their proliferating sites with the enhanced permeation and retention (EPR) effect has already been explained.[254;255] Furthermore, AuNP functionalization with specific antibodies allows the active targeting of CD-bearing cells, which was also discussed previously. Thus, the simultaneous conjugation of effector molecules and antibodies to AuNPs could enable a directed delivery.

In the last decade, gold nanoparticles have actually been determined to be efficient transport vehicles for the intracellular delivery of effector molecules such as under-expressed substrates,[256] oligonucleotides and siRNA[257] or drugs.[258] In addition, the receptor-specific delivery using bivalent gold nanoparticle bioconjugates functionalized with a pharmaceutic and an aptamer was also presented.[259]

However, to obtain maximal effectivity it may be necessary for the cargo to be separated from the AuNP transport vehicles once the area of interest is reached, especially if gene modifications within the condensed nucleus are the target.

For this intent, the delivery process may be combined with the switchable, light-induced release of cargo at the place of destination by separating a photocleavable linker or by melting gold nanoparticles via irradiation (**Figure 2.14a**).[210;260]

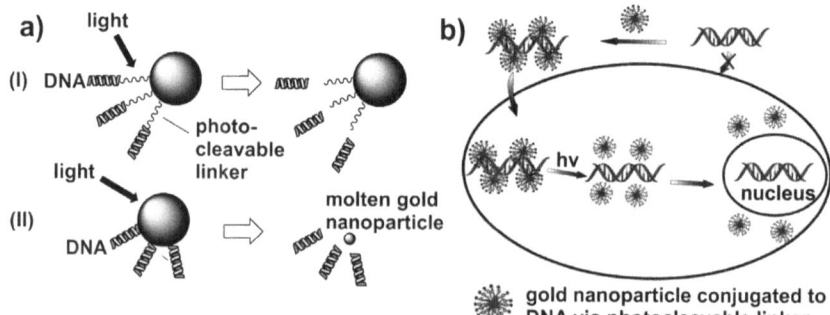

Figure 2.14. Directed delivery of effector molecules using AuNPs as transport vehicles. a) Scheme of light-induced ligand separation from AuNP bioconjugates via photo-cleavage of a linker (**I**) or by particle melting (**II**). **b)** Non-invasive delivery of DNA into the cell nucleus by AuNP bioconjugates. Adapted from Han et al.., copyright 2007 by Springer Science + Business Media.[210]

In this context, Han et al. presented the non-invasive delivery of DNA into cell nuclei by ensuring that gold nanoparticles were safely transported through the cell membrane and by using a photocleavable linker on the AuNPs for light-induced separation of the ligand molecules (**Figure 2.14b**).[261] Furthermore, Poon et al. showed the controlled denaturing of Au-S bonds due to photothermal dehybridization in response to pulsed laser irradiation.[262]

Actually, a selective release of cargo from gold nanorods has previously been demonstrated by Wijaya et al.[260] In their experiments, nanorods with different aspect ratios were selectively melted by irradiation with adequate laser wavelengths, which led to a controlled delivery of the DNA ligands.[260]

In summary, AuNP bioconjugates are widely used in biomedical and reproductive research and their application prospects are growing rapidly. The combination of selective targeting/delivery, therapy and sensing/diagnosis has established a new field of research termed *theranostics* which are smoothing the way for the personalized medicine of the future.[263] However, in order to achieve the proper combination of specific functionalities, the properties of gold nanoparticles must be modified for each individual application and thereby specific design criteria need to be considered.

2.6. Design criteria of gold nanoparticles

How to achieve biological functionality for specific demands

Discussing gold nanoparticles, one may classify their properties as being intrinsic or caused by additional functionalization. Particles' intrinsic properties cover primary particle size, shape and charge while additional properties may be configured via conjugation with functional molecules. Due to a high affinity of sulphur to gold surfaces and a strong thiol-gold bonding, thiolated biomolecules attach nearly covalently to the particles, resulting in stable gold conjugates. Since specifications of AuNP bioconjugates need to be adjusted to meet the individual demands of biomedical requests, 6 regulative design criteria may be defined. Biocompatibility **(I)** of conjugates is the main facet for applications regarding biomedical science, but also aspects covering selective receptor coupling **(II)**, cellular penetration **(III)**, effect initiation **(IV)**, imaging **(V)** and *in vivo* resistance **(VI)** must be matched with regard to the individual objective (**Figure 2.15a**).

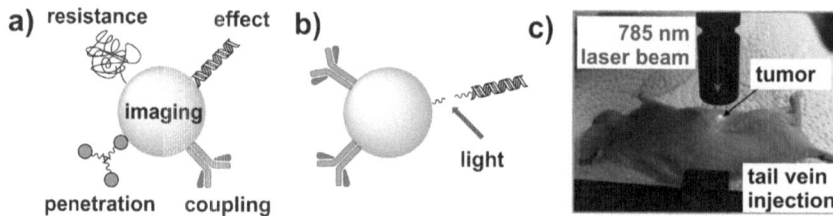

Figure 2.15. Customized AuNP design and effectivity. a) Scheme of an individual particle design with four specific attributes. **b)** Scheme of effect initiation with a potential stimulus, such as e.g. light irradiation. **c)** Tumor detection and treatment by laser excitation of accumulated AuNP bioconjugates *in vivo*. Adapted with permission from Qian et al., copyright 2008 by the Nature Publishing Group.[232]

For instance, the *in vitro* detection of tumor cells requires biocompatibility, nanoparticle coupling to a tumor specific receptor and bio-imaging of particles, while the photothermal therapy of cancer depends on additional *in vivo* resistance and cellular penetration. In this context, the customized design of gold nanoparticle bioconjugates is not trivial and should be carefully considered, with special regard to the multivalent functionalization of particles.[264] Thus, with controlled coupling of distinct ligands to a single gold nanoparticle, all three required design criteria for tumor cell detection can be matched with a single multivalent nanobioconjugate.

In addition to a customized design, a second aspect regarding the functionality of gold nanoparticle bioconjugates involves the utilization of switchable and mainly light-induced stimuli (**Figure 2.15b**). With external irradiation of intracellular particles, effects such as the aforementioned ligand separation[261] or particle aggregation[265] are triggered. Particle aggregation may be necessary to induce a higher extinction cross section of nanoparticles for detection or therapeutic issues within the NIR therapeutic window at the accumulation area (**Figure 2.15c**).

Thus, the tunable structure-function relationship of gold nanoparticle bioconjugates as well as their strong ligand binding and ability to carry diverse functional classes simultaneously, characterize them as perfect candidates for biomedical research applications.
To review the design criteria of gold nanoparticles, effect initiation (**IV**) may be induced by stimulating the plasmon-coupling for photothermal therapy or with the conjugation with effector molecules. The conjugates are mainly applied for the delivery of pharmaceutical drugs to inflammation areas or cancer cells. These aim on a local, stimuli-induced activation of ligands at the area of interest in order to avoid full-body medication.[259;258;266] Whereas, a second issue implies the transport of regulative moieties like oligonucleotides,[218;267;39] siRNA[257] or glutathione[256] for therapeutic applications on the gene and protein levels. With regard to receptor coupling (**II**), recognition molecules such as aptamers[268;269] or antibodies[270;36;271;221] are generally attached to gold nanoparticles to

achieve specific binding of the nanoconjugates to targeted cells or intracellular fragments. With regard to *in vivo* resistance (**VI**), masking molecules such as poly(ethylene) glycol[118] or albumin[272] are often applied to camouflage nanoparticles from immune system recognition and inflammation reaction.

However, these described design criteria mainly depend on conjugated moiety, whereas imaging (**V**), penetration (**III**) and biocompatibility (**I**) criteria of gold nanoparticle conjugates are more often connected with the particles' intrinsic characteristics and will not be discussed in detail within this thesis.

In summary, the appropriate design of nanobioconjugates is crucial for their biological functionality and highly dependent on their task and field of application. Six regulative criteria are defined and should be adjusted carefully for each bioconjugate, with respect to the biomedical request.

2.7. Fabrication of gold nanoparticles and AuNP bioconjugates

2.7.1. Conventional fabrication methods

Nanoparticles are generally defined as spheres between 1 and 100 nm in diameter.[273] Their common states of appearance are solid powders, gaseous aerosols and colloidal dispersions in water or organic solvents. Among those, colloids are often preferred for research due to their safe and stable handling form, which will reduce the risk of particle inhalation.

Focusing on the fabrication of gold nanoparticles, a multiplicity of fabrication methods have been established in the last decades, which are typically grouped into chemical (bottom-up, precursor-based) and mechanical/physical (top-down, precursor-free) synthesis approaches. In addition, there are also some exotic generation techniques e.g. the synthesis in plants and yeast.[274;275]

The mechanical top-down generation of AuNPs may be performed by grinding gold powder to a nanoscaled dimension[276] using e.g. a planetary ball mill. Although a high volume may be processed at once, the drawbacks of this approach are a high polydispersity of nanoparticle sizes, a limitation of minimum size, extremely long grinding times of days to weeks and contaminations that arise by abrasion of grinding gears. In general, the grinding method is more common for the size-reduction of e.g. carbonate nanoparticles than for AuNPs.[277]

Thermolysis has been used as a top-down physical method to produce alkyl-group passivated AuNPs.[278;279] However, heat-treated AuNPs often form 2D superlattices which may limit their biological application.[280]

The most economic method for AuNP generation is the chemical bottom-up synthesis, which can be carried out in a solid, liquid or gaseous state, implying the configuration of structures based on an atomic level. In addition to the chemical fabrication of AuNPs using microemulsions,[281] (copolymer) micelles[282] and seed growth,[283] the most common chemical technique is the wet-chemistry chemical reduction method (CRM), involving the nucleation, growth and agglomeration of atoms into nanoclusters. CRM was originally discussed in 1857 by Michael Faraday, who prepared gold hydrosols by reducing an aqueous solution of chloroaurate with phosphorus dissolved in carbon disulfide.[60] During the next century, various standard protocols were developed.[284;37;38;285;286] In the general CRM principle, Au^{3+} ions of a gold salt such as chlorauric acid ($HAuCl_4$) are reduced with reduction agents such as sodium borohydride[37] or sodium citrate[38] to zero-valent gold atoms (nucleation) (**Figure 2.16** and **Figure 2.17a**).

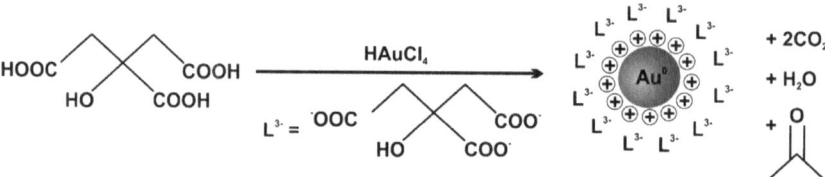

Figure 2.16. Reaction scheme of the Turkevich method for citrate-stabilized AuNP fabrication. Adapted from M. Noyong, copyright 2005 by Michael Noyong, dissertation.[287]

These gold atoms collide with other atoms or ions in the solution and grow into stable *seed nuclei*. As more and more of these nuclei form, the solution becomes supersaturated and the gold begins to precipitate in the form of (sub)nanometer particles. Further growth and agglomeration of the seed nuclei is then controlled with stabilizing agents such as sodium citrate or thiol ligands that protect the particle surface and allow for a precise and monodisperse nanoparticle size control (**Figure 2.17a**).

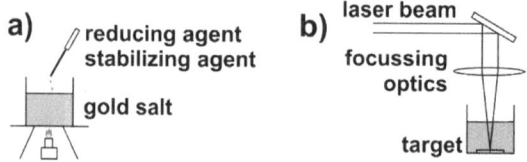

Figure 2.17. Chemical and physical synthesis approach for the generation of AuNPs. a) Scheme of chemical reduction method. **b)** Scheme of pulsed laser ablation in liquids process.

However, the biocompatibility of stabilizers is often restricted, e.g. as shown by Massich et al. that citrate may cause cell death by apoptosis.[121]
The functionalization of CRM-fabricated gold nanoparticles with biological active moieties is commonly facilitated by a successive substitution of the stabilizing agent with the

functional ligand molecules in a so-called *thiol-mediated ligand-exchange reaction*.[288-290] Unfortunately this procedure is known to be inefficient, due to a significantly high ligand excess and low degree of functionalization of the fabricated AuNP bioconjugates.[291]

In summary, the bottom-up chemical synthesis of AuNPs is the most economic fabrication route compared to conventional top-down or biological synthesis approaches. However, although a controlled monodisperse colloid may be gained by CRM, the AuNPs bear stabilizers on their surface which may induce cytotoxic effects and which limit the functionalization efficiency of exchange reaction. Thus, a method to fabricate stabilizer-free AuNPs is required.

2.7.2.   Pulsed laser ablation in liquids

The physical AuNP synthesis approach with pulsed laser ablation in liquids (PLAL) has become a reliable alternative to the conventional CRMs. The principle refers to the removal as well as the nucleation and growth of nanoparticulate material with a complex physico-chemical processes during laser irradiation of a solvent-immersed target.[40;292] Thus, it cannot be clearly assigned to either the top-down or the bottom-up approaches.

The basic experimental set-up was pioneered by Patil et al. in 1987, when they ablated an iron target in water to produce an iron oxide coating on a surface.[293] Six years later, Henglein applied that method for the fabrication of colloidal gold nanoparticles.[294] To date, many research groups have adopted this technique for AuNP generation (**Table SI 14**)) using lasers with femtosecond (fs), picosecond (ps) and nanosecond (ns) pulses at visible or near infrared wavelength. Among them, infrared wavelengths are preferred, since most solvents are transparent in this spectral regime and spherical metal nanoparticles usually do not feature extinction in the NIR regime.

The PLAL process offers specific advantages compared to conventional fabrication methods:
(I) The precursor-free environment allows the fabrication of highly pure (100 %) gold nanoparticles with surfaces free from any contaminations and without the requirement for purification in water and organic liquids. A selective overview of recent publications is found in **Table SI 14**. Because no chemicals are involved and no waste is produced, PLAL may be termed as a *clean* fabrication technology.
(II) There is no requirement for stabilizing additives because the PLAL-generated nanoparticles are usually charged, which results in strong particle repulsion and colloidal stability. Moreover, charge delivery by micromolar anions improves stability und monodispersity of fabricated nanoparticles.[295]

(III) The fabricated particles provide a high degree of occupational safety because they cannot cross the liquid-gaseous interface into the air. Thus, they are not inhalable and do not cause health risks compared to the ablation in air.[296]

(IV) The process is highly variable because there are only few limitations with the materials to be ablated. The aforementioned nanoparticle fabrication of metals,[297;298;292] alloys[299;300] and ceramics[301;302] adds to the variability of the process. Likewise, various liquids such as water, organic solvents and saline media that are transparent to the laser wavelength may be adopted, which leads to a significant number of possible material-solvent combinations.[303;41]

(V) No complex experimental arrangements (e.g. vacuum chambers and nanoparticle collectors) are required for PLAL. The general set-up (**Figure 2.17b**) consists of a pulsed laser system, a set of beam guidance and focusing optical components and a vessel that contains a solid material plate at the bottom, covered with a liquid layer of ablation medium. Furthermore, the process may be accomplished on the time-scale of seconds to minutes, while production up to milligram scale for target ablation and up to the gram scale for wire ablation has already been achieved.[302;304]

(VI) The functionalization of laser-generated gold nanoparticles with biomolecules can easily be achieved with *in situ* or *ex situ* conjugation. *In situ* conjugation involves the direct addition of the functionalization agent to the ablation medium prior to the laser process, thus enabling a simultaneous nanoparticle generation and functionalization in a single step.[39] In contrast, during *ex situ* conjugation, the functionalization agent is mixed with the particles in a second synthesis step.[305]

Despite these outstanding advantages, the current limitations of the PLAL process should not be neglected:

(I) Due to limitations of available ultrashort-pulsed laser systems regarding the combination of a high repetition rate with high pulse energy and ultrafast scanning speed, the yield of nanoparticles by target ablation is limited on the milligram scale and is on focus of current yield enhancement research.[301;302] Thus, commercial fabrication still lags behind the productivity of chemical synthesis.

(II) Monodisperse nanoparticle sizes in *MilliQ* water may not be gained by the PLAL process without the addition of stabilizers, biomolecules or inorganic salts. Instead of that, particle size distributions in the range from ten to one hundred nanometers are developed in ultrapure liquids. Likewise, the particle shape is confined to spheres, while other shapes have only been produced in exceptional cases.[306]

(III) In order to control the phase and structure of PLAL-produced nanoparticles, it is strongly recommended, that the manufacturer has a deep understanding of the ablation mechanism and the physicochemical processes.

Ablation mechanism – laser-matter-interaction

Until recently, there has been a lack of knowledge concerning the physical and chemical processes during PLAL and the nanoparticle formation, that may take place with mechanisms such as nucleation or target-ejection of hot drops and solid fragments.[307] While the action of plasma processing is at least widely acknowledged,[308;309;292;310;40;303] the more complex reaction mechanisms between plasma species and the liquid media are not yet fully understood and different particle generation models have been proposed. Amendola et al.[311] classified six temporal stages of ablation process (**Figure 2.18**):

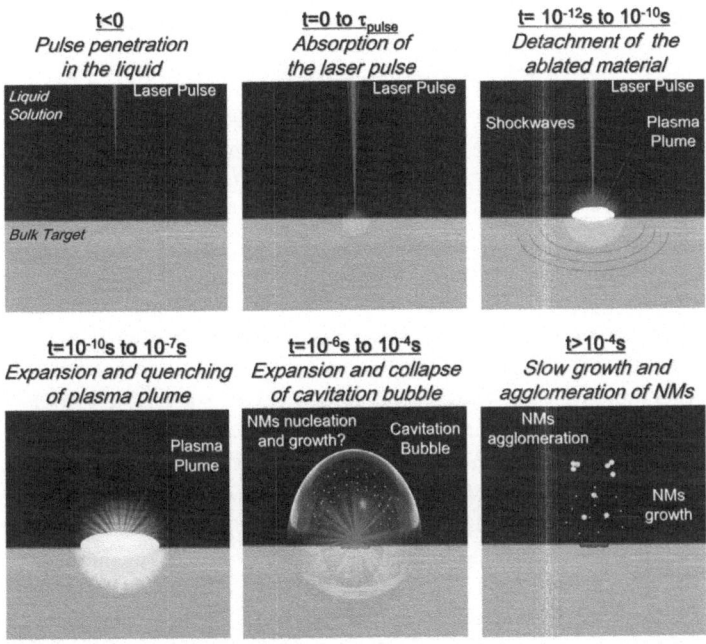

Figure 2.18. Illustration of PLAL process. Presented on the example of ns-PLAL with fluence above the ablation threshold and hypothesized time ranges. NM = nanomaterials. Reprinted with permission from Amendola et al., copyright 2013 by the Royal Society of Chemistry.[311]

(I) Pulse penetration in the liquid
In an ideal scenario, the liquid solution should be transparent at the adopted laser wavelength to enable delivery of laser energy to the target without solvent absorption. To avoid liquid breakdown effects when working with pulses of picoseconds or longer durations, defocused conditions should be followed. Whereas, to avoid nonlinear optical effects like self-focusing when working with femtosecond pulse durations, the liquid layer thickness must be reduced to maintain the defocusing conditions. In all cases, to avoid

nonlinear optical effects like multiphoton absorptions from the liquid, the critical fluence threshold (F_{th}) must not be exceeded.[312;311]

(II) Absorption of the laser pulse
Using nanosecond or long picosecond laser pulses for ablation, photonic energy of the laser beam is coupled to the electrons of the bulk material, which start to oscillate rapidly. The interplay of fast moving electrons with the stationary atoms through electron-phonon interactions transfers the energy and cools the electrons. This transfer causes energy vibrations in the lattice and consequently, phonon-phonon interactions that create lattice waves in the target material and enable thermal material removal by melting or thermal vaporization.[313;311]
On the contrary, laser pulses with high fluences and pulse durations shorter than the time needed to couple the electronic energy to the lattice (femtosecond or short picosecond) cannot initiate electron-phonon and phonon-phonon interactions in the target material. For this reason, there is no or very little thermal damage to the target surface.[313;311]

(III) Detachment of the ablated material
Using nanosecond or long picosecond pulses, the propagation of shockwaves in the material damages a thin layer of the target surface beyond the focal spot.[313;311] If laser irradiance is above the melting threshold of the target but below its vaporization threshold, surface melting of the target and formation of liquid droplets of target material occurs. These liquid droplets can be termed *molten globules*. On the contrary, if laser irradiance is higher than the vaporization threshold, vapors are generated from the target material at the solid-liquid interface. The front part of the laser pulse ionizes this material vapor to create hot laser plasma called *plume*. The plume absorbs and screens the last part of the laser pulse from the bulk material surface (plasma shielding). Thus, a comparatively smaller amount of energy is transferred to the target surface for materials removal.[313;311]
Conversely, for femtosecond or short picosecond pulses, the comparatively higher irradiance than the vaporization threshold causes the delivery of the maximum part of laser energy in a very short duration and the homogeneous, non-thermal, explosive ablation and fragmentation of material on the target surface. Furthermore, the laser pulse terminates before the energy is completely redistributed in the solid and no laser-plasma interaction will occur.[313;311]

(IV) Expansion and quenching of plasma plume
The plasma plume forms at or near the surface and propagates back up the laser beam. It is an expanding, thermodynamic state of high temperature, high pressure and high density.[314] During expansion, the plasma plume cools down and releases energy (heat) into the liquid solution. Thus, the laser-induced plasma (which contains metastable species from the target material) heats the neighboring liquid layer at the plasma-liquid interface to

temperatures that are greater than the boiling temperature of the liquid. This occurs in conditions of standard pressure and thus generates plasma of liquid species, which may be termed *plasma-induced plasma*.[313] Because the plasma-induced plasma is sandwiched between the expanding plasma plume and the liquid, it develops strong pressure (confinement) which results in an explosive ejection of metastable atomic or ionic species from the laser-induced plasma into the plasma-induced plasma.[313;311]

These species are quickly cooled and become clustered into elemental, embryonic noble metal particles such as Ag, Au or Pt, or they react with species of the liquid media to form e.g. oxide, nitride or carbide compound nanostructures of active metals such as Fe, Si, and Al.

(V) Expansion and collapse of cavitation bubble

Due to high pressure differences after the rapid expansion of the laser plume and because there is an energy release into the surrounding liquid, the formation of a vapor bubble termed *cavitation bubble* is initiated at the laser-target interface.[313] The cavitation bubble expands in the liquid with supersonic velocity up to a millimeter radius, while its temperature and internal pressure drop to a value below the surrounding liquid. Thereafter, the bubble collapses on a time scale of hundreds of microseconds with the emission of another shockwave and accompanied by an energy release that is large enough to cause a secondary ablation of the material.[311;313]

(VI) Slow growth and agglomeration of NPs

After shockwave generation, the system reaches physical and chemical steady state again. At this stage, the NPs can experience a secondary growth due to a coalescence with ablated clusters that are still in the solution and because there is an attachment of free atoms and ions.[311;313] These collisions occur as result of the diffusive mobility and thermodynamic instability of the cluster states and the particle growth finishes on the milliseconds to seconds scale when all of the surrounding clusters and atoms are consumed.[315] This secondary growth and eventually the ejection of molten globules were assumed to be the main reasons for the broad size distribution of PLAL-generated NPs.[316;298] However, recently Ibrahimkutty et al. proposed that even in the cavitation bubble two different particle species may be distinguished (**Figure 2.19**).[317] These species are primary particles of approximately 8 to 10 nm average diameter and secondary particles with 45 nm average diameter that result from collisions of the primary particles.

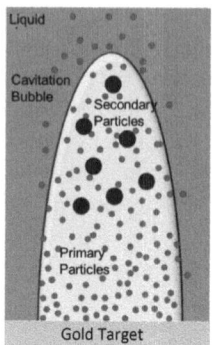

Figure 2.19. Illustration depicting two nanoparticle species found in the laser-induced cavitation bubble. Primary particles with a diameter of approximately 8 to 10 nm and secondary particles with an average diameter of 45 nm can be distinguished. Adapted with permission from Ibrahimkutty et al., copyright 2012 by AIP Publishing LLC.[317]

If the colloidal dispersion is not stable, agglomeration will begin, which results in particle precipitation on a timescale of minutes to days.[311;315] However, surfactants may be adopted to interact with the nanoparticles during condensation (or even within the cavitation bubble[295;318]) while preventing them from further coalescence and agglomeration through efficient stabilization and reduction of particle size distribution.[297;319]

2.8.   Parameters affecting NP formation during PLAL

Points of consideration

During the PLAL process, there are several criteria that may affect the particle formation regarding the ablation medium, the presence of stabilizing agents as well as diverse laser and process parameters.

For instance, the particle size distribution (PSD) of AuNPs is generally broad after ablation in *Milli-Q* water (size distribution range: 5–140 nm[69]), compared to ablation in organic solvents[320;321] such as n-alkanes, dimethyl sulfoxide, acetonitrile and tetrahydrofuran, featuring size distribution that range on average from 2 to 10 nm. Moreover, AuNPs that were fabricated in nonpolar, organic solvents are often subject to aggregation on a time scale of days or weeks.[321]

The stability of AuNPs is a crucial aspect of their biofunctionality and to prevent agglomeration by van der Waal's attraction, the particles need to be stabilized by either electrostatic or steric repulsion.[322] In detail, three types of stabilization may be distinguished (**Figure 2.20**).

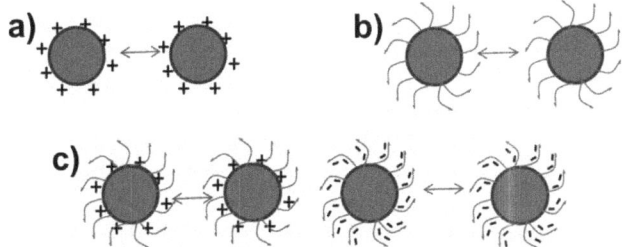

Figure 2.20. Three types of particle stabilization. a) Electrostatic stabilization. **b)** Steric stabilization. **c)** Electrosteric stabilization. Reprinted with permission from Liu et al., copyright 2014 by the Royal Society of Chemistry.[323]

If the NP surface is covered with charge carriers, such as ions, the equally charged particles will electrostatically repel each other in the solution. In 2004, Sylvestre et al. revealed with X-ray photoelectron spectroscopy, that the PLAL-fabricated AuNPs in *Milli-Q* water are not exclusively formed by metallic Au^0, but that they also contain Au^+ and Au^{3+} oxidation states that are mainly due to Au-O compounds.[324;325] They assumed those Au-O compounds are resulting from chemical reactions between the highly excited ejected Au atoms/ions or clusters and oxygen-containing species present in the laser-generated plasma or at the plasma/water interface. In addition, also carbonato complexes were detected ($Au-OCO_2^-$ and $Au-OCO_2H$), which are most likely resulting from the reaction of water-solved atmospheric CO_2 (as HCO_3^-) with the gold surface or after exposure of the dried sample to air.[324]

As a function of pH, either Au-OH or $Au-O^-$ is dominant; this yields a negatively charged surface with high zeta potential and electrostatic repulsion between the particles. This process is termed *electrostatic stability* (**Figure 2.20a**). Unfortunately, electrostatically stabilized particles are affected by high salt concentrations which screen the surface charges by reducing the electrical double layer thickness and result in particle agglomeration.[326;327]

In contrast to the AuNP fabrication in MilliQ water, the generation in organic solvents yields heterogeneous results. In polar solvents like acetone and alcohols, stable and non-aggregated AuNPs are obtained, due to adsorption of anions like enolates or alcoholates.[328] In contrast, in nonpolar solvents like n-hexane and toluene which are unable to give rise to anionic species, the resulting AuNPs are completely unstable and subject of aggregation.[328]

These particles need to be additionally stabilized with conjugation to ligands such as (charge-neutral) polymers or surfactants which prevent the close inter-particle contact with their long molecular chains. This effect is termed *steric stabilization* (**Figure 2.20b**). This effect was demonstrated by Compagnini et al. when they applied thiolated ligands for the steric stabilization of AuNPs in n-alkane, yielding long-term stable colloids without aggregation behavior.[329] Moreover, the structure of the nanoparticles was varied according to the chain length of the sodium alkyl sulfate molecules and the shape could be turned from spherical to elongated through the appropriate molecular mass of the alkane molecules.[329] Different from electrostatic stabilization, the steric stabilized particles are not affected by high salt concentrations.

A combination of electrostatic and steric stabilization is termed *electrosteric stabilization* and defines ligand-stabilized AuNPs that bear and overall conjugate charge (**Figure 2.20c**). It does not matter thereby, whether the charge is located on the particle surface or on a polyelectrolyte ligand. This is the most common stabilization type for PLAL-fabricated AuNPs, yielding an optimized colloid stability and enhanced zeta potential. In this context, Muto et al. demonstrated the conversion of AuNP surface charge as a function of ligand concentration using a cationic surfactant and resulting in positive zeta potential values.[77]

Stabilization agents and other ligands have also been applied to limit the growth of nanoparticles during the nucleation process.[319;330-332] In this context, Mafuné et al. found that the particle size distribution of laser-generated gold nanoparticles shift to smaller sizes with the addition of sodium dodecyl sulphate (SDS).[319] The adjustment of particle size distribution by ligand addition was further reported by Besner et al., and Kabashin et al. who described a significant narrowing of size distribution as a function of increasing ligand concentration (**Figure 2.21a**).[330;332]

Moreover several groups reported that AuNP fabrication by PLAL in electrolytes with low salinity yield highly stabilized particles with small PSD.[324;318;295;333] The oxidized particle surface reacts efficiently with anionic species such as OH-, Cl- and Br- in order to augment its net surface charge and to increase the electrostatic repulsion between particles.[318] The particle surface screening described earlier is not dominant in the applied micromolar concentration ranges and the anionic electrostatic stabilization is resulting from the Hofmeister effect[334] which defines stabilization and precipitation tendencies in proteins. Efficient size control was achieved with these means.[324;318;295;335]

Another possibility to narrow the particle size distribution of AuNPs is the re-irradiation of the colloid with a laser wavelength that is close to the SPR of gold (typically 532 nm laser). The size reduction can be a photothermal (ns, long ps pulses) and/or a photofragmentation (short ps, fs pulses) process, caused by multiphoton ionization. When there is a photothermal effect, the NP size is reduced as a consequence of their increased temperature up to the boiling point. Whereas, the reduction that occurs with photofragmentation has three steps; namely, the electron ejection form NP (leading to surface charging), the formation of a transient state in the NPs and the Coulomb explosion of surface atoms (fragmentation).[336]

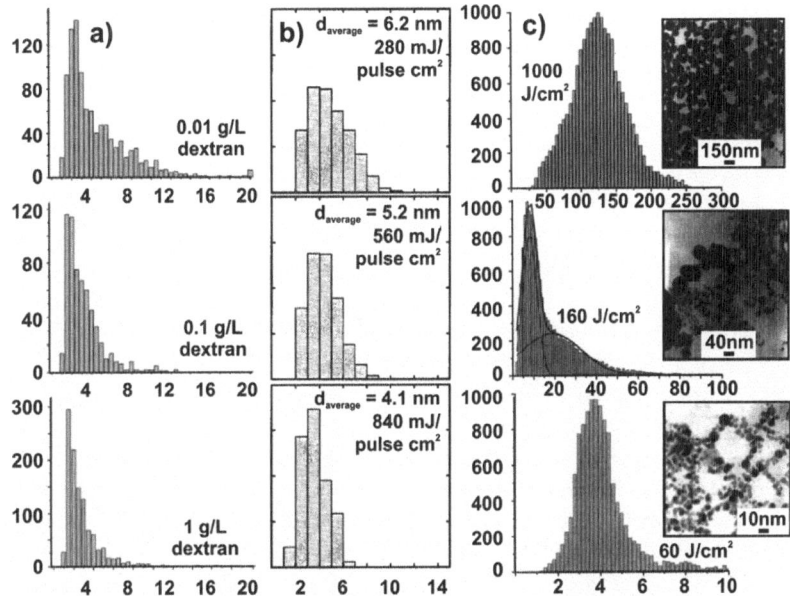

Figure 2.21. Examples on the modulation of PSD of AuNPs. a) PSD modulation as function of surfactant concentration. Adapted with permission from Besner et al., copyright 2009 by the American Chemical Society.[330] **b)** PSD modulation as function of re-irradiation fluence. Adapted with permission from Mafuné et al., copyright 2001 by the American Chemical Society.[316] **c)** PSD modulation as function of fabrication fluence during fs-PLAL with corresponding transmission electron micrographs.[298] Adapted with permission from Kabashin et al., copyright 2003 by AIP Publishing LLC.

The fragmentation efficiency is a function of laser fluence and Mafuné et al. presented that average 8 nm-sized AuNPs were fragmented to a 6.2, 5.2 and 4.1 nm average size for fluences of 280, 560 and 840 mJ pulse^{-1} cm^{-2}, respectively (**Figure 2.21b**).[316] Werner et al. further determined the threshold laser fluences of fragmentation to be 3.4 mJ cm^{-2} for fs laser induced fragmentation and found the process to be dominated by Coulomb explosion.[337]

The photofragmentation of *parent* nanoparticles causes gold atoms and small aggregates to be dispersed in solution, which can re-form smaller-sized *product* particles after condensation or which may be attracted by present NPs in the solution while growing them (coalescence).[338] The size, shape and phase of the nanoparticles might be changed during this process. Thereby, the coalescence rate increases with the concentration of small fragments until they are all consumed. Thus, a competition between fragmentation and coalescence takes place and the minimum particle diameter is only realized when the rate of fragmentation is equal to that of coalescence.[316]

Several studies have already been accomplished on a laser-induced photofragmentation effect on AuNPs after re-irradiation of the colloid with a focused or unfocused laser beam.[339-344;316;345;346;97;337]

Laser parameters such as pulse duration, wavelength, fluence and repetition rate may also influence the nanoparticle formation and should be considered carefully.

For instance, Riabinina et al found the ablation rate to be a function of pulse duration (40 fs–200 ps) with a maximum NP concentration at a pulse duration of 2 ps (using 5 mJ/pulse laser energy).[347]

Regarding wavelength dependency, Giorghetti et al. outlined an AuNP ablation with a 1064 nm wavelength that was more efficient than an ablation with a 532 nm wavelength. This was due to enhanced multiphoton absorption and photofragmentation of AuNPs using 532 nm wavelength.[342]

Kabashin et al. reported on the fluence-dependent particle size distribution of fs-PLAL generated AuNPs. While the thermal-free ablation occurs at low fluences (< 100 J cm^{-2}) and leads to very small and almost monodispersed colloids (3–10 nm), the plasma-induced ablation takes place at high fluences (< 100 J cm^{-2}) and results in much larger particle sizes and broader size distributions (5–70 nm; 25–250 nm) (**Figure 2.21c**).[298] Similar results were determined by Sobhan et al.[348] However, the low fluence regime is related to low production efficiency, which makes the higher fluence regime more attractive in practice. In this instance, the nanoparticle formation is more complex and may result in a two-component size distribution as reported by Sylvestre et al.[309]

Sobhan et al. and Ménendez-Manjón et al. discovered that a significant narrowing of NP size distribution occurred when the repetition rate was increased from 0.1 to 5 kHz.[349;344] Moreover, Sobhan et al. determined that narrowing was a function of irradiation time due to competition between ablation and photofragmentation.[349] Ménendez-Manjón also found that the narrowing was a function of liquid temperature that ranged from 283 to 353 K and which is directly related to a higher compressibility of water.[350] In another publication Ménendez-Manjón et al. showed that the flat-top (homogeneous) beam intensity profile yielded narrow, monomodal size distribution in the fluence range from 0.6 to 4.4 J cm^{-2}, while the Gaussian (inhomogeneous) beam intensity profile resulted in a bimodal size distribution. This was due to different thermalization pathways that formed during laser ablation.[351]

In summary, by tuning the fabrication parameters during the generation process, a flexible size adjustment of broad or narrowed size distributions with different distribution maxima can be provided for PLAL-generated AuNPs. In addition, the choice of ablation medium and the addition of stabilization agents will influence the formation of NPs and modify their size, (shape) and stability. By these means, a precise modulation of intrinsic particle parameters is achieved.

2.9.   Bioconjugation of AuNPs *in situ* during PLAL

Decoration and function

The term *bioconjugation* describes the linkage of biologically active molecules to nanoparticles by chemical or biological means, resulting in combinations of useful properties such as imaging and biological functionality.

Gold nanoparticles fabricated by CRM were conventionally bioconjugated using the thiol-mediated ligand exchange method as explained in **Chapter 2.7.1.** However, this procedure suffers from low surface coverages and it has difficulty equipping the AuNPs with more than one functional moiety due to affinity competitions between the different biomolecules to the gold surface.

Concerning the PLAL method, the linkage of NPs and biomolecules may occur after fabrication (*ex situ* bioconjugation). With simple mixing and a 24 h incubation of colloid and biomolecule solutions, the subsequent ligand coordination on the particle surface takes place either by thiol-gold bonding or electrostatic interactions (**Figure 2.22a**). In this manner, Salmaso et al. achieved the cellular internalization of laser-generated gold nanoparticles, conjugated to a thermosensitive polymer using the *ex situ* process.[117] However, the fabricated nanobioconjugates featured the typical broad PSD of laser-generated AuNPs, making them highly interesting for size screening experiments but inappropriate for size-limited applications

In another approach, Gamrad et al. fabricated size-quenched AuNPs (small PSD) with PLAL in micromolar salinity electrolyte (see **Chapter 2.8**) and conjugated the particles *ex situ* with CPPs in controlled ligand-per-particle ratios.[335] The presence of unbound ligands in the sample had to be considered, because no purification step was performed. However, the size quenching of nanoparticles during PLAL and the adoption of defined ligand-per-particle ratios, enabled the fabrication of highly controllable and reproducible nanobioconjugates.

Another approach was presented by Mafuné et al. in 2001.[319] They demonstrated that the addition of SDS surfactant to *Milli-Q* water prior to ablation leads to the formation of stable gold clusters with sizes that are smaller than those obtained in pure *Milli-Q* water, as a function of SDS concentration (see **Chapter 2.8**).[319]

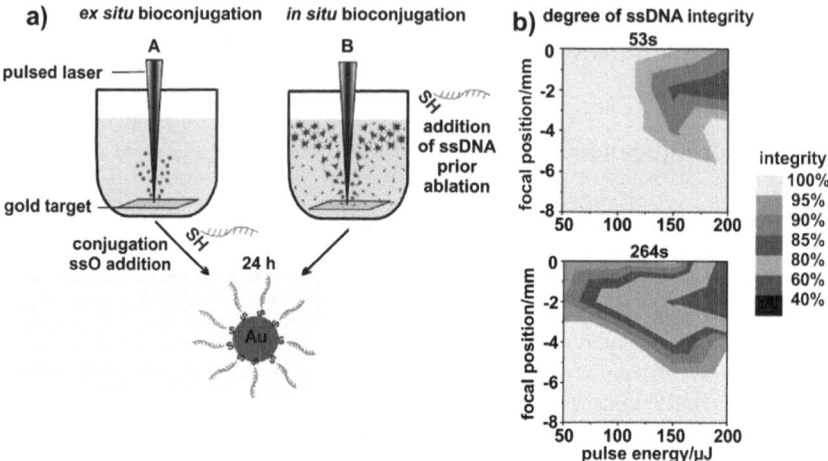

Figure 2.22. Bioconjugation methods of NPs and examples on the integrity of *in situ* bioconjugated AuNP bioconjugates. a) Scheme of *ex situ* bioconjugation (**A**) and *in situ* bioconjugation (**B**) methods with thiol-functionalized single-stranded DNA (ssDNA) that result in fabrication of functional AuNP bioconjugates. Reprinted with permission from Petersen et al., copyright 2009 by the American Chemical Society.[305] b) Integrity of ssDNA after 53 and 264 s laser ablation for different focal positions and pulse energies using a laser power of 0.5 W. Adapted with permission from Petersen et al., copyright 2009 by John Wiley and Sons.[39]

Obviously, the added surfactant competes with the particle growth processes during condensation (see **Chapter 2.7.2**) by decreasing the diffusion rate of small fragments and by covering the nanoparticles' surface while confining their size.[352] By this means, smaller NPs tend to be produced in a concentrated surfactant solution.[353] This observation is commonly termed *size quenching effect*[39] and because the coordination occurs during the fabrication process, the procedure was described as *in situ* bioconjugation (**Figure 2.22a**). The entire *in situ* process is highly sensitive to process parameters such as fluence, ablation time and focal position, since biomolecules are damaged easily by heat or physical degradation (**Figure 2.22b**).[39]

However, the AuNPs of Mafuné et al. were only stabilized by the SDS but did not feature any functionality due to conjugation. In 2009, the *in situ* conjugation of laser-generated AuNPs with biopolymers was demonstrated by Besner et al. for dextran-coated particles intended for biosensing of lectins.[330] In addition, the *in situ* conjugation of laser-generated AuNPs with functional biomolecules was enabled by Petersen et al. with single-stranded oligonucleotides.[39] Later, biological functionality proof of *in situ* functionalized gold-aptamer nanobioconjugates was demonstrated by Walter et al., by the efficient staining of prostate cancer tissue (**Figure 2.23a**).[269]

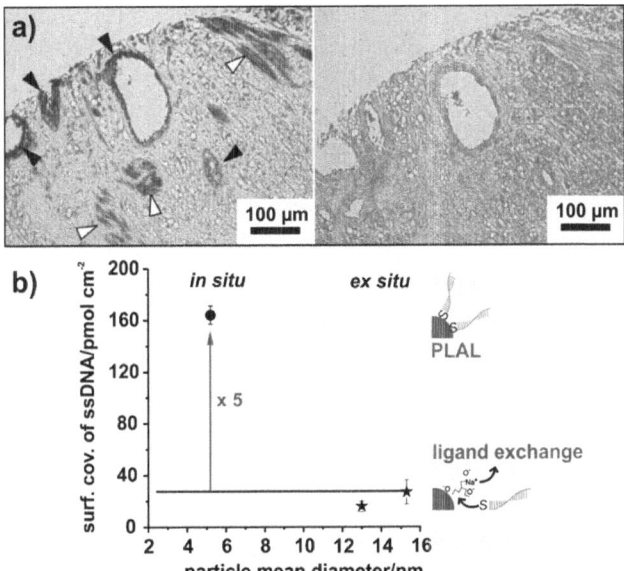

Figure 2.23. Functionality proof of *in situ* bioconjugated AuNP bioconjugates and example on obtained surface coverage values. a) Detection of PSMA in human prostate cancer tissue by immuno-histochemical staining using anti-PSMA aptamer-gold nanobioconjugates (left image) and miniStrep ap-tamer-gold nanobioconjugates as negative control (right image). Black arrows = specific staining, white arrows = unspecific staining. Reprinted with permission from Walter et al., copyright 2010 by Walter et al., licensee BioMed Central Ltd.[269] b) Comparison of surface coverages (attached ligands per nanoparti-cle) obtained by *in situ* bioconjugation of laser-generated AuNPs (circle) and by *ex situ* ligand exchange on chemically synthesized AuNPs (stars). Adapted with permission from Petersen et al., copyright 2009 by the American Chemical Society.[305]

Since that time, research concerning the *in situ* bioconjugation of laser-generated AuNPs has gained more attention due to an enhanced biomedical demand for AuNP bioconjugates and because conjugation is easily accomplished in a single-step process.

Moreover, Petersen et al. determined that the *in situ* functionalization degree of la-ser-generated AuNPs with thiolated biomolecules is up to 5 times higher (**Figure 2.23b**) than for conventional *ex situ* conjugation of CRM-fabricated AuNPs in which the stabili-zation agents on the particle surface are exchanged in a thermodynamic manner with the ligands.[305] This enhanced cargo load is highly attractive for delivery applications (see **Chapter 2.5**). Thus, AuNP *in situ* functionalization with a wide range of biomolecules including oligonucleotides,[39;354;355] aptamers,[269] proteins,[356;357;40] antibodies[358] and cell-penetrating peptides[197] has been reported to date.

When looking at biological functionality, the thermal impact on the biomolecules during *in situ* conjugation must be fully understood. Takeda et al. determined that a significant lysozyme degradation took place as function of laser power and ablation time.[356] Addressing this issue, Petersen et al. reported a detailed study on biomolecule integrity and AuNP bioconjugate yield during the fs-pulsed laser ablation process. In this study, they assessed the ssDNA integrity as function of laser and process parameters such as focal position, pulse energy and ablation time (**Figure 2.22b**).[359] Using this model, optimal parameters can easily be characterized to ensure high biomolecule integrity and AuNP bioconjugate yield.

Alternatively, based on the approach of AuNP generation in liquid flow,[360] the *in situ* conjugation of AuNPs with fluorophore-labelled oligonucleotides was executed by Sajti et al. in a flow system.[354] Due to the flow-associated removal of molecules from the irradiation zone, they found a significant decrease in biomolecule degradation by a factor of 4 even at high pulse energies. These energies facilitate enhanced ablation efficiency and nanoparticle yield regarding economic cost-effectiveness[354] (**Figure 2.24a**).

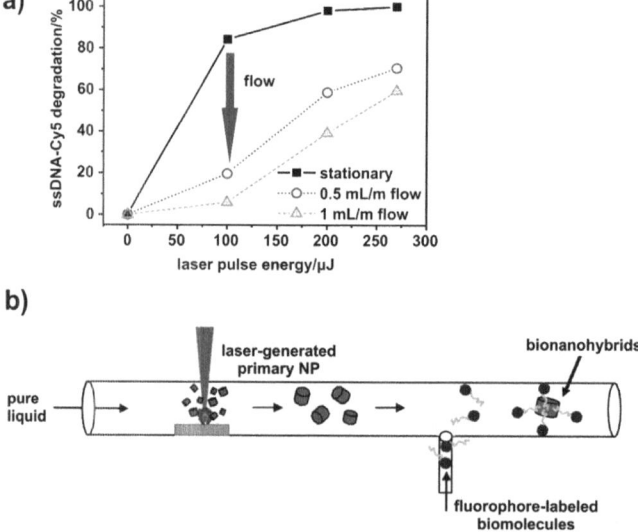

Figure 2.24. Bioconjugation of NPs in liquid flow and example on the biomolecule degradation during flow conjugation. a) The degree of biomolecule degradation in stationary liquid (black solid line) and liquid flow (0.5 mL/m flow: red dotted line, 1 mL/m flow: green dashed line) as function of laser pulse energy. Adapted with permission from Sajti et al., copyright 2010 by Springer Publishing Group.[354] **b)** Time-delayed, fast *ex situ* biomolecule conjugation to NPs in a liquid flow system. Adapted with permission from Sajti et al., copyright 2011 by the American Chemical Society.[361]

In a continuing study, biomolecule degradation was completely avoided with the delayed addition of ligands using a fast *ex situ* bioconjugation method in liquid flow (**Figure 2.24b**).[361]

In summary, PLAL-generated AuNPs may be functionalized with biomolecules during laser fabrication in a single-step, *in situ* process. The resulting AuNP bioconjugates feature higher surface coverage values than obtained by the chemical ligand exchange method. In addition, the AuNP bioconjugates show overall biological functionality if the biomolecule degradation is avoided by implementing adjusted PLAL parameters or adopting a liquid flow system.

2.10. Parameters affecting nanobioconjugate formation and function

Challenges and points of consideration

The bioconjugation of nanoparticles is a highly sensitive process, as various disfigurations of the nanobioconjugates may easily occur (**Figure 2.25a**).[362]
For instance, an incorrect ligand orientation can prevent the exposure of recognition epitopes and can therefore significantly reduce the biological functionality of the nanobioconjugates (**Figure 2.25a 1**). This is a main issue for the attachment of molecules which feature an active center such as antibodies, because their antigen-binding fraction (Fab) has to point outward from the NP surface to allow for the specific interaction and targeting of complementary ligand/analyte epitopes (**Figure 2.25b, Figure SI 1**).

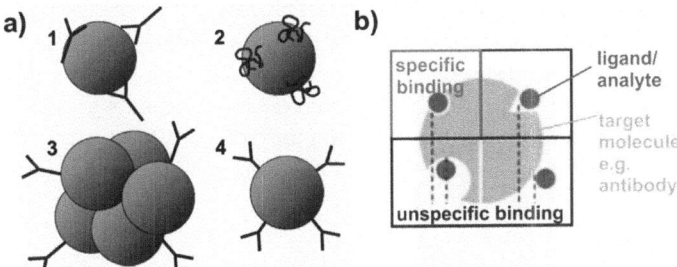

Figure 2.25. NP design requirements regarding ligand orientation and specificity. a) Different NP-antibody bioconjugate formations with **1** = wrong ligand orientation, **2** = ligand degradation, **3** = NP aggregation, **4** = ideal, functional NP bioconjugate. Reprinted with permission from Szymanski et al., copyright 2013 by Elsevier.[362] **b)** Scheme of specificity, illustrating different receptors binding options of a target molecules to a ligand/analyte. Adapted with permission from Yan et al., copyright 2012 by Macmillan Publishers Limited.[363]

The conventional adsorption of antibodies on AuNP surfaces with electrostatic interactions[229] generally yields a random ligand orientation that results in functional and non-functional molecules on the gold surface. High antibody concentrations are required for this approach and the attached ligands can easily be replaced by other molecules because of the weak electrostatic interactions. In a similar manner, also the EDC/NHS (1-Ethyl-3-(3-dimethylaminopropyl)carbodiimide/N-Hydroxysuccinimide) conjugation features the problem of orientation, since the terminal lysine amine groups used for attachment to a carboxyl group of a reaction partner are located at varying positions on the surface of the antibody.[364] Moreover, this method is generally used for nanoparticles that were chemically synthesized and equipped with a carboxyl function on their surface.

A covalent conjugation of antibodies can be achieved with different approaches such as the application of an orthopyridyl-disulfide-polyethylene(glycol)-N-hydroxysuccinimide (OPPS-PEG-NHS) linker or with an Avidin-Biotin system[365;366]. However, both linker systems are not specifically attaching to the non-targeting Fc region and thus they do not allow for a directed orientation of the antibodies on the AuNP surface.[366] Alternatively, the conjugation may be achieved with a heterobifunctional hydrazide-PEG-dithiol linker, which can specifically attach on the Fc portion of glycosylated antibodies while leaving the Fab portion unhindered.[270]

Another topic of disfiguration covers ligand degradation (e.g. by denaturating conditions) which can disable the biological functionality of the biomolecules (**Figure 2.25a 2**). Biomolecule degradation generally occurs with harsh environmental conditions such as inappropriate pH and high salt concentration in the solution or increased temperature that can denature the molecule irreversibly. However, disruption of the molecular structure (integrity reduction) as result of laser irradiation may also occur, especially if inappropriate laser parameters are used (see **Chapter 2.9**).

Finally, aggregation of nanoparticles can occur if the particles are insufficiently stabilized in solution or transferred from *MilliQ* water into high-concentrated salt solutions such as buffer media. This undesired particle aggregation can lead to *in vivo* issues like ineffective cellular uptake or renal clearing hindrance (**Figure 2.25a 3**; **Chapter 2.4**). Thus, ideal conjugation conditions are required to develop functional nanobioconjugates (**Figure 2.25a 4**).

Therefore, when considering PLAL and the *in situ* functionalization method, specific actions must be performed to avoid the formation of non-functional AuNP bioconjugates (**Figure 2.25a 1–3**).

- Correct ligand orientation is achieved by equipping the ligands with linkers which bind specifically to a molecule part at the opposing site of the active unit and which contain a thiol or a disulfide function for the covalent attachment to the gold surface.

- To avoid biomolecule degradation, either *ex situ* conjugation or the *in situ* conjugation has to be executed within a defined parameter window as presented in **Chapters 2.9** and **4.1.1.4.**
- NP aggregation is prevented by the PLAL process itself because partially oxidized and thus electrostatic stabilized particles are fabricated in *Milli-Q* water. However, if the conjugation must occur in non-polar solvents then the adoption of additional stabilizing agents might be required.

Another general issue that should always be considered is that due to local hydrophilic/hydrophobic patterns,[189] the nanoparticle surfaces will progressively and selectively adsorb biomolecules when they come into contact with complex biological fluids (e.g. cell culture media, blood), forming a biomolecule *corona*.[367;368] It is supposed, that the corona forms within 30 s of the nanoparticle's exposure to the culture medium and attaches to the particle surface irreversibly.[369] This corona is in many cases the part that interacts with biological systems and could affect or cover the functionality of former *in situ* attached ligands. Thus, it is recommended to have a stable (optimal: covalent) ligand attachment and complete coverage of the surface to reduce or control the corona formation. If this cannot be provided by the functional ligand itself (due to cost effectiveness) a PEGylation should be considered to cover the empty area on the particle surface. However, although surface modifications (such as PEGylation, or controlled presaturation e.g. with bovine serum albumin) block adsorptions spots and reduce the binding of additional biomolecules, some associations may still occur.[370;371] According to these recommendations, *in situ* bioconjugation of AuNPs is generally performed with thiolized ligands and surface saturation is enabled by PEGylation with a thiolized PEG that is smaller than the functional ligand.

In the previous chapters, CLSM and optimal cellular uptake were discussed. Studies have shown that the mean size of primary nanoparticles must exceed a threshold limit, which according to several groups of researchers, appears to be 50 nm in diameter.[93;82] However, also smaller particles sizes may also be applied if controlled aggregation is achieved afterwards.[372;197] In addition, a positive particle charge strongly enhances particle internalization, while negative zeta potential is an intended property for colloidal stability, biocompatibility and low cytotoxic effects during particle uptake.[105]

These requirements are provided by PLAL-generated gold nanoparticles that feature a negative zeta potential and a partially positive particle charge due to the fabrication process. The particle distribution is tunable during the generation process or with subsequent photofragmentation and a positive net charge can be achieved with additional bio-functionalization, e.g. with cationic peptides.[335]

In addition to these basic design aspects, the customized design of AuNP bioconjugates may further be required to match specific biomedical applications. This customization generally covers the attachment of particular biomolecules, the defined surface coverage values or the multi-valent functionalization of particles with more than one functional ligand to match the combined medical demands and current research trends.[264]

In 2009, Petersen et al. highlighted the flexible adjustment of AuNP surface coverage with oligonucleotide molecules and found a saturation function with increasing biomolecule concentration.[305] In detail, using different nanoparticle to ligand ratios, the surface coverage of 5 nm laser-generated AuNPs with oligonucleotides was tunable from 10 pmol cm^{-2} up to 140 pmol cm^{-2}, which aligned with previous calculations of coverage data perfectly. Furthermore, the attractive issue of multi-valent functionalization may be addressed with different procedures including ablation and *in situ* bioconjugation in biomolecule mixtures or the *in situ* bioconjugation with a first ligand followed by subsequent *ex situ* conjugation with a second ligand.

In summary, the basic aspects of AuNP bioconjugate design including correct ligand orientation and surface masking should be considered prior to fabrication and are provided by *in situ* bioconjugation, if specific actions are followed.
Thus, with features such as a fabrication-related size-flexibility, good biocompatibility, imaging ability and conjugation potential with various functional biomolecules, PLAL-generated AuNP bioconjugates enable an individual and tunable design that can satisfy specific biomedical requirements and that may be used by biologists for a multitude of prospects.

3. Experimentals

3.1. Experimental techniques and adopted laser systems

3.1.1. Pulsed laser ablation in liquids (PLAL)

Ligand-free nanoparticles and nanobioconjugates were generated using the PLAL technique with a focused laser beam that removed material from a water immersed target. Detailed information on the technique and an extensive discussion of the pros and cons can be found in **Chapter 2.7.2.**

For NP fabrication with PLAL two different Ti:Saphire femtosecond-pulsed laser systems (**a**: Spitfire Pro, Spectra Physics, **b**: Legend Elite, Coherent) and a Yb:YAG picosecond-pulsed laser system (Tru Micro 5250, Trumpf) were used. While the fs-pulsed system had a vertical setup, the ps-pulsed system benefited from a horizontal installation, which allowed for a constant water level and facilitated bubble removal from the target surface. By this means, yield and reproducibility were significantly increased. Furthermore, the Tru Micro 5050 system featured a variable pulse energy and repetition rate. The parameters of all laser systems have been summarized in **Table 3.1–3.3.**

Table 3.1. Laser parameters of the fs-laser system Spitfire Pro.

Laser Parameters fs-Spitfire Pro	
Wavelength	800 nm
Pulse Duration	120 fs
Beam Diameter	4 mm
Maximum Power	2.5 W
Maximum Pulse Energy	500 µJ
Maximum Repetition Rate	5 kHz

Table 3.2. Laser parameters of the ps-laser system Tru Micro 5250.

Laser Parameters ps-TruMicro 5250	
Wavelength	1030 nm
Pulse Duration	~ 7 ps
Maximum Power	50 W
Maximum Pulse Energy	250 µJ
Maximum Repetition Rate	200 kHz

Table 3.3. Laser parameters of the fs-laser system Legend Elite with Micra oscillator.

Laser Parameters fs-Legend Elite	
Wavelength	800 nm
Pulse Duration	100 fs
Beam Diameter	8 mm
Micra Oscillator	
Maximum Power	500 mW
Maximum Repetition Rate	80 MHz

3.1.2. Nanoparticle bioconjugation

As mentioned previously (**Chapter 2.7.2** and **2.9**), the bioconjugation of PLAL-generated NPs can either be accomplished *in situ* by ablation in a biomolecule solution or *ex situ* by mixing the generated NPs with biomolecules.

Pros and Cons

The *in situ* method is clearly confined by a distinct process-parameter window that must be adhered to in order to avoid biomolecule degradation. This window also limits the accessible nanobioconjugate yield (**Chapter 4.1.1**). In contrast, the *ex situ* approach can be implemented with an up-concentrated NP colloid, which yields a high nanobioconjugate concentration (**Chapter 4.1.1**).

For this thesis, both conjugation approaches were performed, depending on the experimental sub-goal. However in general the *in situ* method was favored.

3.2. Analytical methods

3.2.1. Ultraviolet-visible (UV-vis) spectrophotometry

UV-vis spectrophotometry, likewise termed *UV-vis spectroscopy* analyzes the interplay between electromagnetic radiation and matter and is routinely used in analytical chemistry for the qualitative and quantitative determination of analytes. In this process, a sample is irradiated with light which usually features 140–630 kJ mol^{-1} energy (wavelength range 900–190 nm).[373] The valence electrons of the sample may absorb the light of specific wavelength/energy while being excited (electronic transition) from a low-energy bonding orbital (single bond = σ, double/triple bond = π, ion bond pair = n) to a high-energy, empty, antibonding orbital (σ*, π*) (**Figure 3.1**).[373];[375]

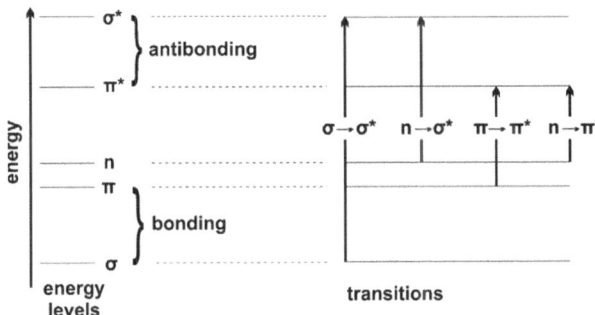

Figure 3.1. Electron transitions in UV-vis spectrophotometry.

Considering a single-bond molecule (e.g. molecular hydrogen, H-H) in the ground state, the σ orbital is termed the *highest occupied molecular orbital* (HOMO), while the antibonding σ* orbital is termed the *lowest unoccupied molecular orbital* (LUMO). After exposure to light of a distinct wavelength/energy, it is necessary to overcome the HOMO-LUMO energy gap (ΔE). In doing so, this wavelength is absorbed and the energy will transfer the electrons from HOMO to LUMO. However, the σ → σ* transitions are often too high in frequency and require UV light < 190 nm. Alternatively, for double-bonded molecules (e.g. ethene, H_2-C=C-H_2) the π orbital is the HOMO and the π* orbital is the LUMO. Because the π → π* energy gap is narrower than the σ → σ* gap (**Figure 3.1**), longer wavelengths may be absorbed and the electrons undergo a π → π* transition.[373] Thus most of the absorptions observed in UV-vis spectrophotometry involve π → π*, n → σ* and n → π* transitions only. In addition, the absorbed photons are not re-emitted as in fluorescence spectroscopy, but the energy is lost in a non-radiative manner as heat.

The amount of light that is absorbed with each wavelength is determined by measuring the light intensity before and after sample interaction (transmittance). It is then calculated by taking the $\log_{10}$ of the value of the absorbance at a given wavelength. The resulting spectrum is then presented as a graph of absorbance/a.u. versus wavelength/nm. A qualitative determination of analytes is enabled with the wavelength of maximal light absorbance and the absorption peak distribution that may be correlated to distinct types of bonds.[373]

In this regard, molecule conjugation can also be identified using UV-vis spectrophotometry when the extinction peak shifts to longer wavelengths. This occurs for instance, if the σ-bonded electrons of a single bond interact with the π bond electrons of a double bond and the energy of the excited state is reduced. Thus, each additional double bond adjusts the absorption maximum.

The conventional double-beam setup of a spectrophotometer consists of 5 components. Light initially stems from two sources: a hydrogen or deuterium lamp (UV) and a tungsten/halogen lamp (visible). That light passes a monochromator (1) and is divided by a prism (2) into its optical spectrum. With a silt diaphragm (3), the individual wavelengths are selected. Crossing a mirror, the light reaches a divisor (4) which produces two beams with identical power. One beam passes the reference sample containing the dilution solvent without analyte (blank), while the second beam passes the colloid. Two detectors (photomultiplier, photodiode) (5) collect the signals and the software subtracts the intensity differences and converts them into an extinction spectrum.

Depending on the material, nanoparticles with element-specific LSPR and with diameters that are > 2 nm may be characterized by UV-vis spectrophotometry. The LSPR results in an intense color for the dispersion and absorbance and scattering of the light also occurs, which is represented by an extinction peak in the UV-vis spectrum. For AuNPs, the LSPR peak is conventionally found at approximately 520 nm, while the exact peak position and band width depends on the NP size and their dispersity.[374] Further a (bio)functionalization may cause the LSPR peak shift by a local increase in the medium refractive index due to biomolecule presence or by a change in the free electron density of the AuNPs due to a strong surface coupling with the biomolecules (Drude model).[375;376]

Pros and Cons

In general, this method is very simple and spectra are collected within minutes, which provide immediate information about NP characteristics and functionalization status. However, the applicability is clearly confined by the detector sensitivity and sample concentration. For AuNPs, the detection limit is ~ 5 μg mL^{-1} and for ssDNA biomolecules it is approximately 0.15 μM. Therefore, as 1 μM biomolecule concentration was a typical concentration that was applied for conjugation experiments and because 100 % conjugation efficiency was rarely reached, the verification of biomolecule attachment to AuNPs by UV-vis spectrophotometry was not always possible. In those cases, the more sensitive infrared spectroscopy was preferred for analysis.

3.2.2.   Fluorescence spectroscopy

The technique of fluorescence spectroscopy is complementary to UV-vis spectrophotometry for the analysis of organic compounds in biochemical, medical and chemical research. While UV-vis spectrophotometry measures transitions from the ground electronic state (low energy) to the excited electronic state (high energy) = absorption, the fluorescence spectroscopy deals with vibrational relaxation from the excited state to the ground state within the timeframe of 10^{-5}–10^{-8} sec = fluorescence (**Figure 3.2**).[377]

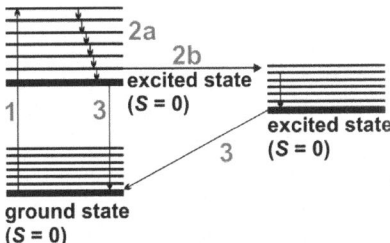

Figure 3.2. The principle of fluorescence. A photon is absorbed by a molecular system (excitation, **1**), then radiation-less relaxation occurs, either through vibrational states (vibrational relaxation, **2a**), through electronic states (internal conversion, **2b**), or both and then de-excites to the ground state takes place by emitting a photon that has a lower energy than the exciting photon (fluorescence, **3**).

Thereby, the wavelength (energy) of the emitted light depends on the energy gap between the ground state and the excited state.

The most striking example of fluorescence occurs when the absorbed photon is in the UV region of the spectrum and is thus invisible and the emitted light is in the visible region. This is termed *Stokes-Shift*.[377]
The excitation and emission spectrum are two characteristic spectra of fluorescent molecules. The excitation spectrum shows the fluorescence intensity as a function of the excitation wavelength at a constant emission wavelength while for emission spectra, the opposite occurs. The emission spectrum provides information for both qualitative and quantitative analysis of fluorescent analytes.

When combined with lifetime measurements dynamic and static quenching mechanisms can be analyzed. In addition, the energy transfer between two fluorophores (donor → acceptor) can be monitored using Förster fluorescence resonance energy transfer (FRET).[377] The efficiency of the process is a function of the molecules' relative distance.

A conventional spectrofluorometer is designed to measure fluorescence spectra, polarization and/or fluorescence lifetime and consists of 5 main components.
A light source (**1**) such as a mercury-vapor lamp, a xenon arc lamp, light-emitting diodes, laser diodes or lasers can be applied. A monochromator (or bandpass filter) (**2**) transmits the light of an adjustable wavelength with an adjustable tolerance to select a specific spectral band. The most common type utilizes a diffraction grating, which means that, collimated light illuminates a grating and exits with a different angle-depending on the wavelength. The emission monochromator can then be adjusted to select which wavelengths to transmit. In order to allow anisotropy measurements the addition of two polarization filters are necessary: one after the excitation monochromator or filter, and one before the

emission monochromator or filter. Photomultipliers (**5**) are used as high-sensitivity detectors to register the fluorescence signals at a 90° angle relative to the excitation light. This occurs in a wide range of wavelengths (200–900 nm) with high sensitivity. A photomultiplier (PMT) is capable of detecting individual photons. Each photoelectron results in a burst of approximately one million electrons, which can be detected as individual pulses. Hence, PMTs can be operated in photon-counting mode or as a current source for which the current is proportional to the light intensity. In the photon-counting mode, the individual anode pulses for each photon is detected and counted. In the current source (analog) mode, the individual pulses are averaged which provides an average anode current. The stability of a PMT in photon counting mode can be increased by operating the PMT at a high constant voltage.

Many fluorescence spectrometers are currently available as plate reader systems, which offer the ability to detect fluorescence and UV-vis absorption with a single device. Such a system was applied for the colorimetric assays.

Pros and Cons

The low detection limit and the high sensitivity of the method must be highlighted. The high specificity, which is due to the specific excitation (absorption) and emission (fluorescence) wavelengths of each fluorophore, is also significant. However, not all (bio)molecules feature a fluorescence effect. Thus, they must be equipped with a fluorophore or they cannot be analyzed with fluorescence spectroscopy. Moreover, fluorescence spectra are highly sensitive to the biochemical environment of the fluorophore and may change as function of concentration or pH modification.

3.2.3. Fourier-Transform Infrared (FT-IR) spectroscopy

Infrared (IR) spectroscopy is a physical method of analysis that is generally applied for the qualitative identification of unknown substances. This method is often chosen because it creates a unique molecular fingerprint of the analyzed sample. In contrast to UV-vis spectrophotometry and fluorescence spectroscopy where electronic transitions take place, the interaction of matter with infrared light excites molecules to undergo vibrational and rotational transitions that are characteristic for each functional organic group and that present themselves as transmission bands in the IR spectrum.[378]

The absorption of IR radiation is only possible for molecule bonds which feature a change in dipole moment with light absorption (vibrational transition). For instance, transitions with $C=O$ and $O-H$ bands are accompanied by a change in dipole moment and thus they absorb strongly in the IR region (IR active bonds), while C-C bonds feature IR inactive transitions.[378];[375] In general, the greater the polarity of a bond, the stronger is its IR absorption and vibration.

A molecule can vibrate in different vibrational modes. Usually diatomic molecules vibrate by stretching the bond between the atoms and polyatomic molecules, which can bend. Moreover, the stretching may be of a symmetric or antisymmetric nature. Exemplarily, all of the vibrational modes mentioned occur on the polar H_2O molecule (**Figure 3.3**).[378]

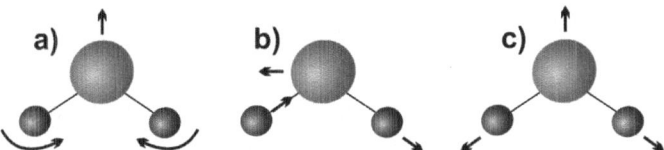

Figure 3.3. The three vibration modes of H_2O. a) The bending/scissoring mode. **b)** The antisymmetric stretching mode. **c)** The symmetric, stretching mode. Red spheres = oxygen atoms, blue spheres = hydrogen atoms.

A specific type of IR spectroscopy is the Fourier-Transform IR spectroscopy (FT-IR), which includes a Michelson interferometer that transforms the polyfrequent radiation of the light source into an interferogram.[378] Hence, a part of the light is reflected within the interferometer with a moveable mirror to generate an optical path difference between the beams. In doing so, interfering light passes the sample and reaches the detector. After the transformation of the detected optical signal into an electric signal, the software can develop a spectrum using mathematical Fourier transformation, presenting the transmission/a.u. versus wavenumber/cm^{-1}. Various regions of the spectrum are correlated to specific types of vibrational bands. For instance, the region from 1450 to 600 cm^{-1} is termed the *fingerprint region* and the region from 4000 to 1450 cm^{-1} is termed the *group frequency region*.[378]

Thus, a conventional FT-IR spectrophotometer consists of 5 components. A black body radiation source (**1**), a beam path with several mirrors (**2**), a Michelson interferometer that incorporates a beam splitter, a motor to constantly change the mirror distance and a HeNe laser that is used as a reference light source to determine the mirror positions (**3**), a black body detector that converts the photons into electric signals (**4**) and the computer software that transforms the electric signals (**5**).

For FT-IR analysis, a solid sample is required. Therefore, the colloid must be lyophilized by freeze-drying. This method is based on the physical process of sublimation and is extremely gentle on the sample. After freezing, a vacuum is established which transforms the ice crystals directly into the gaseous phase without occurrence of a fluid phase. Finally, a solid powder sample is received which must be inserted into a transparent support (usually potassium bromide). The sample is mixed with KBr and pestled thoroughly. Then, the mixture is compressed under high pressure into the form of a pellet. The pressure is required to make the KBr plastic, which becomes transparent with the cold flow.

Thereafter, the IR beam may pass the pellet and the molecule vibrations can be detected. The KBr vibrations are subtracted from the sample by recording a background (baseline) of a pure KBr pellet.

Pros and Cons

FT-IR spectroscopy is a sensitive technique that is used to identify substances in very low concentration that are much lower than it is possible to detect with UV-vis spectrophotometry. Moreover, the modification of functional groups due to molecule transformation and the formation of novel bonds during bioconjugation can be examined. However, the overall sample preparation using lyophilizaton and KBr pellet compression is not trivial and requires a long processing time. In addition, the operator must have a high level of experience. As interesting option, an FT-IR device may be equipped with an attenuated total reflection (ATR) unit, which allows measurements directly in liquid samples without special preparation.

3.2.4. (Surface-enhanced) Raman Spectroscopy (SERS)

Raman spectroscopy analyzes the inelastic scattering of light upon interaction with matter and is used in chemistry and pharmacy for the qualitative and quantitative characterization of substances.

When the sample is illuminated with monochromatic light, an incident photon hits a molecule and raises it for a short period, from a vibrational, ground electric state to a so-called *virtual state* that is located between the ground and the first excited electronic state.[379] Depending on how the molecule relaxes after excitation, several types of scattering can occur[380;379] (**Figure 3.4a**):

I)   The photon is scattered elastically with its original energy (**eq 3.1**) and keeps the same frequency/wavelength. In this case, the molecule relaxes back to its original ground state. This scattering is termed *Rayleigh scattering*.

$$E = h\nu$$

eq 3.1

E = energy of the photon, h = Planck's constant, v = frequency of the radiation.

II)   The photon is scattered inelastically with energy $E = h\nu - \Delta E$. The molecule relaxes back to a higher, more energetic vibrational state than it originally had and therefore the photon is shifted to a lower frequency and longer wavelengths so the system can remain balanced. This scattering is termed *Stokes scattering* and the shift is known as *Stokes shift* (**Figure 3.4b**).

III) If a molecule from a vibrationally excited state is excited to a virtual state, the photon is scattered superelastically with energy E = hv + ΔE. The molecule then relaxes back to a lower vibrational state than its initial state and therefore the photon is shifted to a higher frequency and shorter wavelengths so the system can remain balanced. This type of scattering is termed *anti-Stokes scattering* and the shift is known as *anti-Stokes shift* (**Figure 3.4b**).

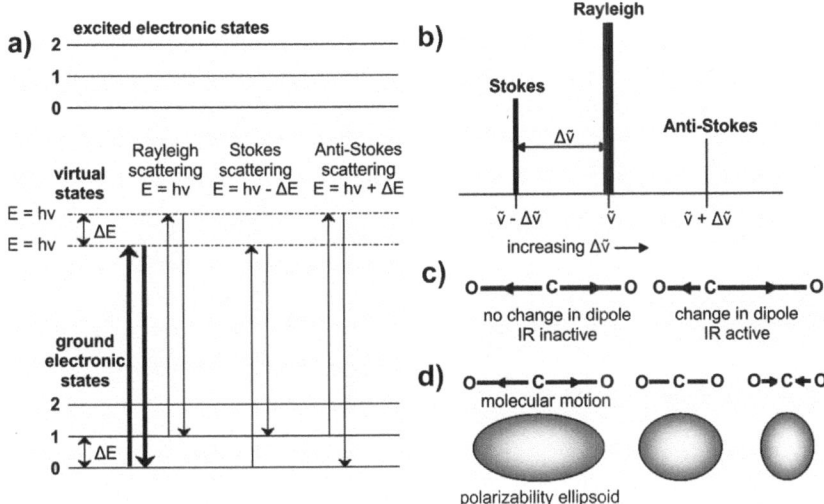

Figure 3.4. **The fundamentals of the SERS method. a)** The three types of scattering of a molecule after excitation. **b)** The shift and relative intensity of the Stokes and anti-Stokes scattering in comparison to the Rayleigh scattering. **c)** Required change in dipole moment for a molecule to be IR active. **d)** Required change in polarizability of a bond to be Raman active. Adapted from Barron et al., copyright June 2010 by Courtney Payne, Andrew R. Barron. Download for free at http://cnx.org/contents/73a1f8a3-32ba-4250-9277-ea639edafb80@1.[379]

Among these types scattering, the Rayleigh scattering is the most common transition and anti-Stokes is the least common (**Figure 3.4b**). However, the Stokes scattering is only used in Raman measurements and the energy change between the incident and scattered photons associated with the Stokes shift is typically measured as a change in wavenumber (cm⁻¹).[379] This is characteristic for a particular bond in its molecular structure (fingerprint). Thus, properties such as the crystallinity, the crystal orientation and the composition of a material/molecule can be determined with high specificity using Raman spectroscopy.

In contrast to IR spectroscopy, which requires a dipole moment or change in charge distribution to be associated with a vibrational mode (**Figure 3.4c**), the Raman measurement depends strongly on the polarizability of a bond in a static electric field.[379] The laser beam can be considered an oscillating electromagnetic wave, which induces an electric dipole moment upon interaction with a sample. Molecules are generally distorted in an electric field: the positive nuclei are attracted to the negative pole of the field, while the electrons are attracted towards the positive pole. This separation of charge causes a temporary induced electric dipole moment (polarization) and due to the deformation (**Figure 3.4d**), the molecules begin to oscillate and emit the light with characteristic frequency.

Because the Raman signal is inherently weak (less than 0.001 % of the source intensity)[379], the target material/molecule is often placed close to a metal surface (Ag or Au) to increase the Raman signal. If the surface plasmons of silver or gold are excited by the laser, this increases the electric fields that surround the metal yielding a significant signal enhancement. This technique is commonly termed *Surface-enhanced Raman spectroscopy* (SERS).[379]

A conventional Raman spectrometer consists of 5 components. The sample is illuminated with a laser beam (**1**). Electromagnetic radiation from the illuminated spot is collected with a lens (**2**) and sent through a monochromator (**3**). Elastic scattered radiation (Rayleigh scattering) is filtered out by a notch or band pass filter (**4**), while the rest of the collected light is dispersed onto a photomultiplier (**5**).

Pros and Cons

Using Raman spectroscopy, all states of matter can be measured, covering solid, liquid and gaseous samples and compared to IR spectroscopy, little or no sample preparation is required. Moreover, many molecules that are inactive or weak in IR spectroscopy will have intense Raman signals, which makes it a complementary technique and as the spectral range reaches below 400 cm^{-1}, both organic and inorganic species can be identified. However, similar to IR spectroscopy, a high level of experience is required of the operator. In addition, for the assignment of Raman signals to particular molecules the access on specific databases is required. The cost of instrumentation for this method is very high.

3.2.5.   X-ray photoelectron spectroscopy (XPS)

XPS, which is sometimes also referred to as *electron spectroscopy for chemical analysis* (ESCA) is a semi-quantitative technique that can measure the elemental composition as well as the chemical and electronic state of elements within a material, based on a three-step process. Thereby, the material is irradiated with a soft, low energy X-ray beam and the photons are absorbed by an atom leading to excitation/ionization of a photoelectron from the inner atomic orbital (photoelectric effect).[381;382] If their binding energy (BE) is lower than the X-ray energy, then they are transported to the material's surface and emitted from the material. The kinetic energy and the number of electrons that escape from the 10 nm surface layer of a material are measured in a high or ultrahigh vacuum and plotted against each other. Corresponding to the electron configuration within the atoms (e.g. 1s, 2s, 2p, 3s), the photoemission spectrum peaks are characteristic for each element and are related to the amount of the element in the sampling volume.[381]

XPS devices apply either a focused beam (20–500 μm diameter) of monochromatic aluminum Kα X-rays or a broad beam (10–30 mm diameter) of polychromatic aluminum or magnesium Kα X-rays. The BE of each emitted electron is determined by the equation from Ernest Rutherford (eq 3.2), using the measured emitted electron kinetic energies and the energy of X-rays with a particular wavelength.

$$E_{binding} = E_{photon}(h\nu) - E_{kinetic} - \phi$$

eq 3.2

$E_{binding}$ = BE of the electron, E_{photon} (hν) = energy of the X-ray photons being used, h = Planck's constant, ν = frequency of the radiation, $E_{kinetic}$ = the measured kinetic energy of electrons, Φ = adjustable instrumental correction factor.

Because the energy of X-ray photons is hν = 1253.6 eV (for Mg Kα) and 1486.6 eV (for Al Kα), the emitted photoelectrons will have kinetic energies in the range of ~ 0– 1250 or 0–1480 eV.[381] The BEs that are characteristic for each atom/orbital the electron derives from (e.g. Al 2s = 73 eV, Al 2p = 117 eV), depend on the chemical state of the sample and represent the strength of the interaction between electrons and the nucleus.[381]

Thus, the nomenclature Au4f implies that the electrons of the gold atom derive from the f-orbital. Due to quantum mechanics, none of the orbital levels except the s level give rise to a single photoemission peak, but they create a closely spaced doublet where the two possible states have different binding energies. This is commonly termed *spin orbit splitting*.[382]

The exact BE not only depends on the orbital from which photoemission occurs but also upon the formal oxidation state of the atom and the local chemical and physical environment and changes that give rise to small shifts in the peak positions.

Only electrons close to the surface can escape without energy loss, while electrons deeper in the surface may transfer a part of their kinetic energy to other electrons within the material through elastic/inelastic processes. These secondary electrons add to the signal background which must be subtracted from the spectrum prior to analysis. One common method is the *Shirley background subtraction*. Furthermore, the effective cross section and the work function/mean free path of the electrons within the material must be considered.

A conventional XPS device consists of 5 components. A 10 keV electron gun (**1**) produces X-rays from an Al or Mg target, a quartz crystal (**2**) monochromatizes the X-ray and focuses it on the sample, an ultra-high vacuum chamber (**3**) avoids surface contaminations with adsorbents from the air and enables the analysis of emitted electrons without interference from gas phase collisions, an electron energy analyzer (**4**) which is often a concentric hemispherical analyzer using an electric field between two hemispherical surfaces to disperse the electrons according to their kinetic energy, and an electron detector (**5**).

Pros and Cons

As XPS detects only the near-surface electrons, the analysis of deeper sample layers is not possible and thus the composition of layered materials cannot be estimated correctly. Furthermore, because an ultra-high vacuum is required, the device is expensive and the analysis and evaluation of data requires qualified technicians. Conversely, the elemental oxidation states of the material can be determined with this method, which is not trivial and is rarely measured with other techniques.

3.2.6.   Dynamic light scattering (DLS)

DLS, sometimes also referred to as Photon Correlation Spectroscopy (PCS) or Quasi-Elastic Light Scattering (QELS) measures the scattered light fluctuations from the colloidal particles and their diffusion rates using Brownian motion and relates these data to the hydrodynamic diameter of the particles with the Einstein-Stokes relation (**eq 3.3**).[383]

$$d_{hyd} = \frac{k\,T}{3\pi\,\eta\,D}$$

eq 3.3

d_{hyd} = hydrodynamic diameter, k = Boltzmann constant, T = absolute temperature, η = viscosity, D = diffusion coefficient.

Thereby, the hydrodynamic size defines the nanoparticle core diameter including the surrounding Helmholtz layer of covalently bound ions.

When a beam of light passes a colloidal dispersion, the particles scatter some part of the light in all directions. If the particles are small compared to the wavelength of the laser used (typically less than d = λ/10, which is around 60 nm for a HeNe laser), the scattering intensity from an illuminated particle will be equal in all directions (Rayleigh scattering), while for larger particles the intensity is angle-dependent (Mie scattering).[383]

If a monochromatic and coherent laser is applied as light source, then a time-dependent fluctuation in the scattering intensity is observed. These fluctuations arise from the random Brownian motion of the particles. Constructive (scattered light has the same phase as the illuminated light) and destructive interference (scattered light has a different phase as the illuminated light) within the illumination zone results in a pattern of bright and dark areas and gives rise to the intensity fluctuation. The analysis of the time dependence of the intensity fluctuation can therefore yield the diffusion coefficient of the particles from which the hydrodynamic particle diameter can be calculated.[383]

A typical DLS system is comprised of 6 main components. A light source (1) illuminates the sample containing cell (2), while a detector (3) measures the scattered light at 173°. The scattered light must be within a specific range, because if too much light is detected, the detector will become saturated. To overcome this phenomenon an attenuator (4) is used to reduce the intensity of the laser source and therefore the intensity of the scattering. On the contrary, it can also increase the laser intensity to detect small particles or samples of low concentration. It covers a transmission range from 0.0003 % to 100 %. The signal is passed to a digital processing board called a *correlator* (5) which compares the scattering intensity at successive time intervals to derive a rate at which the intensity varies. After a correlation function has been measured for different particle sizes this information is passed to the computer where the software (6) will determine an intensity distribution. This intensity distribution can then be transformed using the Mie theory into a volume and a number distribution.[383]

Pros and Cons

It is important to consider, that the hydrodynamic diameter is not the actual metal core diameter of the nanoparticles but that it includes the Helmholtz layer of covalently bound ions. Therefore, the ionic strength of the solvent can affect the measured data as it modifies the particle diffusion speed by changing the thickness of the electric double layer called the *Debye length* (K^{-1}). Due to this, a low conductivity medium will produce an extended double layer of ions around the particle, while reducing the diffusion speed and resulting in a larger, apparent hydrodynamic diameter. Conversely, higher permittivity media will suppress the electrical double layer and the measured hydrodynamic diameter. In addition, any change to the surface of a particle e.g. a biofunctionalization, may affect the diffusion speed and will correspondingly change the apparent size of the particle.

Another point of consideration is that DLS data are generally weighted by intensity or volume and have to be adapted to number-weighted values prior comparison to other data, especially to data derived from number-weighted electron microscopy.
Furthermore, the difficulty with measuring polydisperse colloids using DLS is explained with the Rayleigh approximation (**eq 3.4**).[383]

$$I \sim d^6 \quad and \quad I \sim \frac{1}{\lambda^4}$$

eq 3.4

I = intensity of scattered light, d = particle diameter, λ = laser wavelength.

In the Rayleigh approximation, the d^6 term tells us that a 50 nm particle will scatter 10^6 times as much light as a 5 nm particle. Thus, it is possible that the light from the larger particles will overlay the scattered light from the smaller particles. Therefore, it is difficult to measure a polydisperse mixture of 10–1000 nm-sized particles with DLS, because the contribution to the total light scattered by the small particles will be extremely small. However, because the measurement is performed within a few minutes, this technique is suitable when a quick overview of the particle size trend during photofragmentation (reduction of diameter) or bioconjugation (enhancement of diameter) is needed. Anyway, in serum-containing media a concentration-dependent serum protein adsorption has to be considered.[384]

3.2.7.   Zeta potential

The stability of a colloidal system can be explained with by the DVLO (Derjaguin-Verwey-Landau-Overbeek) theory.[385] This theory suggests that the stability depends on the sum of the van der Waals (vdW) attractive (V_A) and electrical double layer repulsive (V_R) (Born)-forces that exist between particles as they approach each other due to the Brownian motion.[385] An energy barrier that results from the electrostatic V_R prevents two particles from adhering together. However, if the particles collide with sufficient energy to overcome that barrier, the physical V_A between polar molecules will pull them into contact where they adhere strongly together. Therefore, if the particles have a sufficiently high repulsion, the dispersion will be stable. On the contrary, if a repulsion mechanism does not exist, then aggregation will occur.[383]

The electrical double layer (termed *Stern layer*)[386] is built up of an inner layer in which oppositely-charged counter-ions from the solvent are strongly bound to the particle surface with a linearly potential (termed *Helmholtz layer*)[387] and an outer, diffuse region (termed *Gouy-Champan layer*)[388;389]. In this diffuse region, are fewer counter-ions firmly associated with an exponential potential (**Figure 3.5a**).[383]

Between both layers there is a notional boundary which is termed *slipping plane*. When a particle moves, the ions within this boundary remain with the particle, while the outer ions are sheared. The potential that is measured at this boundary is the zeta potential (ζ), which is zero with an electroneutral Stern layer (charge equalization at the isoelectric point).[383] As a consequence, the particles are completely unstable. However, the zeta potential is also proportional to the local environment around the nanoparticles and the electrophoretic mobility of the particles and thus strongly correlates to the strength of particle surface charge.[383]

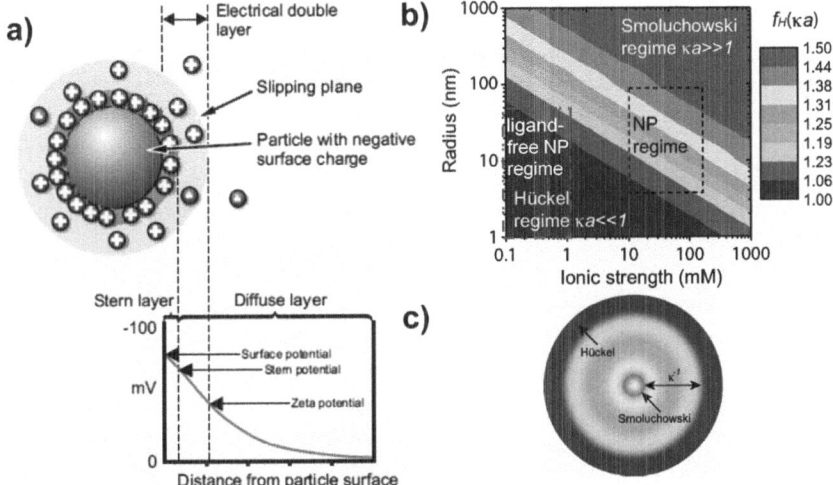

Figure 3.5. Fundamentals of zeta potential and the effects of the nanoenvironment. a) Schematic illustration of the ion distribution around a charged particle and the definition of the zeta potential. Reprinted from Zetasizer Nano Series – User Manual, copyright 2004 by Malvern Instruments Ltd.[383] **b)** Scheme depicting the Henry's factor $f_H(\kappa a)$ as function of solution ionic strength and geometrical nanoparticle radius which is changing with distance from the NP surface. Red coloration = Smoluchowski regime [$f_H(\kappa a) = 3/2$], blue coloration = Hückel regime [$f_H(\kappa a) = 1$], black dash-lined box is identifying particle radii and ionic strengths that are predominant for nanoparticles fabricated by wet chemistry, red dash-dot-lined box is identifying particle radii and ionic strengths that are predominant for ligand-free nanoparticles. **c)** Illustration of the two-dimensional distribution of the Debye layer thickness κ^{-1} with coloration according to the accompanying Henry factor $f_H(\kappa_c)$ in **b)**. Red coloration = Smoluchowski regime, blue coloration = Hückel regime. Images **b)** and **c)** were adapted with permission from Donae et al., copyright 2012 by the American Chemical Society.[390;391] Adaptation was performed in the style of Pfeiffer et al. 2014.[390]

The thickness of the Gouy-Chapman diffuse layer is highly dependent on the ionic strength of the solution and is characterized by the Debye parameter (κ, whose reciprocal value κ^{-1} is termed the *Debye screening length*).[390] According to the DLVO theory, κ^{-1} is decreasing with increasing ionic strength of the solution. Thereby, the surface charge screening by ions is causing reduced electrostatic stability and nanoparticle agglomeration by van der Waal's attraction in consequence (see **Chapter 2.8**). Another factor influencing κ^{-1} is the geometrical radius of the nanoparticles and the correlation between these values, the electrophoretic mobility of nanoparticles (μ) and the zeta potential (ζ) is described by the Smoluchowski equation (for large nanoparticles in solvents with high ionic strengths: $\kappa a \gg 1$)[392] and the Hückel formula (for small nanoparticles and non-polar solvents with low ionic strengths: $\kappa a << 1$)[393]. Thereby, the Henry factor $f_H(\kappa a)$[394] is inversely proportional to the zeta potential and defining the intermediate regime between Smoluchowski and Hückel ($1 \leq f_H(\kappa a) \leq 3/2$), with up to 50 % higher ζ values at given mobility in the Hückel regime. The three regimes are presented on **Figure 3.5b** and the distribution of the Henry factor around a nanoparticle is symmetric under ideal conditions (**Figure 3.5c**). Nanoparticles that were synthesized by wet chemistry are typically ranging between the Smoluchowski and Hückel regime **Figure 3.5b**, right box)[391], while nanoparticles that were fabricated in *MilliQ* water, in particular by PLAL, are usually located in the Hückel regime (**Figure 3.5b**, left box)[41] and are therefore much more sensitive for local changes or fluctuations in the ionic environment as result of the larger Debye length κ^{-1}.[390]

When an electric field is applied across an electrolyte, charged particles suspended in the electrolyte are attracted towards an electrolyte with an opposite charge (electrophoresis). Viscous forces acting on the particles tend to oppose this movement, while shearing the diffuse ion layer. When equilibrium is reached between these two opposing forces, the particles move with constant velocity.[383] This velocity of a particle in an electric field is commonly referred to as its electrophoretic mobility and can be used to calculate the zeta potential according to an equation by Henry (**eq 3.5**)[394]:

$$\mu = \frac{2\varepsilon\zeta f(Ka)}{3\eta}$$

eq 3.5

μ = electrophoretic mobility, ε = dielectric constant of the solvent, ζ = zeta potential, $f(Ka)$ = Henry's function, η = viscosity of the solvent.

Henry's function generally has a value of 1.5 (Smoluchowski approximation) or 1.0 (Hückel approximation).

A zeta potential measurement system is comprised of 6 main components. A monochromatic, coherent laser is conventionally used as light source (**1**), which is split to provide an incident and a reference beam to measure the electrophoretic mobility of the particles using laser Doppler velocimetry (LDV). The incident beam passes the center of a sample cell (**2**) and the scattered light from the particles is detected at an angle of 13° by a detector (**3**). When an electric field is applied to the cell, any particles moving through the measurement volume will cause the intensity of detected light to fluctuate with a frequency proportional to the particle velocity. This information is first passed to a digital signal processor (**4**) and then to a computer where the software (**5**) produces a frequency spectrum from which the electrophoretic mobility and the zeta potential are calculated. The scattered light must be within a specific intensity range for detection, otherwise an attenuator (6) is used to adjust the signal intensity. To correct for any differences in the cell wall thickness and dispersant refraction, compensation optics are installed to maintain optimum alignment.

Pros and Cons

Particles with a zeta potential of ± 30 mV are considered stable (see **Table 3.20**).[383] However, if the particles have a density that is different from the dispersant, they may sediment anyway.

It should be always considered, that the zeta potential describes the potential at the slipping plane (electrokinetic charge) and is not synonymous with the surface-localized charge (surface charge) which is determined behind the slipping plane.[395]

The thickness of the double layer depends strongly upon the concentration of ions in the solution and can be calculated from the ionic strength of the medium. The higher the ionic strength, the more compressed the double layer becomes. The valency of the ions will also influence the double layer thickness. A trivalent ion will compress the double layer to a greater extent than a monovalent ion. Furthermore, the specific adsorption of charged molecules onto a particle surface, even at low concentrations, can have a dramatic effect on the zeta potential. Thus, interpretation of zeta potential values is a delicate challenge.[395]

3.2.8. Electron microscopy, energy-dispersive X-ray spectroscopy (EDXS) and high-angle annular dark-field imaging (HAADF)

Electron microscopy applies a beam of electrons to create an image of a specimen with a much higher magnification than a light microscope, which can provide information about the specimen size, morphology, composition and crystallography.

Using conventional light microscopy for specimen visualization, the resolution limit (RL), defined as the smallest distance between two distinguishable points, is quickly reached which allows for magnifications of approximately 2,000x and 200 nm resolution.[396]

$$RL = \frac{\lambda}{2n \, sin\beta}$$

eq 3.6

RL = resolution limit, λ = wavelength, n = refraction index, β = half opening angle of the objective, $2n \, sin\beta$ = numerical aperture (NA) ≈ 1.4.

According to **eq 3.6**, the resolution limit may be enhanced by increasing the refraction index or the half-opening angle of the objective.[396] The higher the opening angle of an objective, the higher the resolution. Assuming theoretically, that the highest possible angle is 180° and the refraction index of air is n = 1, then the resolution limit would be half of the adopted wavelength (e.g. λ = 500 nm, NA = 1.4, R = 0.5 μm). Therefore, higher resolution can only be achieved with the adoption of even smaller wavelengths as UV light. This demand resulted in the construction of the electron microscope, since electrons feature a wavelength of ~ 4 pm at an acceleration voltage of 100 kV, which is a factor of 10,000 smaller than UV light and allows for magnifications of up to 10,000,000X and a resolution of approximately 50 pm.

After a high voltage electron beam interacts with the atoms in a sample, the electrons may either transmit the sample without being scattered or they may undergo elastic or inelastic scattering.[396;397] During elastic scattering, the electrons are backscattered at the angle of incident and the kinetic energy and velocity remain constant. Whereas in the case of inelastic scattering, some electrons will collide with other electrons from the atom shells, which would result in the emission of X-rays and secondary electrons (SE) with reduced kinetic energy and the emission of backscattered electrons (BSE) at varied exit angles.[396] SEs are produced when the incident electron excites an electron in the sample and loses most of its energy in the process. The excited electron moves towards the surface undergoing elastic and inelastic collisions until it reaches the surface where it can escape if it still has sufficient energy. The production of SEs is very topography related. Due to their low energy only SEs that are very near the surface (< 10 nm) can escape from the sample and be examined. When the SEs are ejected from the atoms, the resulting electron vacancy is filled by an electron from a higher shell and an X-ray is emitted to balance the energy difference between the two electrons.[396] BSEs are beam electrons that are reflected or back-scattered out of the specimen. Their intensity is strongly related to the atomic number (Z) of the specimen, causing elements with high Z to appear brighter than elements with low Z.[396]

There are two main types of electron microscopy; transmission electron microscopy (TEM) and scanning electron microscopy (SEM).

In **TEM,** the electrons that are transmitted through the specimen are detected (**Figure 3.6a**).[397] The transmission of unscattered electrons is transversely proportional to the specimen thickness. Areas of the specimen that are thicker will have fewer transmitted electrons and appear darker, while the thinner areas will have more transmitted electrons and appear lighter. The scattered electrons are then transmitted through the remaining portions of the specimen. The image is magnified and focused on a fluorescent screen or it is detected by a CCD (charge-coupled device) camera.

In **SEM** the surface of a sample is scanned with an electron beam in a raster scan pattern (**Figure 3.6b**).[397] Usually, the SEs are detected using an Everhart-Hornley detector and the image delivers a three-dimensional copy of the specimen with a field of depth. How-

ever, in order to yield information about the distribution of different elements in the
sample, the Z-related BSEs should also be detected.

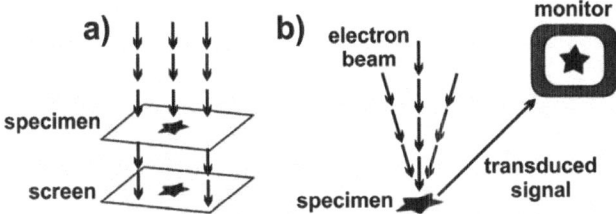

Figure 3.6. Fundamentals of transmission and scanning electron microscopy. a) Setup of
a transmission electron microscope. **b)** Setup of a scanning electron microscope. Reprinted from John J.
Bozzola, Lonnie Dee Russell, copyright 1999 by Jones & Bartlett Learning.[397]

A specific type of electron microscopy is the scanning transmission electron microscopy (**STEM**), with which a standard transmission electron microscope is modified to create a system that scans the highly focused electron beam across the sample.[396] The scanning method allows for connected analysis techniques such as energy dispersive X-ray
spectroscopy (EDXS), in which the high-resolution imaging data and the quantitative data
are recorded simultaneously.[397]
High-angle annular dark-field imaging (**HAADF**) is a method that is used for sample imaging in STEM. Thereby, an annular dark field detector is adopted to collect far more
scattered electrons than can pass through an objective aperture.[398] Compared to annular
dark-field imaging, the HAADF image can only be formed at a very high angle, it has in-
coherently scattered electrons and is highly Z-sensitive.[399]
However, STEM analysis is highly susceptible and exceptionally stable room environment
conditions are required with a limited amount of vibrations, electromagnetic/acoustic
waves and fluctuations in temperature.

EDXS is an analytical technique that is used for the elemental determination of a sample.
As explained previously, the high-energy electron beam is focused on the specimen to
stimulate the emission of characteristic X-rays from the sample. Because the energy difference between the two interacting electron shells and the atomic structure of the element is correlated with the energy of the emitted X-rays, specific elemental classification
can be made.[400] The number and energy of the emitted X-rays is therefore measured
with an energy-dispersive spectrometer. However, it should be considered, that many elements have overlapping peaks, which make multi-material analysis a complex issue.[400]

A conventional electron microscope consists of six components. An electron beam is
thermionically emitted from a hot electron gun (**1**) which is fitted with a tungsten filament
cathode. A Wehnelt cylinder and electromagnetic condenser lenses (**2**) focus and acceler-

ates the electron beam in one direction. The beam passes through the electron column (3) where an ultra-high vacuum is generated by a vacuum system (4) and is focused by several electrostatic and electromagnetic condenser lenses (5) onto the sample. After interaction with the sample detectors (6), the electrons are collected and converted into a signal that is sent to a screen.

For SEM and STEM, the electron beam is additionally passed through pairs of scanning coils or deflector plates in the electron column, which deflect the beam in x and y axes to raster the specimen before it reaches a SEs or BSEs detector.

Conventional transmission electron microscopes work at an acceleration voltage of > 100 kV and a specimen of 5–100 nm thickness can be seen. However, high-resolution transmission electron microscopes work at an acceleration voltage of up to 3 MV and allow for a resolution in the pm regime.

Conventional scanning electron microscopes work at an acceleration voltage of up to 40 kV and allow for a resolution that ranges from less than 1 nm to 20 nm.

Electron microscopy runs in a high vacuum to avoid interplay between the atoms and molecules from the air. Therefore, the sample must be prepared in such a way that it is vacuum-consistent. For TEM the sample needs to be cut in ultrathin sections (e.g. with a microtome), stained for contrast enhancement (e.g. with uranyl acetate), embedded into a polymer if indicated and fixed on a support grid. Whereas for SEM, the sample is fixed or dropped on a carbon-coated disc and the electric conductivity of samples with low Z is enhanced by using conductive heavy-metal silver or gold sputtering.

Pros and Cons

Electron microscopy allows for the visualization of a specimen far below the resolution limit of conventional light microscopes. However, resolution limit of SEM is around 5 nm, while high-resolution TEM can resolve specimen that are smaller than 1 nm in diameter. Moreover, analytical techniques such as EDXS or HAADF can be performed simultaneously if the device features the required equipment, and three-dimensional images can be obtained with SEM. However, electron microscopes are very expensive and should only be handled by an experienced user. The sample preparation is not trivial and the examination of biological samples is restricted due to the high vacuum and requires the use of special environmental chambers with low vacuum conditions, which also require a high level of handling experience.

Moreover, the evaluation of electron micrographs suffers from insufficient statistics.

3.2.9.   Fluorescence & confocal laser scanning microscopy (FluM, CLSM)

As a special type of optical microscopy, the FluM allows for the visual detection of fluorescence and is mainly implemented in life sciences for the specific imaging of cell compartments or components using fluorophores. It is sometimes also termed *epifluorescence microscopy*. In addition to the fluorescence-*in-situ*-hybridization (FISH) for the labeling of DNA and chromosomes, the antibody-targeted staining of proteins/molecules of interest (immunolabeling) is the most frequent application of FluM.[401]

The principle of FluM is the same as for fluorescence spectroscopy. The specimen is fist illuminated with light of a specific wavelength. Thereafter, the light is absorbed by the fluorophore and it emits light of longer wavelengths.[402] Using specific excitation/emission wavelength filters, the cellular distribution of the stained molecules of interest can be detected and recorded with a camera. However, if several, diverse molecules are stained simultaneously, each fluorophore must be analyzed singularly and a multi-color image will be overlaid by the single images with software post-treatment.

There are a wide range of biological fluorescent stains. The most common examples are nucleic acid stains such as DAPI or Hoechst which are excited by UV light and Cy3, Cy5, Texas Red, and several Alexa Fluors that are excited in the visible range and that are generally linked to antibodies.

A typical fluorescence microscope has the basic setup of a reflected-light microscope (**Figure 3.7a**) and consists of 4 main components.

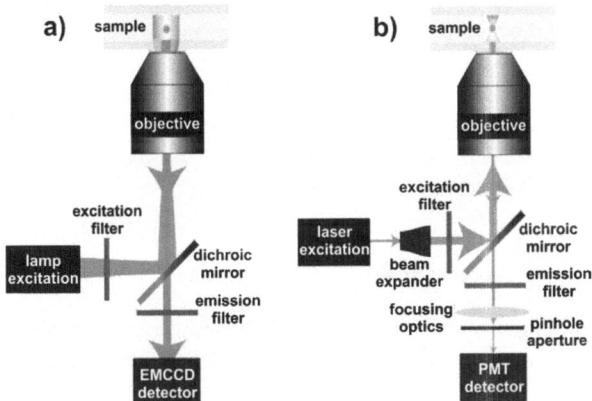

Figure 3.7. Configurations of a fluorescence microscope and a confocal laser scanning microscope. a) The configuration of a fluorescence microscope. **b)** The configuration of a confocal laser scanning microscope. Adapted with permission from J. M. Moran-Mirabal, copyright 2013 by Jose M. Moran-Mirabal, licensee InTech. Download from free at http://www.intechopen.com/books/cellulose-fundamental-aspects/advanced-microscopy-techniques-for-the-characterization-of-cellulose-structure-and-cellulose-cellula.[403]

Either halogen, xenon-arc or mercury-vapor lamps are used as a light source (**1**) because they need to cover the visible and UV spectral range. Two optical filters (**2**) or a filter set (excitation filter, emission filter) are used to choose the excitation and emission wavelength ranges. A dichroic beam splitter or mirror (**3**) guides the light onto the sample, by which fluorescence is established and re-guided by the beam splitter to the ocular (**4**) and a connected detector.

CLSM is an advanced type of fluorescence microscopy that uses optical sectioning to gain better resolution of fluorescent images.[404] The configuration varies slightly compared to the fluorescence microscope (**Figure 3.7b**). Several slits and lenses in a confocal arrangement result in a shared, movable focal plane at which the sample is scanned vertically to the optical axis. Instead of a standard light source, a laser is used to make punctual beam focusing possible. The point scanning allows for the analysis of different sample layers and enables a three-dimensional presentation of the cell. A photomultiplier is used as a detector.

Pros and Cons

Due to the light emitting effect, it is possible to visualize objects with FluM that are below the resolution limit of a light microscope with high contrast and because only the emitted and slightly reflected incident light reaches the objective, a high signal-to-noise ratio is achieved.

However, it is known, that fluorophores lose their ability to fluoresce during constant light illumination. This process is called *photobleaching* and limits the observation time of FluM if it is not reduced with specific chemical treatment. In addition, each fluorophore channel must be recorded separately and then overlaid to obtain a multi-color image.

In contrast, CLSM can be operated at a quasi-theoretical resolution, but the handling is not trivial and demands some experience. Moreover, confocal microscopes are much more expensive than standard fluorescence microscopes.

3.2.10.   X-ray diffraction (XRD)

XRD is a characterization method that has conventionally been applied to study the crystal structures (regular atomic arrangements) of materials by using monochromatic X-rays.

X-rays can be considered electromagnetic waves. When X-rays interact with atoms, they are elastically scattered on the electron shells. The scattered X-rays interfere either constructively or destructively with each other, depending on the optical path difference.[405] The path difference thereby depends on the lattice spacing (gap between the atoms, d) and on the structure of the elementary cell. Due to the periodic structure of the crystals, the constructive interference occurs only for specific angles (θ) and yields a characteristic and regular diffraction pattern (**Figure 3.8**).[405]

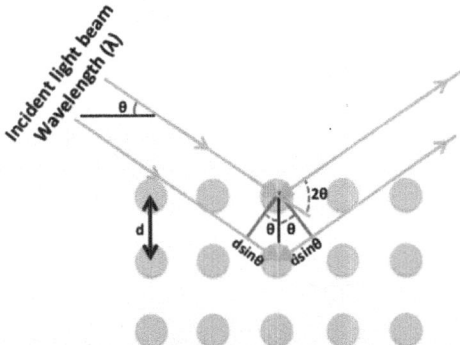

Figure 3.8. The schematic description of Bragg's deflection law. Reprinted with permission from Barron et al., copyright May 2014 by Andrew R. Barron. Download for free at http://cnx.org/contents/ba27839d-5042-4a40-afcf-c0e6e39fb454@20.16.[406]

X-rays feature a wavelength that is in the same order of magnitude (1-100 Angstrom) as the lattice spacing between the crystal planes.[406] Thus, by scanning the sample through the angles at which the diffraction reflectance occurs (2θ), the lattice spacings of the sample can be determined with the Bragg equation (**eq 3.7**)[406] (**Figure 3.8**).

However, because the angles are not only dependent on the elementary cell structure, but also on the atomic number, bond lengths and the atomic position, each material features a characteristic diffractogram.[406]

$$q\,\lambda = 2d\sin\theta$$

eq 3.7

q = an integer number, d = lattice spacing in a crystalline sample, θ = diffraction angle.

A conventional diffractometer consists of 5 main components. Electrons are generated by a cathode ray tube (**1**) and accelerated in an electric field onto a target where they produce X-rays. When the electrons have sufficient energy to dislodge the inner shell electrons of the material, characteristic X-ray spectra containing Kα and Kβ radiation are produced. The X-rays are then filtered with foils or a crystal monochromator (**2**) to produce monochromatic radiation needed for diffraction. The monochromatic X-rays are collimated (**3**) and directed onto the sample where they are diffracted. Because the sample and detector are gradually rotated by a goniometer (**4**), the intensity of the diffracted X-rays is registered using a detector (**5**) and converted to a count rate.

Two common types of X-ray diffraction can be distinguished: powder XRD and single-crystal XRD. As single-crystal XRD is very time-intensive and single crystals are difficulty to obtain, therefore powder XRD is the more popular method.

Pros and Cons

XRD analysis offers the (unique) possibility to characterize the specific composition of a material with highest accuracy. However, the diffraction patterns that are obtained must be matched with XRD databases which require a login access fee. Moreover, samples with a high mass of a few milligrams are generally required for the analysis if the qualified technician is not familiar with the technique of measuring liquid samples between caprolactone foils.

3.2.11.   Mössbauer spectroscopy

Mössbauer spectroscopy is a non-destructive, sensitive analysis method that is able to characterize the nuclear structure of solid materials. It is based on the repulsion energy-free re-emission or resonant absorption of gamma rays from atomic cores and uses a combination of the Mössbauer effect and Doppler shifts to probe the hyperfine transitions between the excited and ground states of the nucleus.[407] Information about the nuclear structure of solid materials can provide very precise qualitative and quantitative information about the chemical, structural and magnetic properties of a material e.g. its oxidation states, spin states and electronegativity. Mössbauer spectroscopy requires the use of solids or crystals that have the ability to absorb the radiation in a recoilless manner. Several isotopes exhibit Mössbauer characteristics but the most commonly studied isotope is ^{57}Fe.[407]

Considering the radioactive isotope ^{57}Co, the excited states $I_e = 5/2$ and $I_e = 3/2$ of ^{57}Fe can be developed by electron capture (**Figure 3.9**).[408] The isotope returns very quickly to its basic state with the emission of gamma radiation. However, the Mössbauer-active transition changes from an excited $I_e = 3/2$ state back to a ground state ($I_g = \frac{1}{2}$) (**Figure 3.9**).[408]

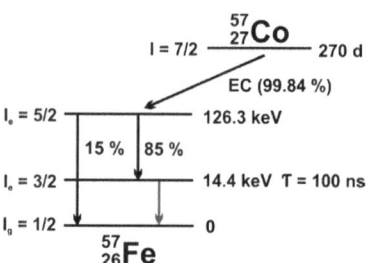

Figure 3.9. The radioactive decay of ^{57}Co. Adapted from M. Guerra, copyright 2012 by Mirjam Guerra, Helmholtz Zentrum Berlin.[408]

The transition-accompanied gamma radiation can be resonantly absorbed in the sample from another ^{57}Fe core, which can undergo a variety of energy level transitions.[407] If the emitting and absorbing nuclei were in identical chemical environments, the nuclear transition energies would be exactly the same. However, depending on the chemical environment of the iron, small shifts can occur in the energetic levels of the excited and the ground states and therefore the required energy for resonance absorption could also shift. The resulting energy difference is balanced with the mechanical Doppler movement of the Mössbauer source. This chemical shift is determined by the electron density at the core and gives information about the spin state and the coordination number. If no other interactions take place, there will be a single line in the spectrum (singlet) which is characterized by the so-called *chemical or isomeric shift* (**Figure 3.10a**).[408;409] Electric or magnetic interactions provoke a further splitting of the states, which results in a resonance line. When this occurs, a doublet is developed during electric interactions, which is defined by the isomeric shift and the quadrupole splitting and gives information about the charge symmetry around the nucleus (**Figure 3.10b**).[408;409] During magnetic interactions, a complete degeneration of the excited state and the ground state occurs. In this case, six different transitions are possible and the resulting sextet is characterized by the isomeric shift and the hyperfine splitting and provides information about the internal magnetic field of a magnetic material (**Figure 3.10c**).[408;409]

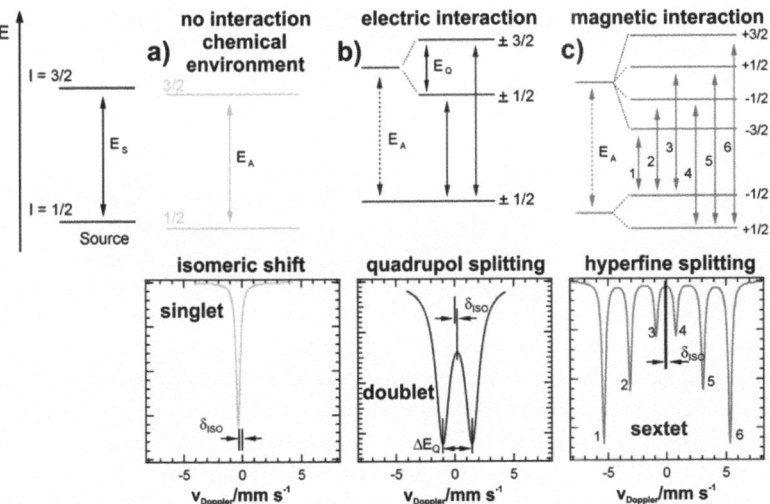

Figure 3.10. Energy level transitions and their modifications that can be determined by Mössbauer spectroscopy. a) The isomeric shift that depends on the chemical environment and that results in a singlet spectrum. **b)** The quadrupole splitting that derives from electric interactions and that results in a doublet spectrum. **c)** The hyperfine splitting that results from magnetic interactions and that results in a sextet spectrum. Reprinted from M. Guerra, copyright 2012 by Mirjam Guerra, Helmholtz Zentrum Berlin.[408]

As a result of these interactions, a material-specific Mössbauer spectrum in extremely fine energy resolution is developed, with the gamma ray intensity plotted as function of the source velocity.[409] At velocities that correspond to the resonant energy levels of the sample, a fraction of the gamma rays are absorbed, which results in a drop of measured intensity (peaks in the spectrum). The number, position and intensity of peaks are used to characterize the sample.

The conventional composition of a Mössbauer device consists of 3 main components. A gamma-ray source (**1**, e.g. $_{57}$Co) which is mechanically movable to generate the Doppler Effect, a collimator (**2**) that filters non-parallel gamma rays and a detector (**3**). In general, there are two methods to measure the Mössbauer effect, which covers the measurement of gamma radiation emission and the measurement of gamma radiation transmission (the latter was used in this thesis).

Because Mössbauer analysis requires a sample that has a mass of several milligrams, the nanoparticles were adsorbed on a tricalcium phosphate support (w/w 5 %) to reach the required mass. The samples were pestled thoroughly prior measurement.

Pros and Cons

Because the method is extremely sensitive, it requires a highly specialized technician for examination and result interpretation. Moreover, Mössbauer devices are rarely found in institutes and although most elements possess isotopes that show the Mössbauer effect, only a few are suitable for practical applications. Among them, the most common isotopes studied are $_{57}$Fe, besides $_{119}$Sn, $_{121}$Sb and $_{151}$Eu. Thus, the main application of Mössbauer spectroscopy is used for the differentiation of bivalent and trivalent iron.

3.2.12.   Flow cytometry & fluorescence-activated cell sorting (FACS)

Flow cytometry is a laser-based technology that allows for the high-speed, multiparameteric analysis of single cells that flow in an electric field. It is often applied in hematology and immunology to determine and quantify specific cell populations, in biology for cell viability analysis and in reproductive biology to separate X-chromosome or Y-chromosome bearing spermatozoa for *in vitro* fertilization.

The cell suspension emerges from a needle and is entrained in the center of a narrow, rapidly flowing stream of liquid (sheath fluid) which moves with great velocity and forces the cells by acceleration to travel one by one through a tiny flow cell (nozzle, **Figure 3.11**).[410] This process is called *hydrodynamic focusing*. If a laser beam is focused onto this flow stream, the cells scatter the light in an extent, which depends on their physical properties such as their relative size, relative granularity and internal complexity.[410]

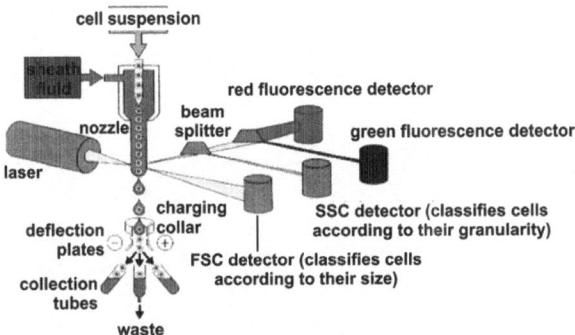

Figure 3.11. Construction of a conventional flow cytometer. Adapted with permission from A. Tabll and H. Ismail, copyright 2011 by A. Tabll and H. Ismail, licensee InTech. Available from: http://www.intechopen.com/books/liver-biopsy/the-use-of-flow-cytometric-dna-ploidy-analysis-of-liver-biopsies-in-liver-cirrhosis-and-hepatocellul.[411]

The forward scattered (FCS) light is mostly diffracted light and is proportional to the volume/size of a cell, while the side scattered (SSC) light is mostly refracted or reflected light and is proportional to the granularity of the cell or its internal complexity. Measuring the scattered light allows for the precise identification of target populations from a heterogeneous population by their phenotype.[412]

An attractive feature of flow cytometry is that the target populations can be separated from the heterogeneous population. Therefore, the fluid is separated with a vibrating transducer into individual droplets, each containing a single cell. The physical character of the droplet cells is measured and the droplet passes an electrical charging collar. A charge is placed on the collar, based on the determined character and the droplet is inversely polarized when it passes the collar. Thereafter, the charged droplets fall through an electrostatic deflection system that diverts them into several collection tubes (**Figure 3.11**).[410]

A common type of flow cytometry is the fluorescence-activated cell sorting (**FACS**, trademark by Becton, Dickinson and Company).[413] In addition to the scattered light, fluorophores can also be measured with the device. Some cells contain natural fluorochromes such as chlorophyll. However, commercial fluorophores such as DAPI and propidium iodide (PI) can be used to stain the cell nuclei and also fluorophore-labeled antibodies are applied for immunolabeling issues. Thus, fluorophore-labeled cells are differentiated and sorted from a heterogeneous mixture by both, their specific light scattering and fluorescent characteristics, which allow for a separation of target populations not only by their phenotype but also by CD markers or viability status.

A conventional flow cytometer with the ability to undergo FACS consists of 5 main components (**Figure 3.11**). A laser system with different wavelengths (**1**) as light source, the sheath fluid jet (**2**) which carries the cells, a flow chamber (**3**) through which the cell flow stream is guided and on which the laser beam is focused, filter systems (**4**) to separate the fluorescence signals on different detectors and a photomultiplier (**5**) for the signal detection with subsequent conversion from light into electric signals and amplification.

The data can be presented in a single dimension as a histogram, in a two-dimensional dot plot or in a three-dimensional image. Based on the fluorescence intensity, a series of subset extractions (gates) can be created to distinguish various regions on the dot-plot. Specific software with defined gating protocols is used to do so.

Pros and Cons

In contrast to fluorescence microscopy, which allows mainly for qualitative labeling information, FACS enables a quantitative and automated, high-throughput screening of labeled cell populations. In detail, it is possible to measure in *real-time*, more than 1,000 cells per second. However, to quantify cells from solid tissues they must be disaggregated before analysis. Thus, only suspended cells from body liquids can be measured without processing.

In general it should be considered, that all of the systems adopted in this thesis vary in precision and sensitivity and feature specific systematic errors. Therefore, most of the measurements that are included in this thesis were performed in triplicate. The mean values and their standard deviations are presented in the results chapter. If the data were insignificant, a detailed explanation of each instance was provided.

3.3.   Experimental procedures

3.3.1.   Nanoparticle fabrication, bioconjugation and processing

Nanoparticle fabrication by PLAL

When the experiments for this thesis started, the Spitfire Pro femtosecond-pulsed laser system with a vertical setup was the only available and enabled nanobioconjugate fabrication by PLAL without biomolecule degradation.[39] Thus, first studies on AuNP and silicon nanoparticle (SiNP) bioconjugation with oligonucleotides were performed with this configuration. However, the TruMicro 5050 picosecond-pulsed laser system with variable pulse energy and repetition rate was evaluated later as optimal alternative to enhance the process productivity. In these terms, the question of maintaining a constant water level height was addressed by changing the direction of laser beam from vertical to horizontal configuration. As both, AuNP properties and nanobioconjugate integrity and functionality for ps-PLAL generation was proven to be nearly identical to the fs-PLAL fabrication (see **Chapter 4.1.1**) the later studies concerning AuNPs and magnetic nanoparticles (MNPs) were all accomplished with the ps-pulsed laser system. Furthermore, some experiments on silicon nanoparticles and their *in situ* bioconjugation were carried out in the framework of an exchange-cooperation with the Italian Institute of Technology (IIT) and a study on SiNP-protein conjugation was performed at the IIT on a Legend Elite fs-laser system, using vertical setup.

Spitfire Pro

If not indicated differently, the process parameters summarized in **Table 3.4** were used for AuNP generation by fs-PLAL with the Spitfire Pro system (**Figure 3.12a**).

Table 3.4. Process parameters for NP fabrication by PLAL with the fs-Spitfire Pro system.

Process Parameters with fs-Spitfire Pro	
Focal Position (Air)	gold: 1 mm behind the target surface (-1 mm); silicon: 4 mm behind the target surface (-4 mm)
Power	0.5 W
Pulse Energy	100 µJ
Repetition Rate	5 kHz
Focusing Lens	40 mm
Ablation Pattern	spiral
Spiral Dimensions	inner radius: 0.4 mm, outer radius: 1 mm, line distance: 0.05 mm
Speed of Axis System	1 mm s^{-1}
Ablation Vessel	48-well plate
Ablation Volume/Well	gold: 500 µL silicon: 800 µL

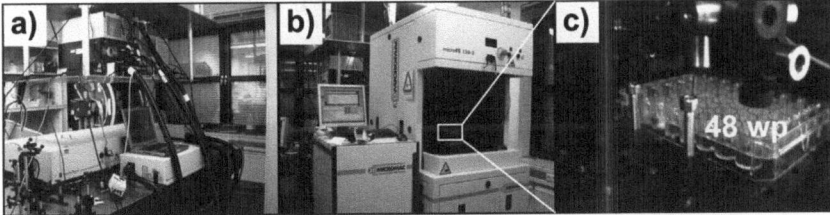

Figure 3.12. Experimental setup for fs-pulsed laser ablation in liquids. a) Photography of the Spitfire Pro System. **b)** Photography of the 3D motorized axis system with ablation vessel. **c)** Magnified view of the process area with the 48-well plate (48 wp) during ablation. Reprinted with permission from C. Sehring, copyright 2010 by Camilla Sehring, Bachelor thesis.[414]

A gold foil (5 x 5 x 0.1 mm; purity: 99.9 %) or a silicon cylinder (diameter: 6 mm, thickness: 10 mm, purity: 99.999 %) was placed on the well bottom of a 48-well plate and covered with *Milli-Q* water or biomolecule solution, resulting in ~ 10 mm water column. The well plate was placed on a 3D motorized axis system (MicroFS, **Figure 3.12b–c**) and the laser beam was focused by a 40 mm lens onto the target.

Tru Micro 5050

If not indicated differently, the process parameters summarized in **Table 3.5** were used for AuNP and MNP generation by ps-PLAL with the Tru Micro 5050 system.

Table 3.5. Process parameters for NP fabrication by PLAL with the ps-TruMicro 5050 system.

Process Parameters with ps-TruMicro 5050	
Focal Position (Air)	1 mm behind the target surface (-1 mm)
Power	0.5 W
Pulse Energy	100 µJ
Repetition Rate	5 kHz
Focusing Lens	56 mm
Ablation Pattern	spiral
Spiral Dimension	diameter: 2.5 mm
Scanner Speed	3.3 m s^{-1}
Ablation Vessel	standard quartz cuvette
Ablation Volume/Cuvette	1 mL

A gold foil (8 x 8 x 0.1 mm; purity: 99.95–99.99 %) or an iron foil (8 x 8 x 0.25 mm; purity: 99.99 %) was fixed by a self-constructed teflon holder within a standard quartz cuvette, filled with 1 mL of *Milli-Q* water or biomolecule solution which resulted in a water column of 10 mm (**Figure 3.13a**).

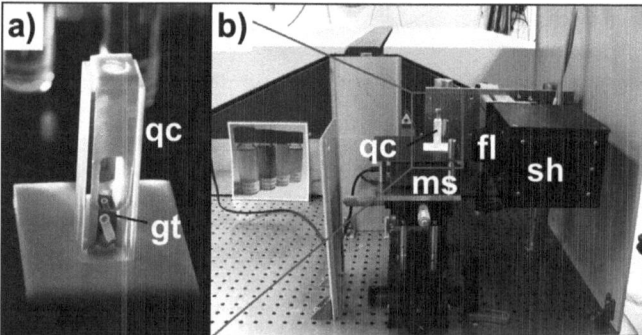

Figure 3.13. Experimental setup for ps-pulsed laser ablation in liquids. a) Photography of quartz cuvette (qc) with inserted gold target (gt). The laser beam and the ablation pattern were visualized by a continuous laser beam (1 mW, 532 nm). Adapted with permission from Barchanski et al., copyright 2015 by the American Chemical Society.[198] **b)** Photography of PLAL setup, including scanner head (sh), focusing lens (fl) and water-filled quartz cuvette (qc) as ablation chamber. Adapted with permission from M. Meißner, copyright 2013 by Marita Meißner, Bachelor thesis.[415]

For gold nanoparticle synthesis, a cylindrical magnet (5 mm x 2 mm) inside the solution and a magnetic stirring plate below the cuvette were applied for continuous liquid homogenization and quick removal of ablated nanoparticles from the process zone. The laser beam was coupled into a HurrySCAN II-14 galvanometric scanner head (Scanlab AG), which allowed deposition of laser pulses with controlled interpulse-distance on the gold target after focusing with a 56 mm telecentric F-theta lens (Sill Optics, S4LFT0055/126) (**Figure 3.13b**).

<u>Legend Elite</u>

The process parameters summarized in **Table 3.6** were used for SiNP-protein *in situ* bioconjugation with the Legend Elite system at the IIT.

Table 3.6. Process parameters for SiNP-protein bioconjugation with the fs-LegendElite system.

Process Parameters with fs-Legend Elite	
Focal Position (Air)	5 mm behind the target surface (-5 mm)
Power	0.12 W
Pulse Energy	120 µJ
Repetition Rate	1 kHz
Focusing Lens	100 mm
Ablation Pattern	Spiral
Speed of Axis System	1 mm s^{-1}
Ablation Vessel	Standard Quartz Cuvette
Ablation Volume/Cuvette	1 mL

For SiNP-Protein A bioconjugation a silicon cylinder (diameter: 6 mm, thickness: 10 mm, purity: 99.999 %) was placed into the quartz cuvette filled with 1 mL Protein A solution, corresponding to ~ 10 mm water layer. The beam was focused by vertical setup using a short 10 cm focal length lens. During ablation the target was moved with a rotation system (T-cube DC Servo motor controller, Thorlabs Inc.) to achieve uniform irradiation from the silicon surface. After each ablation the target was mechanically polished and washed with *Milli-Q* water to remove impurities from the surface.

Gold nanoparticle photofragmentation

For the photofragmentation study, the ps-PLAL-fabricated colloidal solutions were re-irradiated with the TruMicro 5050 laser system. During re-irradiation the gold foil was removed from the cuvette and the laser beam was focused directly into the liquid volume. The solution was stirred constantly and strongly to enable thorough NP recirculation. The used process parameters are summarized in **Table 3.7**.

Table 3.7. Process parameters for photofragmentation of NPs with the ps-TruMicro 5050.

Process Parameters for Photofragmentation	
Focal Position (Liquid)	centered in the liquid volume
Power	varied from 0.25–5 W
Pulse Energy	varied from 100–250 µJ
Repetition Rate	varied from 1–20 kHz
Focusing Lens	56 mm
Ablation Pattern	spiral
Spiral Dimension	diameter: 4 mm
Scanner Speed	3.3 m s^{-1}
Ablation Vessel	standard quartz cuvette
Re-Irradiated Volume	1 mL

Increasing the concentration of gold nanoparticles

To increase the concentration of AuNPs, both an unstabilized and an electrosterically stabilized, ps-PLAL generated colloid were applied for comparison. Ligand-free AuNPs were generated by ablation in *Milli-Q* water and stabilized AuNP bioconjugates were produced by ablation in thiol-functionalized methoxyl (poly)ethylene glycol (mPEG-SH, 5 kDa, 1 µM). An ablation time of 2.5 min mL^{-1} was applied and a total volume of 15 mL per solution was produced. The NP concentration for each solution was precisely set to 250 µg mL^{-1} by UV-vis spectrophotometry.

Increasing the concentration using ultrafiltration

To increase the concentration of nanoparticles efficiently, two filter systems (500 µL volume) which varied in membrane material (*Vivacon®* system: *Hydrosart®* regenerated cellulose; *Nanosep®* system: Omega® modified polyethersulfone) and different molecular weight cut offs (*Vivacon®* system: 10 kDa, 30 kDa, 50 kDa; *Nanosep®* system: 3 kDa, 30 kDa, 300 kDa) were tested. To equilibrate the membrane and to remove preservation agents, all filter tubes were run twice with *Milli-Q* water prior experimentation.

In accordance with the technical data sheets, the *Vivacon®* and *Nanosep®* tubes were all run at 14,000×g. The recommended centrifugation time varied for different molecular weight cut offs (MWCOs) from 3 to 15 min and was adapted individually until the whole liquid had passed the membrane. However, for some MWCOs even an extremely prolonged centrifugation time was not able to filter the whole liquid.

After centrifugation, the filter tubes and membranes were photographed for documentation and the NP concentration of both, the retentates and the permeates were determined by UV-vis spectrophotometry and compared to the initiate concentration prior to ultrafiltration.

Increasing the concentration using controlled evaporation

Two glass vessels with diameters that allowed for a high liquid surface area were both filled with 10 mL of ligand-free AuNPs and placed under the extractor hood for evaporation. During evaporation, one colloidal solution was agitated by magnetic stirring, while the other represented a constant, static system. During six hours of experimentation, 100 µL of the liquids was taken every 30 min for UV-vis spectrophotometry and placed back to the vessel after recording. After six hours, the remaining liquid volume was determined and compared to the start volume.

The same procedure was repeated for the electrosterically stabilized AuNP-mPEG-SH.

Size class separation of AuNPs by successive centrifugation

For the separation of particle size classes, 500 µL of PLAL-generated AuNPs were placed in a 15 mL falcon tube and centrifuged at 1,000 rpm for 10 min. Then the supernatant was carefully separated from the pellet and transferred into a new falcon tube for another centrifugation at 2,500 rpm for 10 min. The same procedure was performed for three more times, while increasing the centrifugation speed stepwise by 2,500 rpm. All separated pellets were taken up in 500 µL *Milli-Q* water and sonicated for 10 min, whereas the final supernatant was applied as present. DLS measurements and SEM preparation was performed and analyzed.

Data were compared with the nanoparticle diameters, which could be theoretically centrifuged at the applied speed and which were calculated using modified Svedberg formula[416] (**eq 3.8**).

$$r^2 = \frac{9\eta_0 s}{2(p_p - p_0)}$$

eq 3.8

r = nanoparticle radius, η_0 = viscosity of the solvent, s = velocity of sedimentation, p_p = density of the particle, p_0 = density of the solvent.

In situ and *ex situ* conjugation of nanoparticles with nucleotides and peptides

For *in situ* bioconjugation of NPs, the biomolecule solution was applied as ablation medium. Suitable biomolecule concentrations ranged from 1 µM to 1 mM, depending on biomolecule type and produced nanobioconjugate yield. After ablation, the dispersion was agitated for 30 min on a tumbling mixer prior to purification by (ultra)centrifugation was initiated.

Ex situ bioconjugation was performed in exceptional cases, e.g. for the bioconjugation of up-concentrated AuNPs. The biomolecule solutions were mixed with the PLAL-fabricated colloids to a final concentration between 1 µM and 1 mM, followed by slow agitation of dispersion on a tumbling mixer for 48 hours and purification.

Antibody conjugation to AuNPs

For the antibody conjugation to AuNPs, three different approaches were used.

For the first approach, the unmodified, native antibody was used directly for *in situ* bio-conjugation during ps-PLAL as aforementioned. The antibodies were attached to the AuNPs only by electrostatic forces in this case.

For the second approach, the antibodies were equipped with an orthopyridyldisulfide-polyethyleneglycol-N-hydroxysuccinimide linker (OPPS-PEG-NHS, 2 kDa) prior conjugation to exhibit a free disulfide function for the gold attachment. Based on an established protocol,[365;417] one part 125 µM OPPS-PEG-NHS in NaHCO₃ (100 mM, pH 8.5) was incubated with 9 parts of antibody solution (1 mg mL⁻¹) at 4 °C. The reaction was allowed to proceed overnight, resulting in a stable amide bond between primary amines on the antibody and carboxyl groups on the PEG chain that are exposed when the NHS terminus is cleaved. The antibody-OPPS-PEG-NHS linker complex was then attached during *in situ* bioconjugation to the AuNPs via the disulfide-containing OPPS group located at the distal end of the PEG linker. This approach featured a covalent, but non-directed conjugation of antibodies to the AuNPs.

For the third approach, the antibodies were pre-treated with heterobifunctional hydrazide-PEG-dithiol linker using an established protocol from Kumar et al. (**Figure SI 2**).[270] Briefly, the antibody solution was diluted to a concentration of 1 mg mL⁻¹ in Na₂HPO₄ buffer (100 mM, pH 7.5). Then 10 µL of NaIO₄ (100 mM) were added and incubated in the dark for 30 min. The reaction was quenched with 500 µL PBS. To control the successful oxidation of the carbohydrate, 20 µL of the antibody solution were mixed with 60 µL of freshly prepared Purpald solution. The presence of aldehydes is indicated by change of solution color from transparent to purple within a few minutes. Then 2 µL of a dithiolaromatic PEG6-CONHNH₂ linker (Sensopath Technologies) was added to the antibody solution and incubated for 1 hour at room temperature. After incubation, 1 mL HEPES buffer (40 mM) were added and the entire volume was filtered with a 10 k MWCO centrifuge filter (Millipore) at 2,000×g (4 °C, ~ 10 min), until approximately 75 % of the solution had passed through the filter. The retained solution was resuspended in HEPES buffer to a final volume of 1 mL, which related to an antibody concentration of 100 µg mL⁻¹. This approach allowed for a covalent and directed conjugation of antibodies to the AuNPs.

In situ bioconjugation was carried out for all approaches using an antibody concentration between 5 and 15 µg mL⁻¹. Following *in situ* bioconjugation, the AuNPs of the second and third approach were incubated with a 10-fold excess of mPEG-SH (5 mM, 5,000 Da) solution overnight at 4 °C to passivate the free spaces on the surface.

Purification of nanoparticles by ultracentrifugation

AuNP & SiNP bioconjugates

If not indicated differently, the unbound biomolecules were removed from gold or silicon nanoparticles by two centrifugation steps (each 15 min on 400 µL solvent) at 10 °C on a Sorvall MTX-150 ultracentrifuge (Thermo Fisher) using a S120-AT3 rotor or on an Optima Max ultracentrifuge (Beckmann Coulter GmbH) at 14,850×g and 59,400×g, respectively. Using Svedberg equation (eq 3.8), these centrifugal forces were calculated to correspond to nanoparticle diameters of 10 nm and 5 nm, which could theoretically be centrifuged.

The use of the Optima Max ultracentrifuge was kindly enabled by the Institute for Biophysical Chemistry (MHH, Dr. Falk Hartmann, PD Ute Curth, Prof. Dietmar Manstein).

MNP bioconjugates

For MNP bioconjugates, a speed of 6,000×g on a MiniSpin bench centrifuge (Eppendorf) for 15 min was sufficient for the separation of unbound molecules. Using Svedberg equation (eq 3.8), this centrifugal forces was calculated to correspond to a nanoparticle diameters of 5 nm, which could theoretically be centrifuged. At least two centrifugation steps were performed for thorough purification.

The nanobioconjugate-containing pellet was resuspended in *Milli-Q* water or buffer for further adoption or storage, while the spare biomolecules-containing supernatant was taken for the quantification of conjugation efficiency and surface coverage values by UV-vis spectrophotometry or fluorescence spectroscopy, respectively.

Transfer of AuNP-nucleotide bioconjugates into saline buffer media

To adapt fs- or ps-PLAL generated AuNP-nucleotide bioconjugates to the ionic strength of saline buffer media, a consecutive addition of salts was carried out. In general 111 µL of a NaH_2PO_4/Na_2HPO_4 (100 mM) solution, followed by 5 increments of 14.5 µL NaCl (2 M) were added consecutively to 1 mL nanobioconjugate solution until final concentrations of 100 mM NaCl and 10 mM NaH_2PO_4/Na_2HPO_4 are reached. Each addition was followed by short mixing on a vortex shaker and 7 hours of agitation on a tumbling mixer.

Immunoblotting with AuNP bioconjugates

Immunoblotting was performed in a *golden blot* format similar to the one described by Walter et al.[269] A low-fluorescent Polyvinylidene fluoride (PVDF) transfer membrane was cut into 1 cm x 0.7 cm pieces, immersed in 100 % methanol for 15 s and rinsed with *Milli-Q* water for 2 min with agitation. The analyte, e.g. IgG (from rabbit) and an unspecific protein as negative control, e.g. penetratin peptide were spotted in three replicates on the membrane, respectively. After ½ h incubation at room temperature, blocking of the membrane was performed with 1 % BSA solution for 2 h. The membrane was rinsed thoroughly and incubated with the staining solution, e.g. AuNP-anti-rabbit-IgG bioconjugates overnight. On the next day, a photography of the membrane was taken for documentation. As positive control, the same procedure was repeated with a commercial labeling agent, e.g. AuNP-anti-rabbit-IgG bioconjugates (Dressed Gold®, Bioassay Works). Photographs of the membrane were recorded and labeling intensities between the commercial and the PLAL-staining were compared by the software Image J.

Removal of amorphous iron-hydroxide from MNPs

By a method from Amendola et al.[418] an ethylenediaminetetraacetic acid (EDTA) solution (5 mM) was added to ps-PLAL-generated MNPs for 1 hour at 35 °C with agitation in order to remove the amorphous iron-hydroxide. Subsequently after reaction time, the solution was centrifuged at 6,000×g for 15 min to remove the EDTA and to protect the particles from being completely dissolved.

EDTA is a six-toothed, metal ion-chelating agent which is used as detergent in washing powder for water softening and acts as Lewis-Base according to eq 3.9.

$$Fe(OH)_2 + H_2EDTA \longleftrightarrow Fe(EDTA) + 2H_2O$$

eq 3.9

3.3.2.   Nanoparticle and nanobioconjugate characterization

Nanoparticle characterization

UV-vis spectrophotometry

The extinction spectra of NP colloids were recorded using a Shimadzu 1650 double-beam UV-vis spectrophotometer, able to measure the wavelength range from 190 to 1100 nm in 0.5 nm step resolution. However, the spectra of silicon nanoparticles were recorded at IIT on a Cary 6000 double beam UV-vis-NIR spectrophotometer (Agilent Technologies) in the range from 200 to 1000 nm and a 0.5 nm step resolution.

For measurement, the colloidal solution was diluted if required and 100 µL were transferred into a low volume quartz cuvette (Hellma Analytics, type 105.201-QS). The baseline was previously recorded on the dilution solvent without analyte (blank).

Determination of colloid concentration

To determine the concentration of SiNPs and MNPs in the colloid, the metal targets were weighed in triplicate on a CPA2P microgram balance (1 µg accuracy, Sartorius AG) prior and after PLAL to obtain the mean ablated mass per volume by arithmetic subtraction. For gold colloid concentration, an improved quantification technique was applied, covering the determination from UV-vis spectra by a method from Muto et al.,[77] who revealed the absorbance intensity at 380 nm to be predominantly contributed by interband transition of AuNPs (isosbestic point). Thus, the ablated target mass of 40 AuNP samples was determined in triplicate by microgram weighing and mean values were plotted against the absorbance intensity at 380 nm and interpolated with a linear fitting (**Figure 3.14**). With this standard calibration the concentration of unknown samples was calculated using the fitting equation and the extinction value at 380 nm.

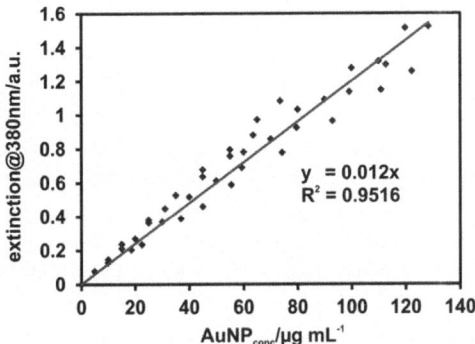

Figure 3.14. Standard calibration of AuNP concentration. The ablated mass of an Au target was plotted against the related extinction values at 380 nm wavelength.

Determination of agglomeration index
The agglomeration index is defined as ratio of extinction intensity at 800 nm (scattering) to extinction intensity at 380 nm (interband absorption) and is equivalent to the share of agglomerated or aggregated primary particles within a specified particle concentration.

Calculation of NP diameter by Haiss et al.
Haiss et al. set up several calculation formulas for the determination of AuNP diameters from 3 to 120 nm.[419] The equations are derived from UV-vis spectra with experimentally determined fitting parameters (B1 and B2) and an average deviation of 11 %. The formula applied within this thesis is summarized in **eq 3.10**.

$$d = exp\left[B_1 \frac{A_{spr}}{A_{450}} - B_2\right]$$

eq 3.10

d = nanoparticle diameter, $B_1 = 3.00$, A_{spr} = absorbance at the SPR peak, A_{450} = absorbance at 450 nm wavelength, $B_2 = 2.20$.

Scanning electron microscopy
For sample preparation, a drop of the diluted colloidal solution was spotted onto a carbon-coated sample disc and the solvent was evaporated at room temperature. Conductive silver was further applied for iron oxide solutions to increase the image contrast.

The sample disc was placed in the vacuum chamber of a QuantaFei scanning electron microscope (Fei Company) and an Everhart-Thornley (ET) detector was used at high vacuum mode (30 kV) to scan the sample surface.

Transmission electron microscopy

To analyze diluted colloidal solutions by TEM, a drop was placed on a hydrophilized carbon-coated, formvar-covered, 300 mesh copper grid and dried at room temperature. The grid was placed into the vacuum chamber of a transmission electron microscope and micrographs were recorded.

TEM images of AuNP colloids were acquired either on an EM 10 C microscope (Carl Zeiss AG), working at an acceleration voltage of 60 kV or on a 400T microscope (Philips, Eindhoven) working at 100 kV. The use of the EM 10 C microscope was kindly enabled by the Institute of Pathology (University of Veterinary Medicine Hannover, Kerstin Rohn, Prof. Baumgärtner). The micrographs derived from the 400T microscope were kindly recorded by Dr. Harald Granzow (FLI Riems, Institute of Infectology).

SiNP colloids were analyzed at IIT using a Jem 1011 electron microscope (JEOL, USA) working at an acceleration voltage of 100 kV.

Determination of NP diameter and size distribution from electron micrographs

Electron micrographs were analyzed using Image J software and the diameter of at least 100 particles was measured to obtain the nanoparticle size distribution. The number of particles with a defined diameter found among the 100 analyzed NP was defined as *particle number frequency* and plotted as function of nanoparticle diameter. If not indicated differently, lognormal fitting was performed.

Dynamic light scattering

The hydrodynamic diameters (d_{DLS}) and the polydispersity index (PDI) of NPs were determined by DLS using a Zetasizer Nano (Malvern Instruments). In general, 100 µL of the solution was added to 900 µL water and the diluted solution was injected into a disposal cuvette (type: ZEN0112, Malvern Instruments). The measurement was run with the parameters that are summarized in **Table 3.8** and the average value of three consecutive measurements was taken for documentation.

For zeta potential measurement, a dip-cell (Malvern Instruments) was inserted into a disposal cuvette (type: DTS1070, Malvern Instruments). Velocity was set between the two electrodes of the dip-cell and mobility of nanoparticles in the electric field was determined with Henry's equation (**eq 3.5**). Measurement was run with the parameters that are summarized in **Table 3.9** and the average value of three consecutive runs was taken for documentation. For the zeta potential measurements presented in this thesis Smoluchowski approximation was generally used, because *i)* almost all nanobioconjugates were fabricated in low salinity solvents such as buffer media: *ii)* ligand-free nanoparticles were fabricated exclusively in polar media such as water; *iii)* the particle size distribution was significantly large for the PLAL-generated nanoparticles when compared to monodisperse distributions of nanoparticles that were synthesized by wet chemistry methods.

Table 3.8. Parameter settings for the measurement of hydrodynamic particle diameter and polydispersity index.

Material			
	Gold	**Silicon**	**Iron**
Refractive Index	0.240	3.50	1.43
Absorption Coefficient	0.290	0.01	1.00
Dispersant			
	Water	**PBS**	**NaCl**
Temperature/°C	25	25	25
Viscosity/cP	0.8872	1.0200	0.97
Refractive Index	1.330	1.335	1.520
Measurement			
Angle/°	173 backscatter		
Duration	automatic		
Number	3		

Table 3.9. Parameter settings for the measurement of the zeta potential.

Material			
	Gold	**Silicon**	**Iron**
Refractive Index	0.240	3.50	1.43
Absorption Coefficient	0.290	0.01	1.00
Dispersant			
	Water	**NaCl**	**PBS**
Temperature/°C	25	25	25
Viscosity/cP	0.8872	0.9700	1.0200
Refractive Index	1.330	1.520	1.335
Dielectric Constant	78.5	5.90	80.0
Measurement			
	Smoluchowski		**Hückel**
f(Ka) Value	1.5		1.0
Duration	automatic		
Number	3		

Fourier-Transform-Infrared Spectroscopy
Colloidal samples were lyophilized in an Epsilon 2-4 LSC freeze-dryer (Martin Christ Gefriertrocknungsanlagen) using the program summarized in **Table 3.10**.

Table 3.10. Freezing program adopted for lyophilization of colloidal samples.

	Freezing	Main-Drying	Final-Drying
Temperature/°C	-25	-25	+8
Vacuum/mbar	atmospheric	0.34	0.001

The powders were pestled in a ratio of 1:50 with dried potassium bromide and mechanically compressed to thin, transparent pellets (3 mm radius, < 0.5 mm thickness) using a KBr press (S.T. Japan). Pellets were analyzed on a Spectrum 100 (Perkin Elmer) with KBr background subtraction and FT-IR spectra were recorded in triplicate with 1 cm^{-1} step width and in a wavenumber range from 400–4000 cm^{-1}.

X-ray diffraction
Colloidal solutions were lyophilized in a freeze-dryer as described above and powders were placed between two sheets of Polyimide foil (Kapton®, Chemplex Industries), which was inserted in a set of specimen holder rings. XRD measurement was carried out on an AXS D8 Advance (Bruker GmbH) with Cu Kα radiation in reflection starting from 30° to 85° = 2θ with a step size of 0.029° and a time per step of 5 sec. XRD measurements were kindly performed by Dr. Olga Wittich (LUH, Institute of Physical Chemistry and Electrochemistry).

X-ray photoelectron spectroscopy
Few sample drops were placed on a highly-oriented pyrolytic graphite substrate (HOPG, ZYA Quality, 1.2 mm, NT-MDT House) and dried in a desiccator overnight. After transferring the samples into the XPS vacuum system no further processing of the samples has been performed. XPS was then carried out using Al-Kα radiation (1486.6 eV, line width: 0.8 eV). The kinetic energies of the photoelectrons were measured with a hemispherical analyzer (radius: 100 mm) in normal emission geometry with pass energy of 20 eV. The spectra were subtracted from Shirley background signals and fitted with Gaussian functions. Deconvolution of peaks was performed by Origin software. XPS measurement was kindly performed by Prof. Christoph Tegenkamp (LUH, Institute of Solid State Physics) and evaluation by Origin was done at LZH.

Photoluminescence measurement
The photoluminescence measurements of SiNP colloids were carried out at IIT on a Fluoromax-4 spectrofluorometer (Horiba-JobinYvon) fitted with a xenon lamp source and a photomultiplier on the excitation and detection sides, respectively.

Mössbauer Spectroscopy
To obtain a sufficient sample amount, 5 wt% of IONP colloid was adsorbed onto a tricalcium phosphate support and pestled thoroughly. The analysis was performed on

a MIMOS-II Mössbauer spectrometer (University of Mainz). The measurement was kindly carried out by Prof. Franz Renz (LUH, Institute of Inorganic Chemistry).

Nanobioconjugate characterization

Calculation of conjugation efficiency

Estimation of conjugation efficiency and surface coverage values was performed by UV-vis spectrophotometry. Initially, a dilution series of the applied biomolecules for nanoparticle attachment was set and measured in the spectral range of 190–400 nm. While proteins and peptides exhibit a characteristic absorbance maximum at 280 nm, the maximal absorbance for nucleotide-based molecules is found at 260 nm. Thus, with respect to the particular biomolecule class, the absorbance values at 280 or 260 nm were plotted against their calculated molar concentration and interpolated by linear fitting (**Figure 3.15**).

This procedure was performed for all biomolecules applied in the experimentations. In the case of fluorophore-labeled biomolecules the specific excitation/emission wavelengths were used to measure the fluorescence intensity with a Fluoroskan Ascent fluorimeter (Thermo Fisher Scientific GmbH) using the solvent without analyte as blank. The fluorescence measurements were kindly enabled by the Institute of Technical Chemistry (LUH, Dr. Johanna Walter, Prof. Thomas Scheper).

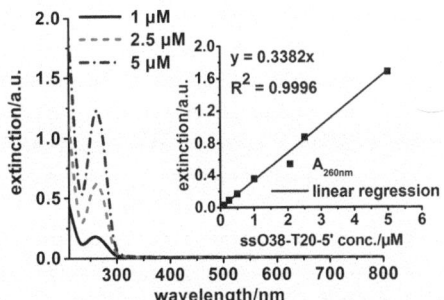

Figure 3.15. Standard calibration of biomolecules. Extinction spectra of dilution series (1 = black solid line, 2.5 = orange dashed line, 5 μM = purple dash-dotted line) and corresponding standard calibration (y = 0.3382x) plotted by the extinction at 260 nm against the molecule concentration; presented on the example of an ssO solution. Adapted from Barchanski et al., copyright 2012 by the American Chemical Society.[420]

According to the developed standard calibrations, the molar concentration of unbound biomolecules [bio$_{unbound}$] was determined from the supernatant of nanobioconjugates after purification by ultracentrifugation. Using the determined value, the molar concentration of nanoparticle-attached biomolecules [bio$_{bound}$] was then calculated by subtraction from the known molar concentration of applied biomolecule [bio$_{input}$] as reference (**eq 3.11**).

$$[bio_{input}]_{known} - [bio_{unbound}]_{determined} = [bio_{bound}]_{calculated}$$

eq 3.11

The calculation was further checked for accuracy, using the pellet UV-vis spectra of nanobioconjugates after subtraction of the AuNP absorbance portion. Again, the molar concentration value $[bio_{bound}]$ was determined from the standard calibrations using the absorbance value at 280 or 260 nm, respectively and used for **eq 3.12**.

$$[bio_{input}]_{known} - [bio_{bound}]_{determined} = [bio_{unbound}]_{calculated}$$

eq 3.12

If the deviation between $[bio_{unbound}]_{determined} = [bio_{unbound}]_{calculated}$ and $[bio_{bound}]_{determined} = [bio_{bound}]_{calculated}$ was less than 5 %, the result was taken for documentation.

Setting $[bio_{input}]$ as 100 %, the conjugation efficiency of biomolecules was finally expressed on a percentage basis by conversion of $[bio_{bound}]$.

Calculation of surface coverage

Surface coverage values were further calculated using distinct parameters of AuNP as density of gold (19.3 g cm^{-3}), molecular weight of a gold atom (197 g mol^{-1}), diameter of a gold atom (0.27 nm) as well as the mean primary particle diameter determined by TEM or SEM, and the AuNP concentration determined from standard calibration by UV-vis spectrophotometry. With these values the total number of particles ($\#_{bio}$) and the total surface area (A) were calculated. Calculated concentration of nanoparticle-attached biomolecules $[bio_{bound}]$ was converted to biomolecule number ($\#_{bio}$) and biomolecule quantity (q) using the Avogadro constant and molecular weight of the biomolecule. Finally, surface coverage values were either expressed as average biomolecule number per nanoparticle ($\#_{bio}$ $\#_{NP}^{-1}$) or as biomolecule quantity per surface area (q_{bio} A^{-1} / pmol nm^{-2}). However, it should be noted that the surface coverage is an approximated and slightly underestimated value, because the particle size distribution gained from electron microscopy is number-weighted and would require a conversion to a surface-weighted size distribution which is typically larger for polydisperse samples.

Determination of diffusion coefficient

The mobility of biomolecules in liquids was determined by calculation of the diffusion coefficient using the Einstein-Stokes relation summarized in **eq 3.13**.

$$D = \frac{k_B T}{6\pi \eta r}$$

eq 3.13

D = diffusion coefficient, k_B = Boltzmann constant, T = temperature, η = dynamic viscosity of the solvent and r = hydrodynamic nanoparticle radius.

Calculation of biomolecule footprint and deflection angle

$$K = \left[\frac{4\pi r^2}{N_r}\right]$$

eq 3.14

$$R = \sqrt[2]{\frac{K}{\pi}}$$

eq 3.15

$$deg = \left[\frac{2R}{r}\right] x \frac{180}{\pi}$$

eq 3.16

K = biomolecule footprint, r = nanoparticle radius, N_r = average number of biomolecules per nanoparticle for given radius, R = radius of footprint approximation on the nanoparticle surface, deg = deflection angle.

The average area which a biomolecule occupies on the AuNP surface is expressed as footprint, while the steric dimensions of the biomolecule define the deflection angle. Both values were calculated based on **eq 3.14-eq 3.16** from Hill et al.,[421] assuming the particles are perfect spheres.

High angular annular dark field and energy dispersive X-ray spectroscopy

HAADF measurements were acquired at IIT via STEM mode using a Jem 2200FS transmission electron microscope (JEOL, USA) working at 200 kV (CEOS spherical aberration corrector of objective lens and an in-column Omega filter) and using a camera length of 50 cm and a probe size of 0.7 nm. The chemical composition of SiNP bioconjugates was analyzed at IIT using EDXS performed in STEM mode with a JED-2300 Si (Li) detector and spherical aberration corrector system (Cs-corrector) for objective lens.

Colorimetric titration assay

To determine the optimal pH range and the minimal antibody concentration for stabilization of AuNP-antibody conjugates, colorimetric titration assays were performed.

For optimal pH identification, equal volumes of AuNP-antibody solution were filled into the wells of a 48-well plate and the pH was adjusted in the range from 5–12 by addition of previously determined K_2CO_3 amounts. The mixtures were set with *Milli-Q* water to a final volume of 700 µL, gently stirred for 20 min and then treated with 1 % NaCl solution. After 10 min and potentially completed color change from red (stabilized colloid) to blue (agglomeration), the absorbance at 525 nm was measured using an Infinite M200 PRO microplate reader (Tecan Group) and compared to a control sample treated with *Milli-Q* water instead of NaCl.

For the evaluation of minimal antibody concentration, equal volumes of ligand-free AuNPs were filled into the wells of a 48-well plate and various amounts of antibodies were added to yield concentrations from 2.5–30 µg mL^{-1}. Then all mixtures were adjusted to the previously determined, optimal pH using K_2CO_3, set to a final volume of 700 µL and stirred gently for 20 min. After treatment with 1 % NaCl solution and 10 min color changing time, the absorbance was measured as described before.

All assays were performed in triplicate for statistical reason.

Surface Enhanced Raman Spectroscopy

The microprobe Raman spectra were recorded on an inVia Raman microscope (Renishaw GmbH) using a 633 nm laser (1.5 mW, 20 sec accumulation time) in backscattering geometry for excitation through a 100X objective (NA = 0.9). The experimental set-up consisted of a grating 1800 lines/mm with spectral resolution of about 1.1 cm^{-1}. The colloid was deposited by drop coating deposition (DCD) technique over the CaF_2 substrate and excess liquid was evaporated, leading to the formation of a coffee ring structure. The measurements were performed at different sample locations to gain a clear picture of the sample content. The recorded SERS spectra were baseline–corrected using maximum third order polynomial with the help of WIRE 3.0 and then normalized to 1.

Integrity determination of ssDNA on AuNP-ssDNA bioconjugates by gel electrophoresis

The integrity of *in situ* conjugated oligonucleotides to AuNPs was determined by gel electrophoresis. Therefore, the oligonucleotides were separated from NPs after conjugation, run on the gel with an untreated oligonucleotide control and band intensities were compared to determine the ratio of biomolecule degradation.

The gel was prepared, by mixing 4 wt-% agarose with Tris/Borate/EDTA (TBE) buffer, boiling the mixture for 10 min and cooling it to 50–60 °C prior casting in a 13 cm x 15 cm gel tray. Then 90 µL of AuNP-ssDNA samples were treated with 10 µL Dithiotheritol (DTT, 100 mM) for 2 h to displace the conjugated oligonucleotides by an exchange reaction and to guarantee analysis of both, unconjugated and conjugated ssO in total. DTT-treated samples were ultracentrifuged for 20 min at 14,850×g. Then 10 µL formamide was added to 50 µL of the supernatants and the mixtures were heated for 2 min at 95 °C. Next, 30 µL of the samples were towed with 7 µL Orange Dye (10x) for visualization and 10 µL of this mixtures was loaded in each well of the polymerized gel. A reference solution containing the same concentration of fresh ssDNA solution was treated equally and run on the same gel. The gel was run for 30 min at 100 V in TBE buffer including 12 µL Ethidiumbromide solution (EtBr, 10 mg mL^{-1}). After running, the gel was illuminated with 250 nm UV light and emission of EtBr at 605 nm was detected. Detection limit for DNA was found to be 10 ng mL^{-1} using a concentration of 0.5 µg mL^{-1} EtBr (150 ng on gel equal to 100 % integrity). Scans of the stained gels were analyzed with Image J software. The intensity of the bands (I_x) was compared to the intensity of the reference (I_0) in order to deduce the degree of degraded biomolecules (D_x) by **eq 3.17**.

$$D_x = 1 - \frac{I_x}{I_0}$$

eq 3.17

Determined values were then indicated in percentage.

3.3.3.   Cellular studies with NPs

Cell culture

For *in vitro* experiments, somatic cell cultures were used, covering adherent human fibroblasts (**Figure 3.16a**), adherent hamster M3E3/C3 pluripotent epithelial cells (**Figure 3.16b**), immortalized, bovine endothelial GM7373 cells (**Figure 3.16c**) and MTH53A canine mammary cells.

Human fibroblasts were kindly donated from Dr. Sabrina Schlie-Wolter (LZH). M3E3/C3 cells were extracted in 1981 from the lung of a Syrian golden hamster (*mesocricetus auratus*) fetus on day 15 after gestation and friendly provided from Prof. Makito Emura (MHH, Institute of Molecular Pathology). MTH53A adherent cells were derived from epithelial healthy canine mammary tissue and obtained from the Cell Culture Collection (University of Veterinary Medicine Hannover, Small Animal Clinic). GM7373 immortalized, bovine (*bos taurus*) endothelial cells were established from a male bovine fetus at 24-week gestation and obtained from the Cell Culture Collection (FLI, Riems, Institute of Infectology).

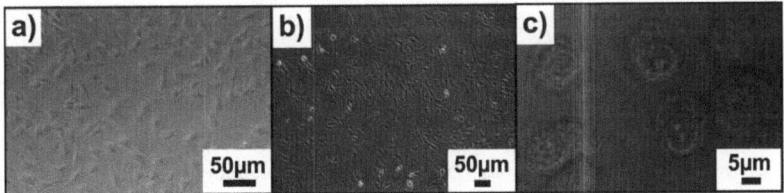

Figure 3.16. Somatic cells that were used for *in vitro* experiments. a) Human fibroblasts. **b)** Hamster M3E3/C3 epithelial cells. **c)** GM7373 cells. Image **c)** was reprinted with permission from Taylor et al., copyright 2010 by the International Society for Advancement of Cytometry, John Wiley and Sons.[122]

Table 3.11. Chemical composition of cell culture media for somatic cells.

Human Fibroblasts	RPMI 1640, 1 % pen/strep, 10 % FCS
M3E3/C3	RPMI 1640, 1 % pen/strep, L-Glutamine, sodium pyruvate, 10 % FCS
GM7373	DMEM, 1 % pen/strep, 10 % FCS
MTH53A	Medium 199, 1 % pen/strep, 10 % FCS

All cells were cultivated at 37 °C in a humidified, 5 % CO_2 atmosphere and the composition of the culture media is summarized in **Table 3.11**.

Cell transfection study

For cell transfection study, $3x10^5$ MTH53A cells were seeded in 6-well plates eight hours prior transfection. Two mammalian expression vectors simultaneously encoding for an expression protein (canine high mobility group protein B1 – rHMGB1 or equine interleukin 12 – eIL-12) and humanized renilla green fluorescent protein (hrGFP) were constructed.[422] hrGFP was used in order to evaluate successful transfection by fluorescence microscopy or flow cytometry. For transfection study, different protocols were applied in triplicate, covering the adoption of conventional FuGENE transfection agent, CRM-fabricated Plano-AuNPs, two PLAL-AuNP colloids with different mean hydrodynamic particle sizes and two magnet-assisted transfection agents (MA Lipofection, MATra-A). Detailed information on transfection protocols are found in the publication of Durán et al.[422] After transfection, the cells were incubated for 24 hours before the plasmid DNA uptake was verified by fluorescence microscopy and flow cytometry. Cell proliferation was evaluated by colorimetric ELISA (Roche Applied Science) and biological functionality of the expressed proteins was analyzed by immunofluorescence directed against eIL-12 and canine HMGB1. Detailed protocols are again found in the publication of Durán et al.[422] Statistical significance was determined using the 1-tailed Wilcoxon-Mann-Whitney test by considering statistically significant differences for $p \leq 0.05$.

Cellular uptake studies

Incubation experiments were accomplished in 24-well plates (M3E3/C3) or 6-well dishes (GM7373) if not indicated differently.

Adherent M3E3/C3 cells were added in a concentration of $1x10^6$ cells per well and grown 24 hours until confluency of approximately 80 % was reached, which is corresponding to a cultivated surface area of approximately 1.52 cm^2. Then cells were transferred into culture medium without fetal calf serum (FCS) and antibiotics, 12 hours prior the nanoparticles were added (5–30 µg mL^{-1}) in full cell culture medium (including FCS) and incubated for 0.5–4 hours. The serum starvation step was performed to synchronize the cell cycle of cells to the quiescent G_0/G_1 phase, thus making the population of proliferating cells more homogeneous.[423] Moreover, the starvation reduces analytical interference and provides more reproducible experimental conditions.[423]

Details on the particle number dose per well and the particle surface dose per cell area are summarized in **Table 3.12**. Data on the particle dose per cell could not be calculated unfortunately, because the exact number of adherent cells was not counted.

For GM7373 cells, no specific cell number was set so that the nanoparticle dose could also not be calculated.

Table 3.12. Number and surface dose of nanoparticles for cellular uptake study into M3E3/C3 cells.

	Particle Diameter/nm	Particles Well^{-1}/#	Particle Area Well^{-1}/nm^2	Particle Area Cell Area^{-1}/nm^2 nm^{-2}
ligand-Free AuNP, 5 µg		6.35E+11	1.69E+14	1.86E+05
ligand-Free AuNP, 15 µg	9.2	1.91E+12	5.07E+14	5.59E+05
ligand-Free AuNP, 30 µg		3.81E+12	1.01E+15	1.11E+06
AuNP-Pen(1 µM), 5 µg		1.65E+12	2.32E+14	2.56E+05
AuNP-Pen(1 µM), 15 µg	6.7	4.94E+12	6.96E+14	7.67E+05
AuNP-Pen(1 µM), 30 µg		9.87E+12	1.39E+15	1.53E+06
AuNP-Pen(5 µM), 5 µg [4]	6.7 – primary	1.44E+09	2.22E+13	2.45E+04
AuNP-Pen(5 µM), 5 µg [4]	particles	4.33E+09	6.66E+13	7.34E+04
AuNP-Pen(5 µM), 5 µg [4]	70 – clusters	8.66E+09	1.33E+14	1.47E+05

After incubation time of 2 hours the supernatant was removed and cells were rinsed twice with serum-free medium and once with PBS to remove membrane-attached nanoparticles. To detach the cells from the cell culture plate, trypsin-EDTA solution was added for 2 min. Then, cells were washed with PBS to remove trypsin-EDTA and prepared for FluM/CLSM (**Table 3.14**).

Immunolabeling of human fibroblasts

Table 3.13. Immunolabeling protocol for FluM and CLSM of human fibroblasts.

Immunolabeling of Human Fibroblasts for FluM and CLSM	
Cultivation	on 10 mm glass cover slips in a 24-well plate tntil 80 % confluency
Fixation	20 min in PFA-solution (4 %) at 4 °C
Rinsing	3 x with PBS at RT
Blocking	1 h with BSA/PBS (2 %) at 37 °C
Rinsing	3 x with PBS at RT
pAb Incubation	**polyclonal rabbit anti-vinculin**, 1:750, overnight at 4 °C
Rinsing	3 x with PBS at RT
sAb Incubation	1 h with **AuNP-sAb**, commercial **sAb-FITC** or **SiNP-Protein A** mixed with Hoechst/0.5 % Tween 20/PBS at 37 °C
Rinsing	3 x with PBS at RT
Mounting of SiNP-Treated Cells	with glycerol
Sealing	using nail varnish

---

[4] The nanoparticle dose of AuNP-Pen(5 µM) bioconjugates is presented for aggregate size, because the dose of primary particles had been the same as for AuNP-Pen(1 µM) bioconjugates.

AuNPs

For the immunolabeling of human fibroblasts with AuNP bioconjugates, the AuNP were functionalized *in situ* with a goat polyclonal secondary antibody (sAb), directed against rabbit IgG (anti-IgG). The nanobioconjugates were purified by a 20 min centrifugation at $14,850 \times g$ and used for immunolabeling in a final NP concentration of 10 µg mL^{-1} by the protocol in **Table 3.13**. After extensive rinsing, the nanoparticle scattering, FITC and Hoechst signals on the cover slips were registered by FluM. Commercial secondary antibodies that were coupled with a Fluorescein isothiocyanate (FITC) fluorophore (sAb-FITC) were applied as positive control.

SiNPs

For the immunolabeling of human fibroblasts with SiNP bioconjugates, no secondary antibody was applied. Instead, NPs were conjugated *in situ* with Fc-specific Protein A (25 µg mL^{-1}), purified by ultracentrifugation and used for labeling in a final NP concentration of 10 µg mL^{-1} by the protocol in **Table 3.13**. After extensive rinsing, the Hoechst and SiNP signal were registered by CLSM.

Human fibroblast labeling and magnetic manipulation with MNP bioconjugates

As described previously, iron nanoparticles were conjugated *in situ* with Alexa 594 fluorophore-labeled BSA and iron hydroxide was removed by EDTA treatment, yielding MNP-BSA-Alexa594 bioconjugates. The fibroblasts were cultivated in a culture disc (35 mm diameter) up to a confluency of approximately 80 % was reached, which is corresponding to a cultivated surface area of approximately 962 mm². Then fibroblasts were transferred into serum- and antibiotics-free culture medium 4 hours prior to nanobioconjugate addition. A particle concentration of 15 µg mL^{-1} was used, which corresponds to a surface dose of 0.043 nm² nanobioconjugates per 1 nm² cell area. After 2 hours of incubation, the cells were detached from the cell culture plate using trypsin-EDTA for 2 min and washed with phosphate buffered saline (PBS) to remove trypsin-EDTA and unbound nanobioconjugates. Then the preparation protocol for FluM was followed to analyze unspecific labeling (**Table 3.14**).

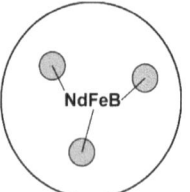

Figure 3.17. Cell culture dish with position of NdFeB magnets for magnetic manipulation assay.

For magnetic manipulation assay, the NP-labeled and PBS-washed fibroblasts were seeded into a new culture disc which was modified with three NdFeB magnets on the outer dish bottom, according to the scheme in **Figure 3.17**. After 48 hours of incubation at 37 °C, magnets were removed and the cell growth at magnet position and in the periphery was documented by photography.

Cell preparation for analysis by FluM/CLSM

To prepare the cells for analysis by fluorescence microscopy or CLSM, the protocol summarized in **Table 3.14** was followed. Cells prepared by immunolabeling protocol were applied for visual examination without further treatment.

For the imaging of human fibroblasts labeled with antibody-FITC-conjugated AuNPs and BSA-Alexa-conjugated MNPs and of transfected MTH53A cells, an AXIO Imager M1/Z1 fluorescence microscope (Carl Zeiss AG) with AxioVision software and the filter sets summarized in **Table 3.15** was used.

Confocal imaging of AuNP uptake into M3E3/C3 cells was performed on an Inverted 2 microscope (Leica Microsystems GmbH), using the excitation and emission wavelengths summarized in **Table 3.16**. The microscope usage was kindly enabled by the Confocal Laser Microscopy Facility (MHH).

Table 3.14. Cell preparation protocol for FluM and CLSM.

Cell Preparation Protocol for FluM and CLSM	
Coating Cells on Polysine-Slides	~ $5x10^5$–$1x10^6$ cells/50 µl PBS + 0.2 % BSA
Cell Adhesion	30 min at 37 °C, removal of excess PBS
Fixation	20 min in PFA-solution (4 %) at RT
Rinsing	3 x 5 min in PBS at RT
Cell Core Staining	10 min with DAPI II counterstain-solution
Mounting	cells mixed 1:4 with Vectashield® GM7373: mounted on a slide within a paper reinforcement ring and subsequently placed coverslip
Sealing	using nail varnish

Table 3.15. The filter sets and specifications from M1/Z1 microscope.

Fluorophore	Filter Set	Excitation λ/nm	Beam Splitter	Emission λ/nm
Alexa594	20	BP 546/12	FT 560	BP 575-640
FITC/GFP	38	BP 470/40	FT 495	BP 525/50
DAPI	49	G 365	FT 395	BP 445/50

Table 3.16. Excitation and emission wavelengths applied for the CLSM analysis of AuNP uptake into M3E3/C3 cells.

Fluorophore	Excitation λ/nm	Laser Intensity %	Emission λ/nm
DAPI	405	60	420-500
AuNP	514	20	520-700

The uptake of AuNPs into GM7373 cells was visualized by an Axioplan 200 apparatus and an LSM510 confocal imaging system (Carl Zeiss MicroImaging GmbH). He-Ne green laser of 542 nm was used to excite the SPR of AuNPs. The imaging was kindly performed by Dr. Sabine Klein (FLI, Institute of Animal Genetics).

Confocal microscopy of SiNPs bioconjugates was performed at IIT using an A1 inverted microscope (Nikon Instruments Europe) equipped with an A1-DUS spectral detector and the excitation and emission wavelengths summarized in **Table 3.17**.

Table 3.17. Excitation and emission wavelengths applied for the CLSM analysis of human fibroblasts which were immunolabeled with SiNP-Protein A bioconjugates.

Fluorophore	Excitation λ/nm	Emission λ/nm
DAPI	350	450–490
SiNP	405	415–550

Cell preparation for TEM

M3E3/C3 cells and human fibroblasts were prepared for transmission electron microscopy using the protocol summarized in

Table 3.18. TEM analysis of M3E3/C3 was kindly performed by Kerstin Rohn (University of Veterinary Medicine Hannover, Institute of Pathology) using an EM 10 C microscope (Carl Zeiss AG) at 60 kV. Human fibroblasts were analyzed at IIT using a Jem 1011 electron microscope (JEOL, USA) at 100 kV.

Table 3.18. Cell fixation protocol for TEM of human fibroblasts and M3E3/C3 cells.

Cell Preparation for TEM	
Rinsing with 0.1 M Cacodylat Buffer	several changes
1 % OsO$_4$ Solution	2 h
Rinsing with 0.1 M Cacodylat Buffer	4 x 10 min
30 % Ethanol	30 min
50 % Ethanol	30 min
70 % Ethanol	30 min or overnight
90 % Ethanol	30 min
100 % Ethanol	2 x 30 min
Glycide Ether + 100 % Alcohol 1:1	30–60 min
Glycide Ether	30–60 min
Glycide Ether + Epon (A + B)	30–60 min or overnight
pure Epon (A + B)	30–60 min
Embedding Samples in Epon with Gelatin	
Warming Cupboard	24 h at 35 °C, 24 h at 45 °C, 4 d at 65 °C

Flow cytometry of transfected MTH53A cells

GFP expression of transfected MTH53A cells was analyzed by flow cytometry. Therefore, cells were trypsinized for 3–5 min, washed with PBS, resuspended in the medium and measured with a FACScan flow cytometer (BD Biosciences) and fluorescence intensities were analyzed with Cell Quest software.

Cell death was identified by PI staining (5 µg mL⁻¹) of trypsinized cells. Cytometry analysis was performed using a FACSCalbur device (BD Biosciences).

3.3.4. Spermatozoa studies with NPs

Spermatozoa

Semen of fertility proven Holstein-Frisian bulls was kindly donated from Masterrind GmbH for experimentation.

Gold nanoparticle penetration studies into spermatozoa

For the co-incubation of spermatozoa with AuNP and AuNP bioconjugates, sperm cells were washed twice by centrifugation for 10 min at 200×g using a buffer for bull semen extension. Then the sperm suspension was diluted to 100×10^6 sperm cells per mL in the same buffer which already contained the nanoparticles and incubated for 2 h at 38 °C. A nanoparticle concentration of 10 µg per mL was used, which can be calculated to a number dose of 1.1×10^5 nanoparticles per sperm cell and a surface dose of 4.4×10^6 nm^2 nanoparticle surface per sperm cell. After another round of centrifugation, spermatozoa were processed for TEM preparation.

Spermatozoa preparation for TEM

To analyze co-incubated spermatozoa by TEM, a gentle and membrane integrity-conserving agarose-based embedment method (**Table 3.19**) was applied.

The sperm-containing Epon bloc was trimmed into 50–70 nm ultra-thin sections using a diamond knife (Diatome, US) on an UltraCut E rotation microtome (Reichert-Jung Leica Microsystems AG). The sections were fixed on 200 mesh copper grids (Plano GmbH) and stained with uranyl acetate and lead citrate. Examination was performed on an EM 10 C electron microscope (Carl Zeiss AG) and the micrographs of approximately 10 different sperm cells per sample were recorded. On a representative sample, a minimum of 30 internalized nanoparticles was counted for the evaluation of the internalized particle size distribution.

The use of the EM 10 C microscope was kindly enabled by the Institute of Pathology (University of Veterinary Medicine Hannover, Kerstin Rohn, Prof. Baumgärtner).

Table 3.19. Agarose-based embedment protocol for TEM preparation of spermatozoa.

Spermatozoa Preparation for TEM	
Fixation	overnight in a mixture of glutaraldehyde (1.5 %) / paraformaldehyde (1.5 %)/PBS at RT
Centrifugation	at 1,000×g
Resuspension	in PBS
Centrifugation	at 1,000×g
Pellet Mixing	1:1 with molten, pre-warmed (38 °C) agarose (2 %)
Centrifugation	at 1,000×g
Sample Hardening	on ice, until hardened
Tissue Excision	by standard protocol
Post-Fixation	OsO₄ (1 %) solution for 2 h
Rinsing	with PBS
Dehydration with 30 % Ethanol	30 min
Dehydration with 50 % Ethanol	30 min
Dehydration with 70 % Ethanol	30 min or overnight
Dehydration with 90 % Ethanol	30 min
Dehydration with 100 % Ethanol	2x30 min
Glycide Ether + 100 % Alcohol 1:1	30–60 min
Glycide Ether	30–60 min
Glycide Ether + Epon (A + B)	30–60 min or overnight
Pure Epon (A + B)	30–60 min

Spermatozoa membrane integrity analysis by FACS

The AuNP-treated spermatozoa and an untreated control were diluted to 1x10⁶ sperm cell per mL and PI was added to a final concentration of 22.5 µM. Flow cytometrical analysis was performed using a FACScan (BD Bioscience) equipped with an argon laser (488 nm, 15 mW) and samples were analyzed in duplicates acquiring 1x10⁴ cells per measurement. PI-positive cells were considered to be membrane-damaged.

3.3.5. Standard terminology in nanotechnology

The terminology in this thesis was adopted from the standard terminology of nanotechnology defined by the American Society for Testing and Materials (ASTM)[424] and the National Institute of Standards and Technology (NIST).[425]
The most important definitions of nanoparticle states are:

Agglomerate: a group of particles held together by relatively weak physical or electrostatic forces (e.g. van der Waals or capillary), that is reversible and may break apart into smaller particles upon processing.
Aggregate: a discrete group of particles in which the various individual components are not easily broken apart, such as in the case of primary particles that are strongly bonded together as a cohesive mass or cluster (e.g. fused, sintered, or metallically bonded particles)

Cluster: a small group of atoms or molecules or an array of bound atoms intermediate in character between a molecule and a solid.

Coalescence: a process in which two phase domains of the same composition come together to form a larger domain, characterized by the disappearance of the boundary between two particles in contact followed by changes of shape and leading to a reduction of the total surface area.

Nanoparticle: a sub-classification of ultrafine particle with lengths in two or three dimensions greater than 1 nm and smaller than about 100 nm and which may or may not exhibit a size-related intensive property.

Further terms within the thesis which are referred to the standard terminology are: dispersion, colloid, monodispersity, polydispersity, hard-aggregate, electrostatic stabilization, steric stabilization, electrosteric stabilization, sedimentation, Brownian motion, interface, electrical double layer, diffuse layer, Stern layer, inner Helmholtz plane, outer Helmholtz plane, shear plane, adsorption, physical adsorption, chemical adsorption, specific adsorption, non-specific adsorption, multilayer adsorption, desorption, isoelectric point, zeta potential, zwitterionic.

Classification criteria of colloidal stability and nanoparticle dispersity

Colloidal stability characterizes the relative ability of colloids to remain dispersed in a liquid and is described by the zeta potential. Whereas, the nanoparticle dispersity stands for the uniformity/heterogeneity of particle sizes in a mixture and is described by the polydispersity index. The criteria to classify colloidal stability and dispersity of this thesis were adopted from Riddick et al. and Müller et al. and are summarized in **Table 3.20**.[426]

Table 3.20. Classification criteria used for data evaluation. **Left table**: Criteria for the stability of dispersion with ZP = zeta potential[426]; **Right table**: Criteria for the dispersity of dispersions.[427]

Stability of Dispersions by Riddick et al.		Dispersity of Dispersions by Müller et al.	
Form of Stability	ZP/mV	Form of Dispersity	Polydispersity Index/a.u.
agglomeration	0 to -11	monodispersity	0.03–0.06
low agglomeration	-11 to -20	narrow dispersity	0.10–0.20
border of agglomeration	-21 to -30	broad dispersity	0.25–0.50
no agglomeration	-31 to -40	unevaluable data	> 0.50
good stability	-41 to -50		
very good stability	-51 to -60		
excellent stability	-61 to -80		

4. Results and Discussion

This chapter is subdivided into three sections. The first section **4.1** deals with the yield enhancement of the PLAL method. For this intent, longer pulse durations than femtosecond pulses were adopted in order to ablate a higher gold mass per time. Moreover, ultrafiltration and evaporation post-processing techniques were applied to further increase the concentration of afore fabricated gold nanoparticles (**Figure 4.1a**).

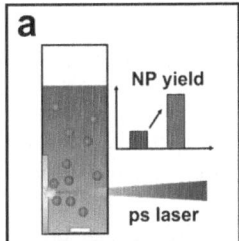 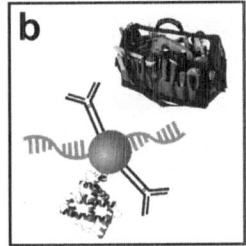 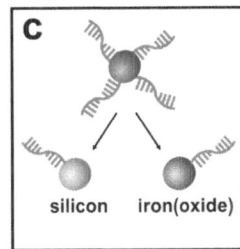

Figure 4.1. Schematic illustration of the subchapter content. a) Yield enhancement of PLAL method. **b)** Adjustment of nanobioconjugate function and design. **c)** Method transfer from gold onto silicon and iron(oxide).

The second section **4.2** focusses on the bioconjugation event and the parameters that were used to adjust a customized design and function of AuNP bioconjugates (**Figure 4.1b**). Finally, in the third section **4.3**, the *in situ* bioconjugation method is transferred onto silicon and magnetic, iron-based materials and critically compared to the gold conjugation (**Figure 4.1c**). Each subchapter begins with in a theoretical introduction followed by a presentation of the results and ends with a concluding summary and discussion.

4.1. Yield enhancement of PLAL method

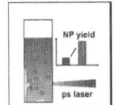 Nanoparticle fabrication by the PLAL technique has become very prominent in the past decade(s), which is proven by the fact that this subject has had more than 350,000 hits on Google search. Curiosity for this method has increased; specifically for nanobioconjugate fabrication, because the idea of using no other additives than the base materials (gold, biomolecules, laser radiation and ultrapure water) makes the fabricated products very attractive for *in vivo* applications. However, the main drawback of this technique is the limited fabrication yield of nanobiohybrids with target ablation on the microgram scale (for laser powers < 5 Watts), which makes the method uncompetitive compared to wet chemical synthesis on the milligram to

gram scale. Thus, considering business economics and cost effectiveness of the fabrication process, efficiency development is unavoidable.

Yield enhancement of PLAL method may be achieved by the ablation of a higher gold mass using longer pulse durations than femtosecond pulses, e.g. with a picosecond-pulsed laser system. Further increase of the concentration of PLAL-fabricated AuNPs and AuNP bioconjugates may be obtained by post-processing methods such as ultrafiltration and evaporation. All these approaches will be presented and discussed in the following chapters.

4.1.1.   Gold nanoparticle fabrication by laser ablation with ps pulses

Barchanski et al. 2015 [1], LZH

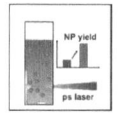

A higher gold mass ablation per time may be achieved by the use of longer pulse duration, e.g. with the adoption of a ps-pulsed laser system instead of a fs-pulsed system.

However, some certain critical questions must be answered: Do the ps-LAL generated particles feature the same intrinsic characteristics as those that were fabricated with fs-pulsed LAL? Furthermore, does the application of longer pulses allow the fabrication of functional nanobioconjugates? The sensitive biomolecules might be damaged by local temperature enhancement and by the establishment of a high-temperature-and-pressure (HTP) region, as was previously demonstrated for ns-pulsed LAL.[357]

To answer these questions, ps-pulsed LAL was applied for the fabrication of ligand-free AuNPs and AuNP bioconjugates and the nanoparticle yield was calculated. The fabricated colloids were thoroughly characterized and compared with those that were generated by fs-pulsed LAL. Finally, the degree of biomolecule degradation during AuNP bioconjugate fabrication was analyzed with gel electrophoresis and the functionality of the nanobioconjugates was tested using immunoblotting and *in vitro* immunolabeling.

4.1.1.1.   Analysis of nanoparticle yield

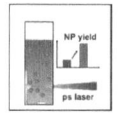

For optimal comparison, gold nanoparticles were fabricated by ps-pulsed LAL using the same settings as for fs-pulsed LAL (0.5 W, 100 µJ, 5 kHz, **Table 3.5**),[39] by changing only the wavelength (1030 nm) and pulse duration (~ 7 ps), compared to 800 nm wavelength and 120 fs pulse duration for the fs-pulsed LAL. Ligand-free AuNPs were produced by ablation in *MilliQ*, while AuNP-DNA and AuNP-protein bioconjugates were ablated in ssDNA (single-stranded DNA, 5 µM) and BSA (bovine serum albumin, 30 µM) solution, respectively. The ablated nanoparticle mass per given time and volume (yield) was determined for a fluence regime of 0.17–1.51 J cm⁻² and the data are summarized in **Figure 4.2**.

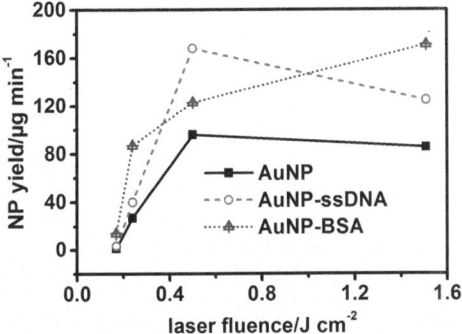

Figure 4.2. Nanoparticle yield obtained by ps-PLAL process for the fabrication of ligand-free AuNPs and AuNP bioconjugates. Yield of ligand-free AuNP (AuNP, black solid line), ssD-NA-conjugated gold nanoparticles (AuNP-ssDNA, green dashed line) and BSA-conjugated gold nanoparticles (AuNP-BSA, red dotted line) for 100 μJ pulse energy, 5 kHz repetition rate and 0.5 W laser power within 0.17 J cm^{-2} to 1.51 J cm^{-2} fluence regime.

Overall, an extremely low NP yield was found for ligand-free AuNPs and AuNP bioconjugates that were fabricated with 0.17 J cm^{-2} fluence. However, a steep increase in yield was determined between 0.17 and 0.5 J cm^{-2} laser fluence. Whereas, between 0.5 and 1.51 J cm^{-2} laser fluence a more stagnating or even regressive tendency was observed.

A maximum NP yield of 95 μg min^{-1} for ligand-free AuNPs was identified for a 0.5 J cm^{-2} fluence, which is a factor of 9 higher than that obtained from ablation with the fs-pulsed laser system[39] and which may be due to optical breakdown phenomena. In addition, the highest obtainable NP concentration was found to be ~ 300 μg mL^{-1} after 5 min ablation time, because self-absorption of the nanoparticles and photofragmentation effects limit a linear development. Interestingly, AuNP bioconjugate yield was generally found to be considerably higher than that for ligand-free AuNPs by a factor of 1.5–3, which is most likely because unstabilized, ligand-free AuNPs tend to form agglomerates at which the loss of laser light appears to be due to scattering issues. Conversely, highly stable nanobioconjugates remain in the solution and increase the particle mass per volume. This result is in agreement with the findings from Petersen et al.[39]

Comparing the maximal yield of AuNP-ssDNA bioconjugates fabricated by the ps-LAL process (168 μg min^{-1}) with the maximum yield for the same AuNP-ssDNA bioconjugates fabricated with a fs-pulsed LAL (11 μg min^{-1}),[39] gives an enhancement factor of 15. Thus in summary, increased yield of approximately one order of magnitude can be reached for ps-PLAL of ligand-free AuNPs or AuNP bioconjugates compared to their fabrication with fs-PLAL. However, for an analytical characterization, it was important to answer the question of whether the ps-PLAL fabricated particles had the same quality and featured the same properties as the fs-PLAL generated ones.

4.1.1.2.   Characterization of ps-LAL fabricated AuNPs

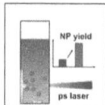 Ablation in the analyzed fluence regime from 0.17 to 1.51 J cm^{-2} resulted in characteristic, red-colored dispersions (**Figure 4.3c**, inset), which is an evidence for the formation of colloidal AuNPs with diameters that are greater than 2 nm and which exhibit the plasmon band absorption. Nanoparticles were found to be of spherical shape (**Figure 4.3a–b**, insets) and size distributions were determined to be dependent on laser fluence with a number-weighted modal nanoparticle diameter of 12 nm for ablation positions near the focal plane (0.7–0.4 J cm^{-2} laser fluence, high NP yield, **Figure 4.3a**). In addition, a number-weighted modal diameter of 34 nm was observed for ablation positions behind the focal plane (0.4–0.1 J cm^{-2} laser fluence, low NP yield), as summarized in **Figure 4.3b**. However, in comparing the size distribution that resulted from ablation near the focal plane with the size distribution obtained for fs-PLAL-generated AuNPs,[39] no significant difference in either the modal particle size or the distribution was registered.

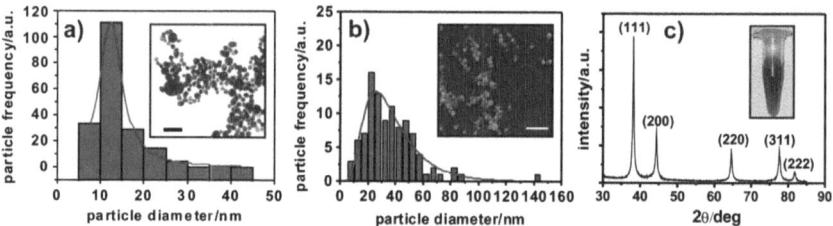

Figure 4.3. The characteristics of ligand-free AuNPs that were fabricated with ps-PLAL. a) Particle size distribution of ligand-free AuNPs fabricated at the focal plane. Corresponding transmission electron micrograph is presented in the inset. b) Particle size distribution of ligand-free AuNPs, fabricated 1 mm behind the focal plane. Corresponding scanning electron micrograph is presented in the inset. c) X-ray diffraction spectrum of ligand-free AuNPs. Scale bars = 50 nm for (a) and 250 nm for (b). Images b) and d) were adapted with permission from Barchanski et al., copyright 2015 by the American Chemical Society.[428]

All fabricated AuNPs exhibited zeta potential values of -20 mV to -33 mV which are characteristic for an electrostatic stable dispersion (**Table 3.20**) and which were in the same range as found for fs-generated AuNPs.[39] Furthermore, X-ray diffraction analysis featured intense diffraction peaks at 2 theta = 38.2°, 44.4°, 64.6°, 77.6° and 81.7° (**Figure 4.3c**) which were indexed to (111), (200), (220), (311) and (222) planes of a face-centered, cubic crystalline structure of three gold modifications (ICSD Collection Codes: 44362 and 53763; NIST M&A Collection code: A 7123 53929).

Thus in summary, a similar quality of ps-PLAL generated AuNPs was found that had equal intrinsic characteristics as exhibited by fs-PLAL generated particles. Thus, switching the PLAL systems from fs to ps pulse duration for highly efficient particle production was possible without comparability issues.

4.1.1.3. Characterization of ps-LAL fabricated AuNP bioconjugates

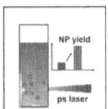 Nanobioconjugates were fabricated with the *in situ* bioconjugation method during ps-LAL, using single-stranded DNA (ssDNA, 5 μM) and bovine serum albumin (BSA, 30 μM) as functional ligands. The electron micrographs of AuNP-ssDNA and AuNP-BSA bioconjugates are presented in the insets of **Figure 4.4a–b**. In contrast to ligand-free AuNPs, a significant narrowing of size distribution to a range from 1 to 50 nm with a number-weighted modal diameter of 9 nm was detected for AuNP-ssDNA bioconjugates due to the size quenching effect (**Figure 4.4a**). Interestingly, for AuNP-BSA bioconjugates, the size quenching effect was not distinct with a size distribution from 1 to 130 nm and a number-weighted modal diameter of 25 nm (**Figure 4.4b**).

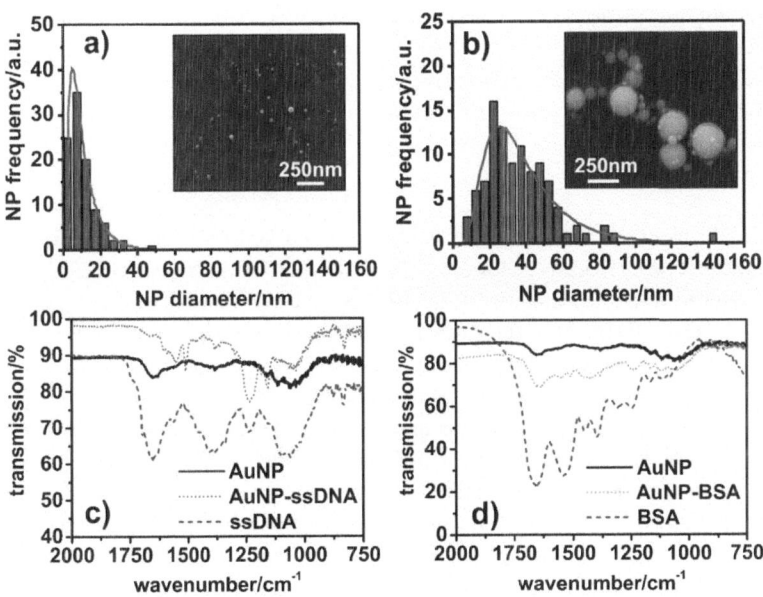

Figure 4.4. The characteristics of ps-PLAL-fabricated AuNP bioconjugates. a) Particle size distribution of AuNP-ssDNA bioconjugates. Scanning electron micrograph is presented in the inset. **b)** Particle size distribution of AuNP-BSA bioconjugates. Scanning electron micrograph is presented in the inset. **c)** FT-IR spectra of ligand-free AuNPs (black solid line), ssDNA (red dashed line) and AuNP-ssDNA bioconjugates (green dotted line). **d)** FT-IR spectra of ligand-free AuNPs (black solid line), BSA (red dashed line) and AuNP-BSA bioconjugates (green dotted line). Images **a)** and **c)** were adapted with permission from Barchanski et al., copyright 2015 by the American Chemical Society.[428]

In this context, it must be considered that BSA was used in a 6-fold higher concentration than ssDNA and that the molarity and thus the density of functional groups (cysteine motives) of BSA were much higher than for ssDNA. The high ligand number could have hindered efficient particle coordination due to enhanced intrinsic molecule interactions. However, calculating the diffusion coefficients for both biomolecules, it was found that the ssDNA molecules featured a 5 times higher mobility in liquids (1.45×10^{-10} m² s⁻¹) than the BSA molecules (3.07×10^{-11} m² s⁻¹). Thus, it is also very likely that the smaller ssDNA molecules are able to diffuse more rapidly, while coordinating the ablated particles and quenching their size to a greater extent than the BSA molecules.

Furthermore, the micrograph of AuNP-BSA bioconjugates (**Figure 4.4b**, inset) featured a significant corona around the nanoparticle's surface, which is most likely an artifact that arises from the interaction of the electron beam with a thick protein-multilayer. Calculating the surface coverage values, a mean number of 6,600 attached BSA molecules per nanoparticle was determined (**Table 4.1**), which supports the assumption of multilayer formation. However, it should be considered that the quite voluminous BSA molecule was dragged into the pellet during ultracentrifugal purification. Although only a single purification step with $14,850 \times g$ was performed, this could be enough force to theoretically centrifuge a particle with 10 nm diameter according to Svedberg relation (**eq 3.8**). Thus, assuming the BSA molecule to have a hydrodynamic size of approximately 14 x 4 x 4 nm as according to Wright et al.,[429] the possibility of undesired centrifugation cannot be excluded.

Because the biomolecule to NP ratio was supersaturated for BSA, a conjugation efficiency of only 20 % was calculated. However, for AuNP-ssDNA conjugates, an efficiency of 80 % resulted in a mean number of 163 ssDNA molecules per NP and 107 pmol cm⁻² surface coverage (**Table 4.1**). These data are highly comparable to the results for fs-based *in situ* conjugation, which were evaluated using a ssDNA with the same length and thiol function and with an additional fluorophore label.[305]

Furthermore, the purified nanobioconjugates were analyzed with UV-vis spectrophotometry and Fourier-Transform Infrared Spectroscopy (FT-IR) to evaluate the conjugation in detail. Distinct UV absorption peaks were observed at 260 nm for ssDNA and at 280 nm for BSA in the nanobioconjugate pellets, while they were absent in the ligand-free AuNP colloid (**Figure SI 6**). This proves that there is a biomolecule presence in the purified pellets. Furthermore, the surface plasmon resonance peak position ($\lambda_{SPR\ max}$) was found to have red-shifted from 525 nm for ligand-free AuNPs to 527 nm for AuNP-ssDNA bioconjugates and to 550 nm for AuNP-BSA bioconjugates (**Figure SI 6**). In accordance with the Drude model, the $\lambda_{SPR\ max}$ strongly depends on the dielectric constant of the surrounding medium as described by **eq 2.4–eq 2.6**. Thus, a local increase in the medium refractive index due to biomolecule presence or a change in the free electron density of the AuNPs because of a strong surface coupling with the biomolecules is most likely the reason of the SPR red-shift.[375;430;374]

Table 4.1. Calculated conjugation efficiencies and surface coverage values for AuNP-ssDNA and AuNP-BSA bioconjugates.

NP Size/nm	Conj. Efficiency/% / Conj. Amount/nmol		Surface Coverage/ # Biomolecules NP-1		Surface Coverage/ pmol cm-2	
	AuNP-ssDNA	AuNP-BSA	AuNP-ssDNA	AuNP-BSA	AuNP-ssDNA	AuNP-BSA
1			0.22	0.42	12	22.36
9[5]	70	20	163	309	107	201
25[6]	/	/	3500	6603	297	560
50	3.5	6	2.8E+04	5.3E+04	593	1118

The FT-IR spectra of purified nanobioconjugates verified successful biomolecule conjugation because they feature particular stretching bands of ssDNA and BSA molecules (**Figure 4.4c–d**). For ligand-free AuNPs, the distinct bands at 1650 cm^{-1}, in the range of 1380–1420 cm^{-1} and at 1050 cm^{-1} are highly comparable to the spectra of AuNPs that are gained by biological synthesis (**Figure 4.4c, Table SI 2**).[431;432] These results indicate metal-carbonato coordination and therefore the presence of Au-O compounds at the AuNP surface and are in line with data published by Sylvestre et al. regarding gold nanoparticles that were fabricated with fs-pulsed LAL.[325]

The ssDNA spectra featured characteristic peaks of the nucleosides (adenine, guanine, cytosine and thymine), ribose sugar and the DNA phosphate backbone and correspond to the natural B-form DNA (**Figure 4.4c, Table SI 2**).[433-435] The conjugation of ssDNA to AuNPs was clearly determined by the distinct peaks at 1560 cm^{-1}, 1510 cm^{-1}, 1156 cm^{-1}, 1234 cm^{-1} and 833 cm^{-1}. The other bands were also observed in weak frequency but they could not be analyzed because they overlapped with bands of the ligand-free AuNP spectra. Because the ssDNA peaks, including the peak that was at 1234 cm^{-1}, were clearly identified for nanobioconjugates, it can be assumed that the secondary conformation of DNA remained in B-form after conjugation and did not change to A-form or Z-form (**Figure SI 7**). However, the intensity enhancement of the bands from aromatic amines and P-O/P=O stretchings may indicate a shift of ssDNA conformation from a coiled structure to a stretched alignment on the AuNP surface, which is most likely due to a close package.[420] Thereby, the phosphate backbone is elongated, which allows for different electrostatic interactions between the aromatic rings of nucleotides.

---

[5] number-weighted modal diameter of AuNP-ssDNA conjugates.
[6] number-weighted modal diameter of AuNP-BSA conjugates

The BSA molecule featured characteristic and very intense FT-IR bands of amide I, amide II and amide III (**Figure 4.4d, Table SI 2**).[436-439] For laser-generated AuNP-BSA bioconjugates, the amide I band could not be clearly assigned since the broad C=O stretching on a ligand-free AuNP overlaps with the amide I band of BSA. However, a small and sharp peak maximum at 1632 cm[-1] was detected, which could indicate a change from a BSA α-helix structure to an extended chains plus β-sheet. In contrast, a peak maximum of the amide II band was clearly determined for nanobioconjugates. Interestingly, another peak at 1511 cm[-1] was found for AuNP-BSA bioconjugates in the amide II region, which could be an indication of a BSA-AuNP bond by NO_2 valence. Moreover, in analyzing the amide III region, it was clearly observed that a peak maximum of BSA at 1311 cm[-1] was no longer visible for AuNP-BSA bioconjugates, while the BSA peak maximum at 1243 cm[-1] was detectable for AuNP-BSA bioconjugates and accompanied by an intensity increase. This confirms a conformation change of the BSA structure on the surface of nanobioconjugates from α-helix to β-sheet as indicated by the peak frequency changes of amid I. This result is in perfect agreement with data from Servagent-Noinville et al., that showed a decrease in α-helix content and an increase in intermolecular β-sheet content after adsorption of BSA on montomorillonite.[440]

In conclusion, the data indicate the covalent attachment of ssDNA and BSA molecules to AuNPs during ps-pulsed *in situ* bioconjugation with comparable surface coverage results as determined for the fs-pulsed approach. However, FT-IR results depict significant conformation changes of biomolecules after conjugation, which raises a question of their integrity and functionality after the laser beam interaction.

4.1.1.4.   *Integrity of biomolecules after in situ conjugation*

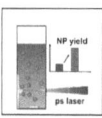 The integrity of ssDNA was analyzed after *in situ* bioconjugation and subsequent separation from the AuNP with dithiotheriol (DTT) using gel electrophoresis. The determined results were compared with an untreated ssDNA sample (**Figure SI 8**). The analysis range covered the ablation positions from 0 mm (defined as a position with a determined focus on the target in air) to +1 mm (defined as a position with a determined focus in front of the target) and to -2 mm (defined as a position with a determined focus behind the target), different laser powers from 0.5 W to 1 W with a variation of the pulse energy/repetition rate combination and ablation times that varied from 15 s to 120 s. The results are summarized in **Figure 4.5a–c** (for 0.5 W) and **Figure SI 9** in the supporting information (for 0.75 W and 1 W, respectively).

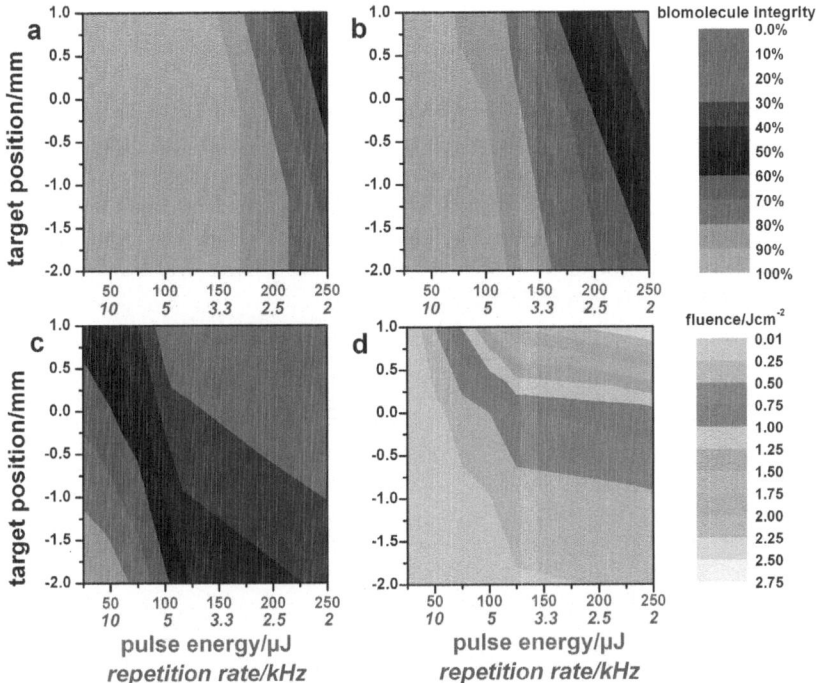

Figure 4.5. Results of DNA integrity study after ps-pulsed *in situ* bioconjugation of AuNPs. a)–c) Integrity of ssDNA after 15 s **(a)**, 45 s **(b)** and 120 s **(c)** of ablation using different target positions and pulse energy/repetition rate combinations for a laser power of 0.5 W. Target position 0 is defined as the position of the determined focal point in air, while positive and negative target positions are defined as positions in front of and behind the 0 position, respectively. **d)** Determined laser fluence for the analyzed parameter regime. Reprinted with permission from Barchanski et al., copyright 2015 by the American Chemical Society.[428]

In detail, a systematic increase in degradation was observed for *i)* ablation time prolongation for each power series; *ii)* power enhancement for each ablation time; and *iii)* pulse energy enhancement/repetition rate decrease. Regarding ablation position, higher integrity values were observed in the range from -0.5 mm to -2 mm, corresponding to a low fluence regime of 0.01 J cm^{-2} to 0.5 J cm^{-2} (**Figure 4.5d**) while higher degradation was found in the range from -0.5 mm to 1 mm, corresponding to a higher fluence regime of 0.5 J cm^{-2} to 2.75 J cm^{-2} (**Figure 4.5d**). These results defined a distinct parameter window (highlighted with a light green color in **Figure 4.5a–c** and **Figure SI 9**, which is highly suitable for the ps-pulsed *in situ* bioconjugation. Overall, the data are in agreement with results published by Petersen et al. regarding the integrity of ssDNA after fs-pulsed *in situ* conjugation.[39] Furthermore, they correlate with results from Takeda et al., who determined that ssDNA decomposition after laser irradiation was fluence-dependent.[357]

Interestingly it was found that the ssDNA decomposition can also be monitored by UV-vis spectrophotometry. During the re-irradiation of ps-PLAL-generated AuNP-ssDNA bioconjugates with a fluence of 165 J cm^{-2}, a significant modification of the nanobioconjugate absorption in the UV region was registered in relation to the irradiation time. In detail, the peak maximum at 260 nm and the SPR peak were both reduced by 35 % for 600 s of irradiation, while the maximum at 208 nm increased by 45 %, respectively (**Figure 4.6, Figure SI 10**).

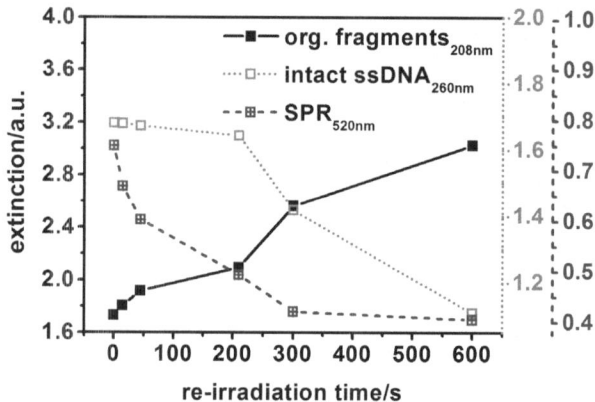

Figure 4.6. Correlation between DNA decomposition and AuNP photofragmentation. Trends concerning the decrement of SPR extinction (at 520 nm, red dashed line) and intact ssDNA extinction (at 260 nm, green dotted line) and the rise of ssDNA-fragment extinction (at 208 nm, black solid line) with progressing re-irradiation time of AuNP-ssDNA bioconjugates. Adapted with permission from Barchanski et al., copyright 2015 by the American Chemical Society.[428]

It is known, that the aromatic bases of DNA nucleotides feature a characteristic absorption of approximately 260 nm in B-form DNA,[441] while an absorption at 208 nm can be assigned to the sugar and phosphate components of the DNA backbone.[441] From this knowledge it appears to be certain that the ssDNA decomposes during re-irradiation since the nucleotide is significantly reduced while the amount of free sugar and phosphate fractions increase. In fact, similar results were also obtained by Giusti et al. and Giorgetti et al. using dendrimer-coupled AuNPs. They discovered a significant photodegradation effect during ps-pulsed laser fragmentation, which was illustrated by increased UV contribution.[342;341] Thus, UV-vis spectrophotometry can be applied for the straightforward qualitative integrity evaluation of AuNP-ssDNA bioconjugates.

However, the sensitivity of the spectrophotometer should be carefully considered for the evaluation. With the spectrophotometer used in this study, a minimum ssDNA concentration of 0.15 μM could be detected which was referring to an extinction of approximately 0.022. Thus, a reduction in ssDNA concentration by more than 0.15 μM was the minimum that could be correlated to biomolecule decomposition in this thesis.

In summary, considering biomolecule integrity after *in situ* bioconjugation, a process parameter window was defined for optimal ps-pulsed fabrication that yields AuNP bioconjugates with nearly 100 % biomolecule integrity at a concentration of ~ 100 μg mL^{-1}. Furthermore, the decomposition of ssDNA can also be monitored on-line during fabrication using UV-vis spectrophotometry for instance.

4.1.1.5.   *Functionality of ps-LAL fabricated AuNP bioconjugates*

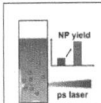 To analyze the functionality of ps-fabricated AuNP bioconjugates, *golden blot* (immunoblotting) assay was performed. Adopting IgG and the cell-penetrating-peptide TAT (negative control) as analytes and PLAL-fabricated, ligand-free AuNPs (negative control), commercially available AuNP-anti-IgG$_{comm}$ bioconjugates (positive control) and the AuNP-anti-IgG$_{PLAL}$ bioconjugates as detection samples, the formation of red spots on the membranes was investigated. The results are presented on **Figure 4.7**.

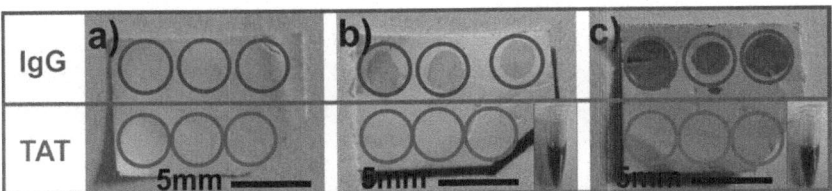

Figure 4.7. Functionality proof of ps-PLAL generated AuNP bioconjugates by a *golden blot* assay. IgG was immobilized on the membrane within the red circles and TAT within the purple circles, respectively. a) The membrane was incubated with ligand-free AuNPs as negative control. b) The membrane was incubated with commercially available AuNP-anti-IgG bioconjugates as positive control. c) The membrane was incubated with PLAL-generated AuNP-anti-IgG bioconjugates. Test tubes including the employed conjugate solutions are shown in the inset. Reprinted with permission from Barchanski et al., copyright 2015 by the American Chemical Society.[428]

No spots at all were detected on the negative control membrane (**Figure 4.7a**), which implies that the ligand-free AuNPs are unable to bind to the proteins. In contrast, on the positive control membrane (**Figure 4.7b**) red spots were clearly indicated, highlighting a significant labeling of the IgG analyte with the commercial AuNP-anti-IgG$_{comm}$ bioconjugates, while no labeling occurred with the non-specific TAT analyte. A very similar result was obtained for the ps-PLAL-generated AuNP-anti-IgG$_{PLAL}$ bioconjugates (**Figure 4.7c**), however the labeling intensity was significantly increased by a factor of 4.4 (approximation by the Image J software). The immobilized amount of analyte on the membrane and the applied concentration of AuNP-anti-IgG bioconjugates solutions were kept constant for all cases. Thus, the enhanced labeling intensity must be associated with an improved *quality* of AuNP-anti-IgG$_{PLAL}$ bioconjugates compared to AuNP-anti-IgG$_{comm}$ bioconjugates. Unfortunately, no details about the surface coverage of commercial AuNPs with anti-IgG antibodies were provided by the manufacturer. However it seems certain that two thing could occur: *i)* according to the high surface coverage of AuNP-anti-IgG$_{PLAL}$ bioconjugates with correctly oriented antibodies, the binding efficiency could be significantly enhanced. In this case, the labeling intensity would be NP number-dependent; *ii)* a higher amount of large nanoparticles (> 50 nm) with an increased antibody-to-particle ration than for small nanoparticles (< 50 nm) could have bound the analyte, which would most likely be due to an exceeding polydispersity of laser-generated AuNP-anti-IgG$_{PLAL}$ bioconjugates compared to the chemically-derived AuNP-anti-IgG$_{comm}$ bioconjugates. These large nanoparticles feature different optical properties with light absorption in the NIR regime and exhibit enhanced scattering characteristics. In this case, the labeling intensity would be NP size-dependent.

With these results, the specific functionality of ps-PLAL fabricated AuNP bioconjugates was verified for a laboratory assay and the nanobioconjugates were further tested on an *in vitro* system using immunolabeling technique. For this intent, the cellular membrane-cytoskeleton protein vinculin of human fibroblasts was targeted by a primary anti-vinculin antibody and then the anti-vinculin antibody was targeted by the ps-PLAL fabricated, fluorophore-coupled AuNP-anti-IgG$_{FITC}$ and conventional anti-IgG$_{FITC}$ secondary antibodies (positive control). The labeling was compared to an untreated negative control and results are presented on **Figure 4.8**.

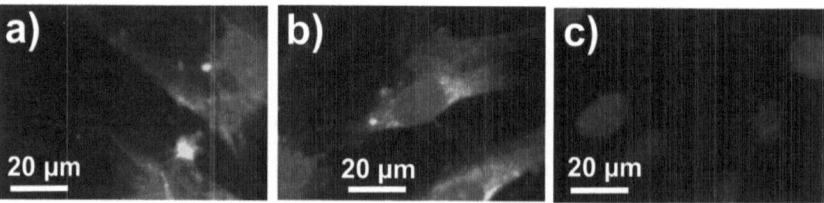

Figure 4.8. Functionality proof of ps-PLAL generated AuNP bioconjugates by immunolabeling. a) Immunolabeling of the membrane-cytoskeleton protein vinculin in human fibroblasts using laser-generated AuNP-anti-IgG$_{FITC}$ bioconjugates. **b)** Immunolabeling of the membrane-cytoskeleton protein vinculin in human fibroblasts using commercially available anti-IgG-FITC secondary antibodies. **c)** The untreated, negative control. Vinculin = green, cell nuclei = blue, scale bars = 20 μm. Reprinted with permission from Barchanski et al., copyright 2015 by the American Chemical Society.[428]

As illustrated in **Figure 4.8a–b**, no significant difference was observed for the vinculin labeling between AuNP-anti-IgG$_{FITC}$ bioconjugates and commercial secondary antibodies, neither in labeling amount nor in labeling intensity. However, no vinculin signal was found in the untreated, negative control sample (**Figure 4.8c**).

In summary, the results of the immunoblotting assay and *in vitro* labeling imply that the biomolecules that are attached to AuNPs with ps-pulsed *in situ* conjugation are fully active and feature specific functionality with the same or even enhanced labeling intensity as conventional labeling markers. Thus, the ps-pulsed fabrication of AuNPs and AuNP bio-conjugates allows for significant yield enhancement compared to fs-PLAL, producing nearly 100 % integrity-preserved and active nanoconjugates.

4.1.2. Increasing the concentration of PLAL-generated AuNPs

4.1.2.1. Increasing the concentration by ultrafiltration

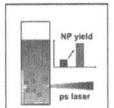 From a biologist's point of view, ultrafiltration is a favorable post-processing method for increasing the concentration of macromolecules in a solution. Due to comparable dimensions, it may be beneficial to use this technique to increase the concentration of nanoparticles as well. In contrast to the centrifugation process, the ultrafiltration is far gentler on the particles because a high-speed is not required for the sedimentation of light-weight fractions. In detail, the solvent is pressurized through a membrane using spin-flow and the macromolecules retain on the membrane (retentate) and can be resuspended in a low volume. Using this method, concentration factors of 30-50 can be reached using commercial filter systems and a 500 μL start volume. However, interactions of AuNPs with the filtration column materials may limit the applicability. The retention property of ultrafiltration is expressed as the *molecular weight cutoff* (MWCO) of the applied membranes, which is correlated to the *molecular weight* (MW) of the filtered molecules/particles. It is recommended to select a MWCO that is 3 to 6 times smaller than the MW of the particles that shall be retained. The MW of AuNPs can be calculated for different diameters, which result in an exponential development (**Figure SI 3**). In this calculation, the smallest MW is 6x10^4 g mol^{-1} for 1 nm particle size and 8x10^8 g mol^{-1} for a 50 nm-sized particle.

Because PLAL-generated AuNPs are polydisperse and feature a broad particle size distribution, it was necessary to screen various MWCOs that ranged from a 3 kDa to 300 kDa pore size in order to achieve optimal retention. The effect of membrane material was analyzed by using two different ultrafiltration tubes. *Vivacon*® tubes feature a *Hydrosart*® regenerated cellulose membrane and a polycarbonate/polypropylene tube body. In contrast, *Nanosep*® tubes exhibit an *Omega*® membrane (polyethersulfone, PES, modified to minimize protein binding) and a low-binding polypropylene filter body.

For analyzing the efficiency of ultrafiltration method to increase the concentration of PLAL-generated nanoparticles, ligand-free AuNPs (generated in *MilliQ*) and stabilized AuNPs that were fabricated by *in situ* conjugation to mPEG-SH (methoxyl polyethylene glycol thiol, 5 kDa) were used in this study. Both colloids featured a similar modal particle diameter of approximately 36 nm (**Table 4.2**).

Table 4.2. Characteristics of ligand-free AuNPs and AuNP-mPEG-SH conjugates that were applied for the ultrafiltration study.

	AuNP	AuNP-mPEG-SH
Concentration/µg mL^{-1}	86	86
d$_{DLS}$/nm	74	76
Mean d$_{SEM}$/nm	38	35
PDI	0.20	0.14
Zeta Potential/mV	-28	-32

However, the PSD was much broader for ligand-free AuNPs (**Figure 4.9a**) than for AuNP-mPEG-SH conjugates (**Figure 4.9b**) because no size quenching occurred in the stabilizer-free *MilliQ*. Moreover, the conjugation with mPEG-SH slightly increased the zeta potential of nanobioconjugates and their hydrodynamic diameter (**Table 4.2**).

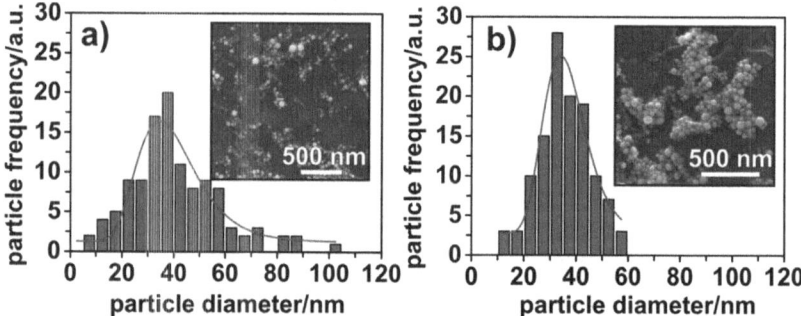

Figure 4.9. Particle size distributions of the adopted samples. a) Particle size distribution with logNormal fitting of ligand-free AuNPs, featuring a number-weighted modal particle core size of 38 nm. b) Particle size distribution with logNormal fitting of mPEG-SH-stabilized AuNPs featuring a number-weighted modal particle size of 35 nm. Scanning electron micrographs of the samples are presented in the inset. Scale bars = 500 nm.

The colloidal solutions were used for the ultrafiltration study with 3 *Vivacon®* and 3 *Nanosep®* tubes of varying MWCOs. A detailed sample overview is shown in **Table 4.3** and the results have been summarized in **Figure 4.10**.

Table 4.3. Sample overview of the ultrafiltration study on the increase of nanoparticle concentration.

	Vivacon®					
MWCO	10 kDa		30 kDa		50 kDa	
Solvent	MilliQ	mPEG	MilliQ	mPEG	MilliQ	mPEG
ID	1	2	3	4	5	6
	Nanosep®					
MWCO	3 kDa		30 kDa		300 kDa	
Solvent	MilliQ	mPEG	MilliQ	mPEG	MilliQ	mPEG
ID	7	8	9	10	11	12

Photographs of the filter membranes and the redispersed retentate after the filtration run are presented in **Figure 4.10a** and **Figure 4.10c**, while the NP concentration data prior to and after ultrafiltration have been summarized on **Figure 4.10b** and **Figure 4.10d**.

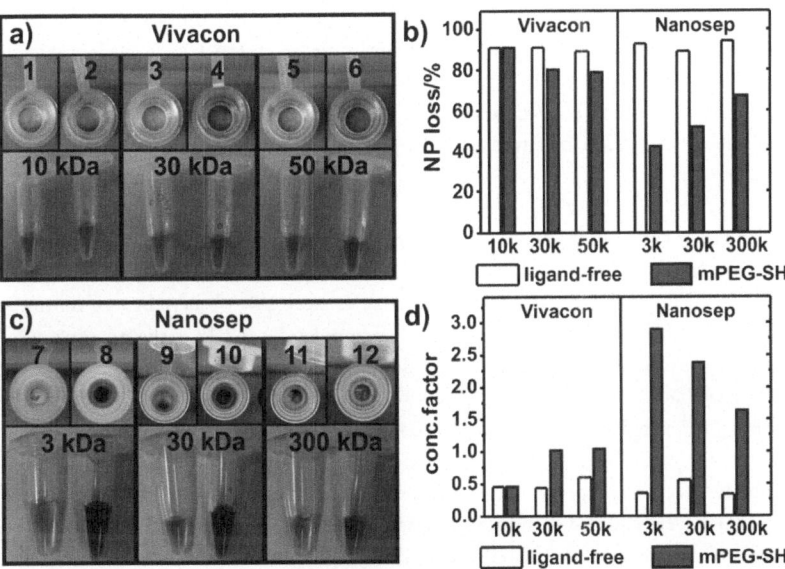

Figure 4.10. Results of the ultrafiltration study for the increase of nanoparticle concentration. a) Photographs of retentates on *Vivacon®* tube membranes and after redispersion. b) The amount of nanoparticle mass loss after ultrafiltration with *Vivacon®* and *Nanosep®* tubes (ligand-free AuNPs: white bars, mPEG-SH-conjugated AuNPs: red bars). c) Photographs of retentates on *Nanosep®* tube membranes and after redispersion. d) The concentration factors that can be achieved with ultrafiltration using *Vivacon®* and *Nanosep®* tubes (ligand-free AuNPs: white bars, mPEG-SH-conjugated AuNPs: red bars).

On all *Vivacon®* tube membranes a dark-pink to purple retentate was observed (**Figure 4.10a**), while a clear liquid was found in the filtrate. It was hardly possible to redisperse the retentates in fresh *MilliQ*. However, the retentate of AuNP-mPEG-SH conjugates was a little bit easier to redisperse than the retentate of ligand-free AuNPs, which adhered extremely to the membrane corners. All applied MWCOs were small enough to prevent particle loss through the membrane, because no NPs and thereby no liquid coloration was found in the filtrates (**Figure 4.10a**). The normalized UV-vis spectra of all retentates featured an extinction shoulder that ranged from 600-750 nm in addition to the SPR peak (**Figure SI 4**). This indicated an aggregated NP subpopulation that could not be disrupted by sonification.

A significant loss of nanoparticle mass of approximately 90 % was determined for ligand-free AuNPs, which was independent from the applied MWCO (**Figure 4.10b**). This result is in agreement with the high number of particles that stuck to the filter membrane and that could not be redispersed. On the contrary, a similar loss of nanoparticle mass was found for AuNP-mPEG-SH conjugates only on the 3 kDa MWCO membrane, while for 10 kDa and 50 kDa membranes, a particle loss of approximately 80 % was determined (**Figure 4.10b**). If the solvent reduction from 500 µL (prior ultrafiltration) to 100 µL (redispersed volume) and the high nanoparticle losses are taken into consideration, then concentration factors of 0.5 to 1 are calculated (**Figure 4.10d**). These values are equal to or less than the colloid concentration prior ultrafiltration. Thus in summary, no increase in AuNP concentration occurred with *Vivacon*® ultrafiltration using a *Hydrosart*® membrane.

Conversely, for the ultrafiltration with *Nanosep*® tubes, a purple retentate was only found for ligand-free AuNPs, while a reddish retentate was discovered for AuNP-mPEG-SH conjugates (**Figure 4.10c**). The stabilized particles were also observed to be more easily redispersable than the ligand-free AuNPs. For samples 11 and 12, a slightly pink-colored filtrate was recovered, which indicated particle loss through the membrane and highlighted the fact that the 300 kDa MWCO was too large to retain the AuNPs (**Figure 4.10c**). However, for samples 7, 8 and 10 it was barely possible to filter the entire liquid volume (even though centrifugation time and speed were increased), which is most likely due to a membrane pore-blockage of large NPs (**Figure 4.10c**). Thus, retentate volumes of 150–200 µL were recovered and no additional liquid was added for redispersion. The differing liquid levels were considered when calculating the concentration factors.
Similar to the *Vivacon*® tubes, high nanoparticle mass losses of approximately 90 % were determined for ligand-free AuNPs with all MWCOs (**Figure 4.10b**). Furthermore, an aggregated NP subpopulation was also found for these retentates (**Figure SI 4**). Interestingly, for AuNP-mPEG-SH conjugates a significantly lower particle loss of 40 to 65 % was determined for 3 kDa to 300 kDa tubes, respectively (**Figure 4.10b**). If solvent reduction and the nanoparticle losses are again taken into consideration, then concentration factors of 1.5 to approximately 3 were determined for AuNP-mPEG-SH conjugates (**Figure 4.10d**). Conversely, for ligand-free AuNPs no increase of concentration was enabled (**Figure 4.10d**). However, for AuNP-mPEG-SH sample 8 that was filtered through

the 3 kDa membrane and that featured the highest concentration factor, an aggregated particle population was also determined in the UV-vis spectra (**Figure SI 4**). Therefore, the AuNP-mPEG-SH sample 10 that was filtered through the 30 kDa membrane without particle aggregation was found to reach the best concentration factor of 2.2.

The different efficiencies for nanoparticle concentration that were found for *Vivacon®* and *Nanosep®* ultrafiltration tubes, can be most likely explained by the difference in membrane material that was aforementioned. In detail, a *Hydrosart®* hydrophilic regenerated cellulose membrane (RC) was used in *Vivacon®* tubes, while an *Omega®* hydrophobic polyethersulfone membrane (PES) with reduced protein binding was used in *Nanosep®* tubes. Polyethersulfone membranes generally feature a zeta potential of approximately – 10 mV at neutral pH with a pI of ~ 3.[442] Moreover, in 1993 Clark and Juncker reported about a less negative zeta potential for regenerated cellulose membranes compared to polysulphone materials.[443;444]

In this study, the water-based colloids featured a zeta potential of -28 and -32 mV, respectively. Therefore, it can be speculated, that they were stronger repelled from the hydrophobic PES *Omega®* membrane with higher negative zeta potential and more attracted from the RC *Hydrosart®* membrane with less negative zeta potential.

However it should be noticed, that the modifications of PES and RC membranes were not indicated by the manufacturer and could result in different zeta potentials for the membranes used in this study.

Thus in conclusion, the *Nanosep®* filtration using a hydrophobic *Omega®* polyethersulfone membrane at neutral pH may results in an approximate increase in AuNP concentration by a factor of 2–3 if stabilized NPs and an optimal MWCO are applied. The optimal MWCO must generally avoid both particle loss into the filtrate and particle shape modification from aggregation and should also feature a good re-dispersability of AuNPs without high particle loss on the filtration membrane.

However, having screened for an optimal ultrafiltration system that would increase the AuNP concentration, the functionality of the up-concentrated nanobioconjugates had to be further verified. Thus, a *golden blot* (immunoblotting) assay was performed using a functional antibody against IgG (anti-IgG) coupled with the PLAL-generated gold nanoparticles using a heterobifunctional hydrazide-PEG-dithiol linker and yielding AuNP-anti-IgG bioconjugates that had a mean size of 9 nm. The bioconjugates featured a calculated surface coverage of approximately 10 antibodies per nanoparticle (**Figure SI 5**). Nanobioconjugate concentration was determined to be 83 µg mL^{-1} after conjugation and 207 µg mL^{-1} after an increase in concentration with a 30 kDa *Omega®* filtration membrane was performed. This concentration increase was corresponding to a concentration factor of 2.5. Applying both IgG and the cell-penetrating-peptide TAT (negative control) as analytes and the ligand-free AuNPs (negative control) and the up-concentrated AuNP-anti-IgG$_{UFconc}$ bioconjugates as detection samples, the formation of red spots on the membranes was examined. The results are summarized in **Figure 4.11**.

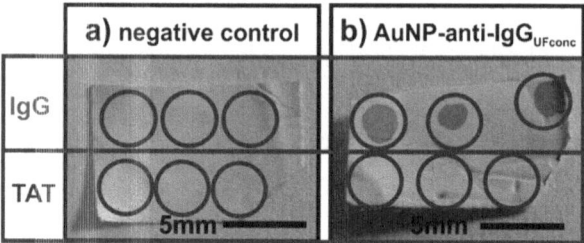

Figure 4.11. Functionality proof of the up-concentrated AuNP bioconjugates using a *golden blot* assay. Immunoglobulin (IgG) was immobilized on the membrane within the **red circles** and TAT within the **purple circles**, respectively. **a)** The membrane was incubated with ligand-free AuNPs as negative control. **b)** The membrane was incubated with PLAL-generated and by 3 kDa Omega® membrane up-concentrated AuNP-anti-IgG$_{UFconc}$ bioconjugates.

Upon analyzing the negative control membrane, no spot was determined, neither for the IgG analyte, nor for the TAT control protein as was anticipated (**Figure 4.11a**). However, for the concentrated AuNP-anti-IgG$_{UFconc}$ bioconjugates, clear dark-red spots on the IgG analyte were visible with the naked eye. The control protein was not labeled, which indicates the efficient specificity of the nanobioconjugates (**Figure 4.11b**). These results verify that AuNP bioconjugates that were up-concentrated using a 30 kDa Omega® filtration membranes are still functional and that they have not been damaged, separated or inactivated by the treatment.

4.1.2.2.   *Increasing the concentration by solvent evaporation*

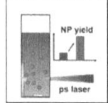

Another interesting approach to increase the concentration of PLAL-fabricated AuNPs is the post-processing reduction of solvent volume using liquid vaporization into a gaseous phase. For instance, heating a solution affects a rapid liquid evaporation. However, controlled heating of small volumes without complete liquid drying is not trivial and the heat may also increase the velocity and therefore the chance of NP collision, which leads to particle aggregation and sedimentation. Thus, steady-state evaporation under an extractor hood without heating could be a gentle alternative. In this case, all variables of the system are kept constant while evaporation takes place under standard conditions as result of increased entropy. The benefit of agitation on the system could be considered promising and should be compared to the steady system.

The same ligand-free AuNPs and AuNP-mPEG-SH conjugates as those used for ultrafiltration experiments were applied for this study (**Table 4.2, Figure 4.9**) and both samples were evaporated without heating but once with and once without magnetic stirring. A detailed sample overview is summarized in **Table 4.4**.

Table 4.4. Sample overview for the increase of nanoparticle concentration using evaporation technique.

Nanoparticles	Treatment	Sample ID
Ligand-Free AuNPs (ddH₂O)	agitated evaporation	1
	steady-state evaporation	2
	prior evaporation	3
Stabilized AuNPs (mPEG-SH)	prior evaporation	4
	steady-state evaporation	5
	agitated evaporation	6

The nanoparticle concentration was constantly monitored with UV-vis spectrophotometry during the experiments and the results are presented in **Figure 4.12**. The increase in concentration due to solvent evaporation was highly effective for the four samples 1–2 and 5–6 (**Figure 4.12a**). An average increase in ligand-free AuNP concentration by a factor of 4 was determined for the homogeneous system, while the concentration rose exponentially for the agitated system, reaching a concentration factor of 10 (**Figure 4.12b**). A similar scenario was observed for the mPEG-stabilized AuNPs with a concentration factor of 8 for the homogeneous system and of 13 for the agitated version (**Figure 4.12b**). Thus, agitation appears to have a positive impact on evaporation with a higher molecular exchange at the air/water interface.

Interestingly, when the evaporation time of the stationary sample exceeded 6 hours and when a NP concentration of 1.5 mg mL^{-1} was reached, flat, golden-colored, spicular organized, macroscopic structures were clearly visible on the air/water interface with the naked eye, for the ligand-free AuNPs (**Figure 4.12c**). However, no sedimentation on the glass bottom was found. Analyzing the macroscopic structures with SEM, a solid gold formation with three-dimensional character was observed (**Figure 4.12d–e**). However, in high magnification some gold nanoparticle aggregates were identified, which indicate a self-assembled coalescence of the particles, yielding a gold hard-aggregated superstructure (**Figure 4.12f**).

The self-assembly of inorganic, nanoscaled materials with capillary force and surface reactions into well-defined one-dimensional, two-dimensional or three-dimensional superstructures has already been discussed in the literature.[445-448] Capillary forces are interactions between particles that are mediated by fluid interfaces. They arise from the Laplace pressure as a result of the overlap of menisci which form from the condensation of liquid around two separate, adhering particles.[449] During rapid destabilization of nanoparticle dispersion, e.g. with heating, the interparticle capillary forces lead to close-packed aggregates because particles quickly adhere to each other and sedimentate from the solution.[450;451]

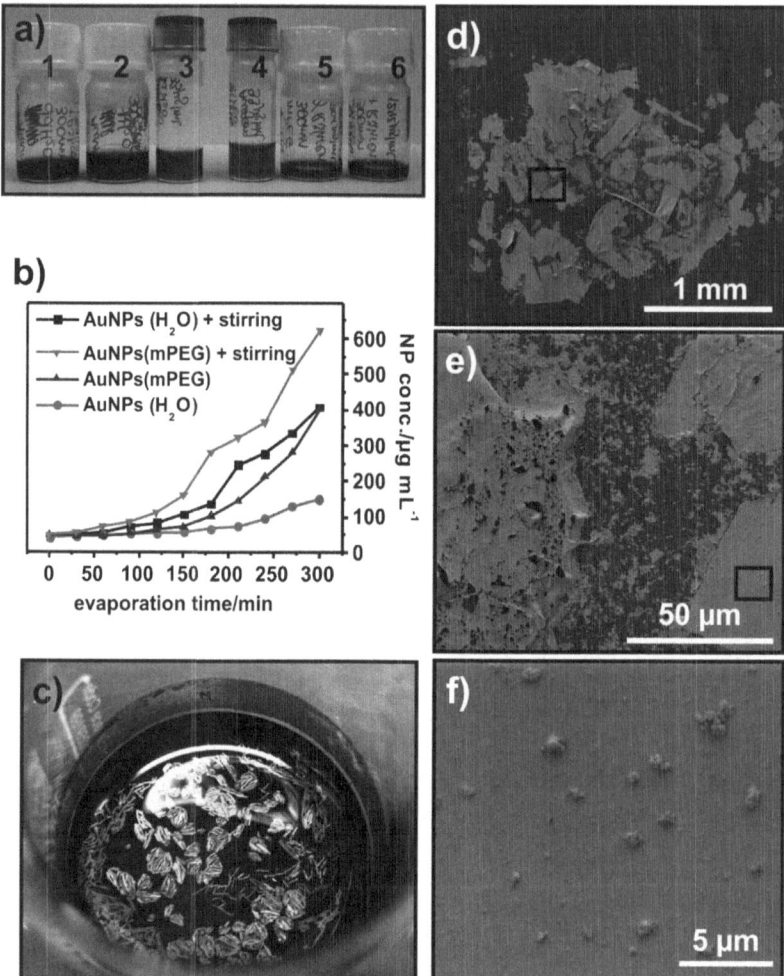

Figure 4.12. Results of the controlled evaporation of AuNPs and AuNP bioconjugates.
a) Photography of samples prior (3&4) and after increase of concentration (1–2 & 5–6) with illustration of
solvent reduction within 6 hours from full bottles (1–2 & 5–6) to lower liquid levels. b) Increase of
NP concentration by evaporation time. c) Macroscopic clustering of AuNPs for AuNPs(ddH$_2$O) sample
of 2 mg mL^{-1} concentration. d)–f) Scanning electron micrographs of macroscopic AuNP formation with
increasing magnification, f) = magnification of boxed area in e) and e) = magnification of the boxed area
in d).

At a slower destabilization, e.g. with evaporation, the high surface energy of the particles,
the contacting cores and the presence of water then catalyze the chemical sintering of the
cores with nanoparticle coalescence reactions. This yields ordered superlattices that ho-

mogeneously nucleate in solution.[452] For example, Nikoobakht et al.[453] showed that capillary forces induce gold nanorods to align parallel to each other. Recently, Mandal et al. proposed the concept of gold cold welding for AuNP assembly including the fabrication of gold networks.[196] Actually, if the electron micrographs of evaporating stationary samples were monitored during the first two hours, a progressive coalescence from chemical sintering or cold welding can be observed (**Figure 4.13a–c**) along with a distinct formation of neck-like contacting joints (**Figure 4.13d**).

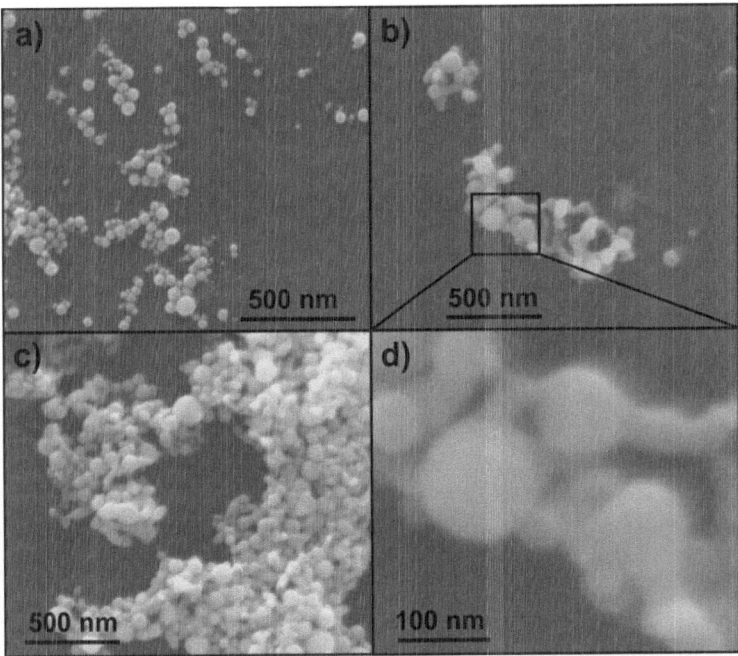

Figure 4.13. **Time-dependent process of macroscopic AuNP formation due to chemical sintering/cold fusion. a)** The progressive coalescence of primary gold nanoparticles by chemical sintering/cold fusion. **b)–c)** Formation of two- and three-dimensional superstructures after 2 hours of sintering/cold fusion. **d)** Magnified section illustrating the contacting cores of nanoparticles during coalescence process.

Interestingly, less assembled structures were observed for the agitated samples of ligand-free AuNPs. This was most likely due to the fast particle motion that worked against the capillary forces and yielded a more homogeneous solution up to a concentration of ~ 4 mg mL^{-1}. However above this concentration, a rapid destabilization and sedimentation of nanoparticles was observed. The formation of gold superstructures was also found for AuNP-mPEG-SH conjugates that were not agitated; however, they appeared at much higher concentrations of approximately 3 mg mL^{-1}. Conversely, for the stirred AuNP-mPEG-SH sample no network formation was registered.

The increase in concentration was indicated by a significant solvent reduction from full bottles (5 mL) to a final volume of less than 1 mL (**Figure 4.12a**). In addition, there was also a darker liquid coloration (**Figure 4.12a**) compared to the initial colloidal solutions (3&4). However, the loss of volume and therefore the increase in concentration was higher for the AuNP-mPEG-SH sample than for the sample which included the ligand-free AuNPs (**Figure 4.12b**). Interestingly, the evaporation of a *MilliQ* water sample without nanoparticles resulted in an even higher volume loss than for the AuNP-mPEG-SH sample. Thus, an increased evaporation in the order of: *MilliQ* > AuNP-mPEG-SH > ligand-free AuNPs seems to have occurred. It can be speculated about the reason for this difference in evaporation.

Due to vapor pressure, the water molecules are able to overcome the surface tension of the water and the ambient pressure and they can cross the liquid-air interface to become water vapor. Considering an open system such as used in the experiment, then the water molecules will evaporate when the vapor pressure is higher than the ambient pressure and the surface tension becomes zero. However, the vapor pressure is also dependent on the temperature and the intramolecular forces of the liquid.
If the system consists of more than one component (e.g. mixture of solvents, solvent and solute), then the partial vapor pressure of each component is equal to the partial vapor pressure of the pure component multiplied with its mole fraction in the mixture.[454] This is defined as *Raoult's law*. Thus, a change in the vapor pressure is dependent on the amount of (solute) molecules and not on their chemical properties.
It is mostly probable, that solute molecules in the solution will take up spaces at the surface of the solution. In consequence, this will limit the number of solvent molecules at the surface, which could evaporate. Thus, if a non-volatile solute is dissolved in a solvent, then the vapor pressure of the final solution will be lower than the one of the pure solvent. To reach the boiling point of the solvent, the vapor pressure has to be raised by the input of more energy. Thus, a non-volatile solute raised the boiling point of the solution. Assuming the transferability of this explanation from non-volatile solutes onto non-volatile colloidal particles, this could explain the slower evaporation of the nanoparticle-containing samples compared to the pure *MilliQ* sample.

The main difference between both nanoparticle samples is the ligand conjugation of the AuNP-mPEG-SH conjugates. These ligands could possibly increase the vapor pressure of the water; especially if they were separated from the AuNPs. However, the AuNP-mPEG-SH conjugates were purified by ultracentrifugation before they were used for the experiments. Moreover, the ligands were attached with a thiol function, which generally should not separate or degrade in solution without heating or an additional treatment of reduction agents. Thus, an effect of separated ligands on the evaporation speed can most likely be excluded.

The colloidal solutions was used for the experiments had the same mass concentration and because of the comparable modal diameters it can be assumed that they also had nearly the same number concentration (**Figure 4.9, Table 4.2**). However, the AuNP-mPEG-SH bioconjugates were highly stable in solution and repelling each other even at increased concentration, while the ligand-free AuNPs were affected by superstructure formation as aforementioned (**Figure 4.12c–f**). This three-dimensional network was covering a significant amount of the surface area and thus most likely hindered the water molecules from evaporation while limiting the total volume loss.

However, due to a required initial volume of > 10 mL and relatively long processing times, the approach of post-processing evaporation for increasing the AuNP concentration is rarely applicable for sensitive nanobioconjugates, which may undergo a structure alteration accompanied with activity reduction. Thus, no functionality analysis of AuNP-coupled antibodies was performed for this method as presented for the sample that was up-concentrated via ultrafiltration (**Chapter 4.1.2.1**).

It can be recommended to use linker-capped AuNPs to examine the increase in concentration with evaporation and to attach the sensitive biomolecules *ex situ* to the concentrated sample.

4.1.3. Summary and discussion

Considering business economics and cost effectiveness of the PLAL fabrication process, yield enhancement of the method is unavoidable. For this intent, picosecond pulse durations were adopted in order to ablate a higher gold mass per time by PLAL than obtained with femtosecond pulses. Moreover, ultrafiltration and evaporation post-processing techniques were applied to further increase the concentration of afore fabricated gold nanoparticles.

Ablation of a higher gold mass using a picosecond-pulsed laser system

As discussion base, Petersen et al. have contributed fundamental data on the correlation between nanoparticle yield and biomolecule integrity during fs-PLAL as a function of process parameters.[39] They determined a particle yield of 11 μg min^{-1} for AuNP-ssO bioconjugates and a yield of 10.5 μg min^{-1} for ligand-free AuNPs for parameters that result in nearly 100 % integrity preservation of nucleotides. Those values will serve as a reference for the comparison to ps-PLAL method. The yield of ps-PLAL may be directly correlated to the yield of fs-PLAL, because in the same manner, the optimal process parameters that enabled nearly 100 % biomolecule integrity preservation were chosen for comparison.

Thus, the determined maximum yield of 95 µg min^{-1} for ligand-free AuNPs and of 168 µg min^{-1} for AuNP-ssDNA bioconjugates by ps-PLAL were found to be factors 9 and 15 higher than the yields obtained with ablation using the fs-pulsed laser system (**Figure 4.14a**).

The highest obtainable NP concentration was found to be ~ 300 µg mL^{-1}. Furthermore, the process was highly reproducible, controllable, could be run with a production speed on the timescale of minutes and could be accomplished under nearly sterile conditions, because the ablation vessels and targets can be autoclaved and pyrogen-free, ultrapure water can be adopted (**Table 4.5**).

The ps-PLAL fabricated AuNPs and AuNP bioconjugates were identified to be of the same high *quality* regarding constitution, surface coverage, integrity and functionality as the fs-PLAL products, which highlights the suitability of both pulse lengths for comparable results.

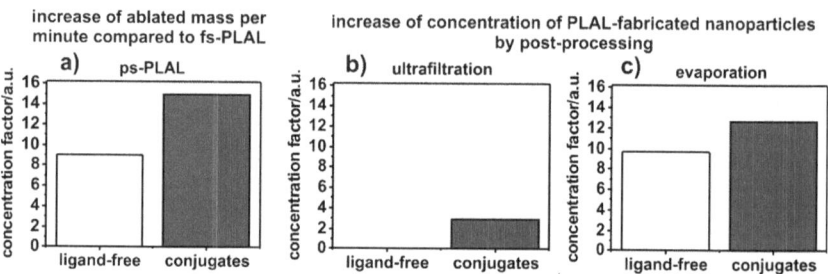

Figure 4.14. Concentration factors of the three adopted approaches. a) Concentration factors of ps-PLAL ablation method. b) Concentration factors of ultrafiltration post-processing method. c) Concentration factors of evaporation post-processing method. Concentration factors are presented for ligand-free AuNPs (white bars) and AuNP-bioconjugates (green bars).

Increasing the concentration of PLAL-fabricated AuNPs and AuNP bioconjugates

Ultrafiltration and evaporation post-processing techniques were applied as post-processing methods in order to increase the concentration of afore fabricated gold nanoparticles.

Using *Vivacon®* ultrafiltration tubes with a *Hydrosart®* membrane made of hydrophilic regenerated cellulose, ligand-free AuNPs and AuNP-mPEG-SH conjugates adhered strongly to the membrane and resulted in a high particle mass loss of 90 %. Thus, there was no significant increase in concentration determined for the screened MWCOs. Conversely, for *Nanosep®* ultrafiltration tubes with an *Omega®* membrane made of a hydrophobic polyethersulfone membrane, a concentration factor of 2.2 was reached for 30 kDa MWCO without particle aggregation. However, only stabilized AuNP-mPEG-SH conjugates could be up-concentrated, while ligand-free AuNPs adhered again on the membrane and particle mass losses of approximately 40 % had to be accepted.

Functionality tests of the up-concentrated AuNP bioconjugates proved, that the nanobioconjugates were still functional and that they have not been damaged, separated or inactivated by the treatment.

In conclusion, the *Nanosep*® filtration using a hydrophobic *Omega*® polyethersulfone membrane that repel the AuNPs can result in an approximate increase in AuNP concentration by a factor of 2–3 if stabilized NPs and an optimal MWCO are applied (**Figure 4.14b**).

When using the post-processing technique of solvent evaporation without heating for 5 hours, an average increase in AuNP bioconjugate concentration by a factor of 9-10 for a steady-state system can be reached. For a continuously agitated system even an increase by more than one order of magnitude (factor 13) is enabled, because of a higher molecular exchange at the air/water interface (**Figure 4.14c**). However, it should be considered that a long process time can damage or inactivate the sensitive biomolecules.

For ligand-free AuNPs in a steady-state system, a maximum concentration of $1.5 \, g \, mL^{-1}$ appears to be reachable without macroscopic gold formation on the air/water interface. The network superstructures were most likely due to a progressive coalescence from chemical sintering or cold welding. The formation of gold superstructures was also found for AuNP-mPEG-SH conjugates, but at much higher concentrations of approximately $3 \, mg \, mL^{-1}$. Conversely, less assembled structures were observed for the agitated samples, because the fast particle motion worked against the capillary forces.

Furthermore, an increased evaporation in the order of: *MilliQ* > AuNP-mPEG-SH > ligand-free AuNPs was determined. The non-volatile colloidal particles raised the boiling point of the solution, which resulted in a slower evaporation of the nanoparticle-containing samples compared to a pure *MilliQ* sample. In addition, the superstructure network of ligand-free AuNPs was covering a significant amount of the surface area and thus most likely hindered the water molecules from evaporation while limiting the total volume loss.

However, both post-processing approaches, the ultrafiltration and the solvent evaporation, suffer from a lack of process control. For instance, the conditions during evaporation are strongly affected by the environment while the adopted vessel size and the speed of agitation will strongly influence the outcome. Moreover, the requirement of a large initial liquid volume and long process times should be considered (**Table 4.5**), which are probably not appropriate for sensitive biomolecules. Conversely, during ultrafiltration not only the filtration volume, the MWCO and the centrifugal speed, but also the applied particle sizes and size distributions will affect the results. Furthermore, high nanoparticle mass losses on the filter membrane are a main drawback of the method (**Table 4.5**).

Table 4.5. An overview of the analyzed methods with the applicable volumes, advantages and disadvantages.

	Volumes	Advantages	Disadvantages
ps-PLAL	volume depends on process time and biomolecule integrity	fast process, accomplishment within minutes, nearly sterile	defined parameter window has to be matched to gain integrity-preserved bioconjugates
Ultrafiltration	low volumes (< 10 mL)	fast process, accomplishment within ½ Hour	MWCO screening required, low concentration factor, high particle mass losses
Solvent Evaporation	high volumes (> 10 mL)	concentration > 4 mg mL^{-1} reachable	long process times which depend on the used volume (hours to days), risk of contaminations

Thus in summary, the adoption of ps pulses for PLAL method was found to significantly increase the mass ablation per volume and time of fabricated AuNPs and AuNP bioconjugates by approximately one order of magnitude compared to the adoption of fs pulses for PLAL. However, if an even higher gold concentration than one order of magnitude is required, then evaporation could be an interesting post-processing technique to yield another order of magnitude increased concentration of stabilized AuNPs. By this means, a total concentration on the mg mL^{-1} scale can be reached.

4.2.   Considerations for the structure-function relationship

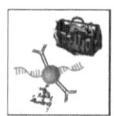 For the customized application of nanobioconjugates, the focus of the design must be set on the structure-function relationship. The demand of the nanobioconjugate function is conventionally given by the applicator and the manufacturer must allow that function with the appropriate nanobioconjugate structure. However, structure setting is modulated by a multiplicity of diverse factors which do not only affect the nanoparticles but which may also influence each other and which may complicate their presentation and discussion. Thus, for a structured orientation, the following chapter will be divided into three consideration areas, while a fourth area, which deals with the (biological) function of the designed nanobioconjugates, will be continuously outlined (**Figure 4.15d**). The 1st consideration area (CA) covers the modulation of the particles' intrinsic parameters such as particle size and charge that may directly affect ligand conjugation and binding amount (**Figure 4.15a**). The other CAs deal with the nanoenvironment and the manipulation of the conjugation process. In this regard, basic, extrinsic issues such as choice of bond type, ligand amount and surrounding medium (2nd area, **Figure 4.15b**) and the ligand characteristics such as their length, dimension, binding orientation or amphiphilic nature and net-charge and the

adoption of diverse ligands for multivalent functionalization (3rd area, **Figure 4.15c**) are discussed.

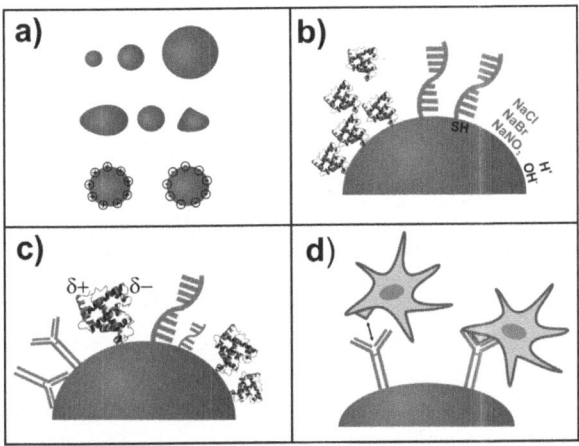

Figure 4.15. Four consideration areas, discussing the nanobioconjugate structure parameters and their functionality proof. a) The 1st area covers the intrinsic nanoparticle characteristics as NP size, shape and surface charge. **b)** The 2nd area focusses on the interaction of the NP surface with the environment and handles basic conjugation issues as the effects of the surrounding medium, the bond type and the ligand amount. **c)** The 3rd area deals with ligands characteristics as their length, their dimension, their binding orientation and their amphiphilic nature and further outlines the issue of multi-valent conjugation with different ligands. **d)** The 4th area is a superior topic and handles the functionality of designed AuNP bioconjugates.

4.2.1.   Intrinsic parameters of AuNPs

The intrinsic parameters of gold nanoparticles are the primary particle size, the particle shape and the particle charge. These parameters have a significant impact on the conjugation efficiency and the surface coverage during *in situ* bioconjugation and they also affect the nanoparticle-cell interactions that regulate particle uptake and cytotoxicity. Therefore, they should be considered carefully and if necessary they should be modified according to the demands of the project. Concerning PLAL-generated AuNPs, there are several methods for the intrinsic parameter modulation, which can be performed either during the ablation process *in situ*, or with a secondary *ex situ* treatment. These modification methods and the impact of the nanoparticles' size and charge on bioconjugation process and biological functionality will be presented in the following subchapters. In addition, a quite speculative chapter on the impact of the nanoparticles' shape and relevant modification methods is found in the Supporting Information in **Chapter 7.1.**

4.2.1.1.   Primary nanoparticle size

Barchanski et al. 2015 [I]*, Cooperation LZH-LUH*
Duran et al. 2011 [XV]*, Cooperation LZH-TiHO*

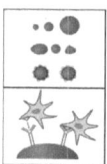 The size of a spherical particle can be quantitatively described by its diameter, since all spatial dimensions are identical. Because PLAL-generated AuNPs are polydisperse colloids that feature a particle size distribution (PSD), it is necessary to average the size of the particle ensemble. There are international standards for presenting the average size of a PSD, including the mean size, the median size and the modal size (**Figure SI 11**).[425] The modal is the highest peak in the differential size distribution curve and represents the most commonly found particle size. The median defines the point according to which half of the size values reside above and half of them reside below. Thus, this is the 50 % size of a cumulative size distribution curve. In the literature, this is often termed *D50* and may be related e.g. to a volume-based distribution (D_v50) or a number-based distribution (D_n50). Finally, the mean size can be expressed according to the measured characteristics which are number, length, surface, volume and weight. The arithmetic mean is calculated by dividing the sum of the diameters from all the individual particles in the distribution by the total number of particles in the distribution. However, the geometric mean is calculated with the nth root of the products of the diameters from the n particles under study.[425] These standard definitions should be followed to avoid confusion of research results.

In general, for symmetric distributions the mean, modal and median sizes are equivalent, while for asymmetric distributions they have different values (**Figure SI 11**).

The most common methods for determination of nanoparticle sizes are TEM, DLS, analytical disc centrifugation or density gradient centrifugation. If not indicated differently, the average size values described in this thesis refer to the mean (Feret) primary particle size if determined by TEM measurement and to the median (D_n50) size if registered by dynamic light scattering.

Furthermore, it is important to distinguish between the average primary particle size, which refers to the individual particle as the smallest unity, and the average global particle size, which includes particle aggregates/agglomerates as single unities. However, as agglomeration is generally triggered by additives or specific treatment of a colloidal solution its effect will be discussed later and the focus of this chapter is set on the primary particle size.

The primary particle size of an object is usually inversely proportional to the surface area-to-volume ratio (SA:V). Considering a perfect sphere with radius r, the formulas for surface area, volume and SA:V ratio are summarized in **Table 4.6**.

Table 4.6. Formulas for particle surface area, particle volume and SA:V ratio calculation and results for

Particle Size/nm	Surface Area (SA)/nm²	Volume (V)/nm³	SA:V ratio/nm⁻¹
	$4\pi r^2$	$\dfrac{4\pi r^3}{3}$	$\dfrac{4\pi r^2}{\frac{4\pi r^3}{3}} = \dfrac{3}{r}$
1	3.14×10^0	5.24×10^{-1}	6.00×10^0
10	3.14×10^2	5.24×10^2	6.00×10^{-1}
100	3.14×10^4	5.24×10^5	6.00×10^{-2}

With increasing nanoparticle size by one order of magnitude, the surface area of a particle increases by a square factor, while the volume increases even by the third power (**Table 4.6**). In consequence, a nanoparticle with primary particle size of 1 nm obtains a SA:V ratio of 6, while it is reduced up to 0.06 for 100 nm particle size (**Table 4.6**). From the literature it is known that small objects with a large SA:V are more reactive than objects with a large size because the stable interatomic bonding arrangements that exist within larger particles are not satisfactory for the increased number of surface atoms of the small particles. A higher surface energy is the consequence of this.[455;456] In these terms, it seems plausible that extremely small NPs of sizes < 2 nm induce strong cytotoxic effects (see **Chapter 2.3**) and that surface functionalization will be more efficient for small particles than for larger ones.

Continuing with this thought process and assuming a constant mass of 50 µg gold and 100 % monodispersity, the particle number and concentration per mL and the total surface area of all particles can be calculated using distinct parameters of gold (19.3 g cm⁻³ = density of gold, 197 g mol⁻¹ = molecular weight of a gold atom, 0.27 nm = diameter of a gold atom). The results are summarized in **Table 4.7**.

Table 4.7. Relationship between particle size, particle number per volume, particle concentration and total surface area of all particles in solution.

Particle Size/nm	Particle Number per mL	Particle Concentration/µM	Total Surface Area of All Particles/nm²
1	4.95×10^{15}	8.22×10^0	1.55×10^{16}
10	4.95×10^{12}	8.22×10^{-3}	1.55×10^{15}
100	4.95×10^9	8.22×10^{-6}	1.55×10^{14}

With an increasing nanoparticle size by one order of magnitude, the particle number and concentration decrease by the third power, which results in a decrease in total surface area by one order of magnitude (**Table 4.7**). These results demonstrate that the primary particle size and the monodispersity of NPs have a deep impact on the available docking area for biomolecule conjugation and that small particles in the range of 1-10 nm with a high degree of monodispersity may yield higher surface loadings than larger sized particles and polydisperse colloidal solutions with broad PSD.

Therefore, CRM appears to be a superior technique for AuNP synthesis, because the desired particle size can be adjusted very precisely depending on the strength of metal-metal bonds and the difference between the redox potentials of the metal salt and the reducing agent applied.[35] The resulting PSD has been found to be highly monodisperse using this method. Solely the purification of particles from chemical reaction by-products confines the method from being optimal.

In contrast, PLAL-generated nanoparticles in *MilliQ* without the addition of stabilizers or ions will always feature polydispersity, which is a process-related characteristic. The PSD may be beneficial for any type of screening experiments, for example, if the particle threshold for cellular entry or cytotoxicity is unknown. However, if a specific size or size class has been evaluated via screening, then the fabrication of the defined, monodisperse particle sizes via PLAL is a complex topic.

Recently, Rehbock et al. reported on the size-controlled, monodisperse PLAL-fabrication of AuNPs in electrolytes with low salinity in a liquid flow system.[318] However, an optimal reproducibility to gain distinct particle sizes was not provided with this method and it only allowed for the fabrication of small primary particles (< 20 nm), while the isolation of larger particles was not the focus of that study.

Thus, two methods to modulate the PSD of PLAL-generated AuNPs will be discussed in the following chapters; the PSD narrowing using *in situ* photofragmentation and the separation of individual size classes using *ex situ* centrifugal processing of the fabricated colloids.

4.2.1.1.1.   *Modulation of NP size distribution by photofragmentation*

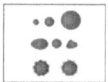 The principle of photofragmentation was already discussed in **Chapter 2.8**, and several workgroups have investigated the technique of laser-assisted size control for the PSD modulation of PLAL-generated AuNPs.[339-344;316;345;346;97;337] Conventionally, either femtosecond- or nanosecond-pulses have been applied thus far with a wavelength of 532 nm, near to the SPR of gold. However, photofragmentation using ps-pulses and NIR-wavelength has rarely been a topic of investigation to date, [342;341] even though a highly efficient second-harmonic generation (SHG) may occur.

Therefore, a small study concerning the ability and efficiency of ps-photofragmentation to reduce the PSD of PLAL-generated AuNPs was accomplished using the process parameters defined in **Table 3.7**. The photofragmentation effect was monitored within 10 minutes of processing using UV-vis spectrophotometry, DLS and SEM analysis and the results are presented in **Figure 4.16**.

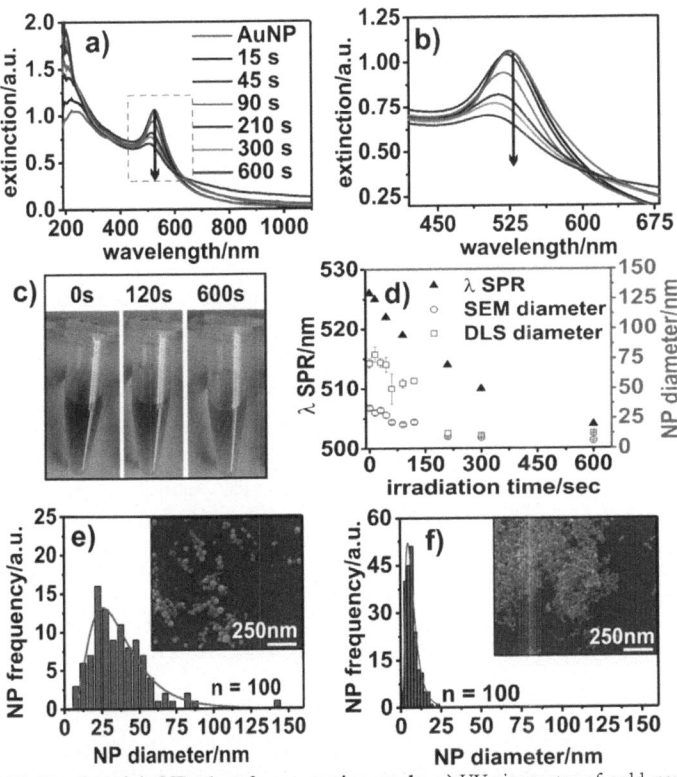

Figure 4.16. Results of AuNP photofragmentation study. a) UV-vis spectra of gold nanoparticles prior to photofragmentation (AuNP) and during photofragmentation time from 15 s to 600 s. **b)** Magnification from the blue boxed area in **(a)**. **c)** Photographic presentation of liquid color fading after 120 s and 600 s photofragmentation in comparison to the untreated AuNP colloid. **d)** Alteration of SPR wavelength (black triangles), Dn50 (DLS, red-framed boxes) and number-weighted modal particle diameter (SEM, red-framed circles) during photofragmentation. **e)–f)** Size distributions with logNormal fitting for the untreated AuNP colloid **(e)** and after 600 s of photofragmentation **(f)**. Scanning electron micrographs of untreated AuNP colloid and after 600 s photofragmentation are presented in the insets of **(e)** and **(f)**. Adapted with permission from Barchanski et al., copyright 2015 by the American Chemical Society.[428]

The UV-vis spectra presented a clear trend. With increasing photofragmentation time the characteristic SPR peak was significantly reduced (**Figure 4.16a–b**). Starting with a sharp peak and peak area of 44.7 nm^2 in the region from 365 to 630 nm for the untreated colloid, only a small peak with an area of 13.4 nm^2 was left after 600 s of irradiation, which corresponds with a decline of 70 %. This trend was accompanied by an SPR wavelength shift from 527 nm to 503 nm (**Figure 4.16d**) and a significant color loss of the corresponding colloidal solution from red to light pink to nearly transparent (**Figure 4.16c**), which indicates a particle shape or size modification. These data are in perfect agreement with results that have been found in other studies.[457;342;343;69;458]

Interestingly, another trend was observed in the UV regime from 190 to 250 nm (**Figure 4.16a–b**). Untreated AuNPs usually feature a plateau between 220 and 250 nm followed by a sharp decline below 220 nm. This extinction around 200 nm is contributed by an interband transition of AuNPs and from transitions of free gold atoms in the solution.[316] For fragmented AuNPs the plateau disappeared after 15 s of treatment and the extinction in the UV increased steeply. With respect to the Mie theory,[68] the decrease in UV absorption and the shifted SPR wavelength are in line with the predicted trend for size-reduced AuNPs through interband transitions of metallic gold and concentration enhancement of free gold atoms in the solution.[342;316]

Particle size measurements with DLS and SEM confirmed the assumption, because a reduction of D_n50 from 72±3.8 nm for the untreated colloid to 13±0.5 nm after 600 s of irradiation was determined, while the modal particle size was reduced from 34 nm to 6.4 nm (**Figure 4.16d–f**). However, it should be considered that SEM resolution is limited and that a size determination below 5 nm scale is defective, and renders the possibility of an even smaller primary particle size for photofragmented AuNPs. For precise measurement, high-resolution TEM should be executed. In addition to the particle size reduction, a distinct narrowing of PSD was also registered (**Figure 4.16e–f**)

When analyzing the NIR regime of UV-vis spectra, the slope decreased slightly for photofragmented AuNPs (**Figure 4.16a**), which suggests that a lower degree of agglomerates were induced by the laser treatment. In fact, in comparing the scanning electron micrographs, the untreated AuNPs were found to agglomerate in small clusters of a few particles, while for AuNPs that were photofragmented for 180 s, the particles were clearly separated from each other (**Figure SI 12**). Interestingly, after 600 s of photofragmentation, the slope was found to be significantly higher than for untreated AuNPs, which indicates particle coalescence (**Figure 4.16a**). The electron micrographs effectively revealed large agglomerated clusters of primary particles that were building networks on the micrometer scale (**Figure SI 12**). Furthermore, the zeta potential dropped noticeably from -20 mV to -12 mV (**Figure SI 13**). This is most likely due to the enhanced total nanoparticle surface that is produced, which is not sufficiently oxidized and result in a loss of stability. The coalescence effect is known in the literature and Eckstein and Kreibig reported in 1993 about the light-induced aggregation of 10 nm gold clusters in a solution by van der Waals-like forces after irradiation with a 514 cw laser.[459] In detail, the mechanical force between the neighboring clusters was indicated to result from Maxwell tensions of the electromagnetic scattering fields of optically excited Mie plasmon resonances. Interestingly, Lau et al. reported recently about the photofragmentation and stabilization of ultra-small AuNP clusters (< 3 nm) in presence of an oxidizing species such as hydrogen per-oxide.[458] The hydrogen peroxide enhances the surface charge density and the electrostat-ic interparticle repulsion by surface oxidation and hinders particle coalescence. This effect is due to a slightly higher redox potential of hydrogen peroxide compared to the one of Au^{3+}. In addition, minute amounts of sodium hydroxide can be added to increase the pH of the solution and to stabilize the colloid, because the low pH that result from the addi-tion of hydrogen peroxide could cause a destabilization of the AuNPs.[458]

In summary, the results clearly indicate a size-reduction from 34 to approximately 6 nm and PSD narrowing of AuNPs after photofragmentation with NIR wavelength and pico-second pulse duration. This is most likely due to a highly efficient SHG and an energy transfer of absorbed laser light during the photofragmentation.

However, the method is only applicable for ligand-free AuNPs, because nanobioconju-gates would be degraded immediately with the focused laser beam. Moreover, it is rec-ommended to add oxidizing species into the colloidal solution prior photofragmentation in order to reduce the coalescence of small particle clusters.

4.2.1.1.2. *Separation of NP size classes by successive centrifugation*

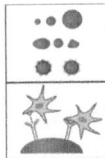 Another possibility to modify the PSD of PLAL-AuNPs is their separation into size classes using successive centrifugation. Because a nanoparticle with a diameter of 100 nm features a higher mass than a nanoparticle with a diameter of 50 nm, the 100 nm particle sedimentates faster due to gravity. Applying centrifugal force, the sedimentation is accelerated and after super-natant removal, the size-classed nanoparticles in the pellet may be re-dispersed in a desired solvent. The supernatant is then used for another round of centrifugation with increased speed to sediment a smaller size class in the next pellet. This procedure can be continued with successively increased centrifugation speed until all nanoparticles are size-classed (**Figure 4.17a**). In addition, Svedberg equation (**eq 3.8**) can be used to calculate the theoretically required centrifugation speed and time to sediment distinct nanoparticle sizes.

For a feasibility study, a standard PLAL-generated colloid was used that featured a D_n50 of 75 nm and a modal particle core size of 42 nm (**Figure 4.17b**). After one minute of centrifugation at 1,000 rpm, the pellet included particles with D_n50 of 100 nm and a modal particle core size of 60 nm. During the next 4 centrifugation steps with constant time and successively increased centrifugation speed, the diameters of nanoparticles in the obtained pellet decreased continuously down to 11 nm, while in the final supernatant nanoparticles with only 7 nm mean size were measured (**Figure 4.17b**). These values correspond quite well to the theoretically calculated data by Svedberg equation (**Table 4.8**). However, for 1,000 rpm a discrepancy between theory and practice was found, most likely due to an imprecise separation of the supernatant and the soft pellet. For higher centrifugation forces the pellet became more condensed, which simplified the separation procedure.

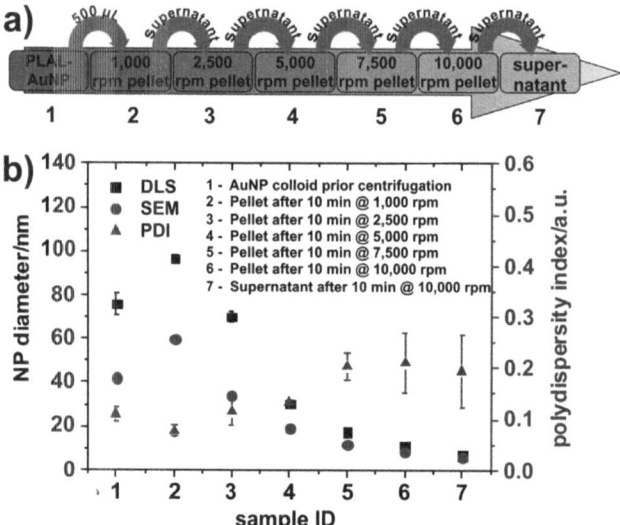

Figure 4.17. Schematic overview and results of centrifugal study. Starting with the untreated PLAL-AuNPs (sample **1**) followed by 5 centrifugation steps (samples **2–6**) and resulting in a final supernatant (sample **7**). Alteration of Dn50 (DLS, black squares), number-weighted modal particle diameter (SEM, red dots) and PDI data (purple triangles) are summarized for 7 samples in the additional graph.

Table 4.8. Nanoparticle diameter that could be theoretically centrifuged using distinct centrifugation speed and time, calculated by Svedberg equation.[416]

Centrifugation Speed/rpm	Centrifugation Time/min	Theoretically Centrifuged Particle Diameter/nm
1000	10	100
2500	10	45-40
5000	10	25-20
7500	10	15
10,000	10	11

Interestingly, the polydispersity index increased from 0.1 to 0.2 after the 5th centrifugation step, which was most likely due to enhanced coalescence of small particles because of increased van der Waals attractions.

The adoption of PLAL-generated and size-class separated AuNPs for biological applications was performed in cooperation with the University of Veterinary Medicine Hannover in the context of a cellular transfection study.[422] AuNPs with a broad PSD from 1 to 180 nm were fabricated (**Figure 4.18a**) and subsequently centrifuged to obtain: *i)* a small PSD from 6 to 26 nm with a modal particle size of 14 nm and a D_n50 of 28.5 nm (PLAL-AuNP-S1, **Figure 4.18b**) and *ii)* a broader PSD from 15 to 90 nm with a modal particle size of 41 nm and a D_n50 of 52.4 nm (PLAL-AuNP-S2, **Figure 4.18c**).

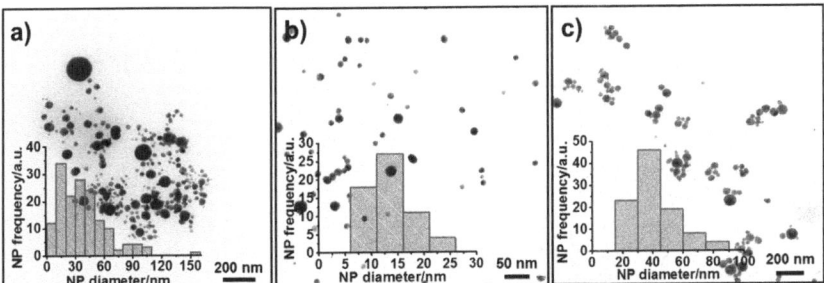

Figure 4.18. AuNP colloid prior and after size separation with successive centrifugation. Scanning electron micrographs and corresponding PSDs of PLAL-AuNP prior and after centrifugal separation into two size classes. a) Untreated PLAL-generated AuNPs. b) Size-classed PLAL-AuNP after first centrifugation. c) Size-classed PLAL-AuNP after second centrifugation. Adapted from Durán et al., copyright 2011 by Durán et al., licensee BioMed Central Ltd.[422]

Both size classes were evaluated as transfection agents to increase the uptake of exogenous plasmid DNA by mammalian cells. The results were compared to other transfection protocols covering the use of a conventional *Fugene®* reagent, commercially available, ligand-stabilized AuNPs with 20 nm diameter (Plano) and two magnetic assays using magnetic nanoparticles with hydrodynamic diameters of 100–200 nm.

Briefly, the application of AuNPs featured significantly higher transfection efficiencies of plasmids for the expression of humanized renilla Green Fluorescence Protein (hrGFP) than conventional Fugene treatment, the adoption of CRM-AuNPs or magnet-assisted transfection methods (PLAL-AuNP-S1: 46 %; PLAL-AuNP-S2: 50 %; Plano-AuNP: 23 %; Fugene: 31 %; magnetic assays mean: 20 %; **Figure 4.19**).

Furthermore, no significant cytotoxic effect was recognized for PLAL-fabricated AuNPs, whereas the chemically derived and ligand-stabilized Plano-AuNPs induced a significant PI % increase and a lower cell proliferation (**Figure 4.19**).

Unfortunately, no size-related transfection effect was determined for the analyzed size classes, although this result is in agreement with the results of a former study from Petersen et al.[460] They demonstrated that a similar transfection efficiency of ~ 50 % was reached for PLAL-generated, size-classed AuNPs with D_n50 values of 24 and 59 nm, while the efficiency was significantly lower (15 % and 8 %) for smaller and larger sized AuNPs with D_n50 values of 14 and 89 nm, respectively.

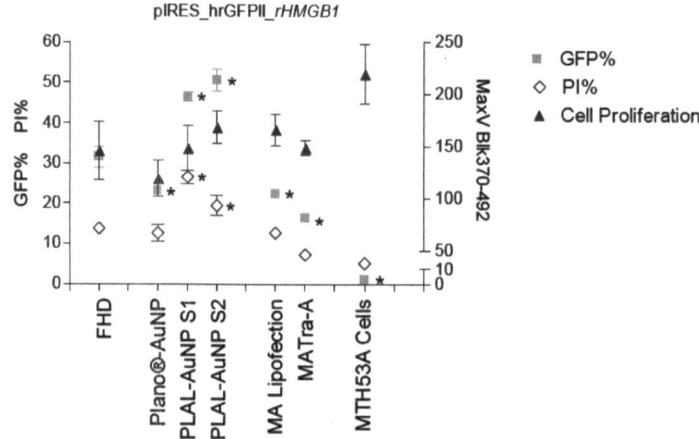

Figure 4.19. Results of AuNP transfection study. The mean cell proliferation (blue triangles) and the amount of GFP-positive (green squares) and PI-positive (red-framed diamonds) cells are presented 24 h after transfection with different protocols. Reprinted from Durán et al., copyright 2011 by Durán et al., licensee BioMed Central Ltd.[422]

However, Petersen et al. fabricated their size-classed AuNPs with a fine adjustment of the laser process parameters, which is an even more complex method than successive centrifugation, due to the requirement of a very broad parameter screening series for each desired size class. In contrast, the favored size class can be narrowed down with a few centrifugation steps using either the supernatant (for small particles) or the pellet (for larger particles) of a centrifuged sample. Moreover, the required centrifugal force and time can be estimated by calculation with Svedberg equation.

One negative aspect of the successive centrifugation method is the high sensitivity to the starting size distribution and volume. However, the separation of nanobioconjugates is also feasible, because no destructive force for size separation is applied.

In summary, both of the presented methods; the photofragmentation and the successive centrifugation, are highly efficient to modify the particle size and the PSD of PLAL-generated AuNPs, if specific size requirements must be met for specific application. However, monodispersity will not be reached with these approaches and chemical synthesis methods should be used to satisfy this demand. Furthermore, it should be considered that the photofragmentation method cannot be performed with nanobioconjugates. Thus, bioconjugation needs to be performed with the photofragmented nanoparticles *ex situ*. Therefore, this process is mainly suitable if size-classed, ligand-free AuNPs are required, while size-separated AuNP bioconjugates can be generated with the successive centrifugation method.

4.2.1.2.   Nanoparticle charge

Barchanski et al. 2015 [I], Cooperation LZH-LUH

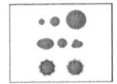 If gold nanoparticles are fabricated in *Milli-Q* water, they usually feature a zeta potential of -20 mV to -30 mV, which indicates a high degree of colloidal stability. As aforementioned in **Chapter 2.8**, Sylvestre et al. determined a partial oxidation of fs-LAL generated AuNPs with the gold oxidation states Au^+ and Au^{3+} with XPS in addition to the one that was elemental gold (Au^0).[324;325] Furthermore, Muto et al. used titration to determine that 3.3–6.6 % of the surface atoms are charged.[77]

In detail, the particles formed by ablation were partially oxidized by the oxygen present in the solution and the Au-OH compounds were further deprotonated, which resulted in an Au-O⁻ surface and the negative zeta potential. Actually, both species are in equilibrium and strongly pH-dependent: Au-OH ⟷ Au-O⁻ + H⁺ (low pH → shift to the left side, high pH → shift to the right side). In addition, carbonato complexes (Au-OCO₂⁻ and Au-OCO₂H) were also detected (**Chapter 2.8**)

The partial oxidation of the AuNP surface enhances the chemical reactivity of the particles, especially for electrostatic interactions with anions, but also for covalent interactions with thiols. Furthermore, the partial oxidation may have an impact on nanoparticle growth, since particle coalescence is limited by the electrostatic repulsion which leads to reduced particles sizes.

Concerning biological applications, it is known, that the charge of nanoparticles strongly influences their cellular uptake behavior and their toxicity (see **Chapter 2.3** and **Chapter 2.4**). Thus, detailed knowledge about the amount of charged surface atoms is essential in order to gain specific biological functionality.

Because no XPS data of ps-PLAL generated AuNPs were presented in the literature to date and because most experiments in this thesis have been conducted with ps-pulsed LAL, an XPS analysis of the AuNPs fabricated in ddH₂O was performed and a high-resolution spectrum of the gold 4f core level is presented in **Figure 4.20**.

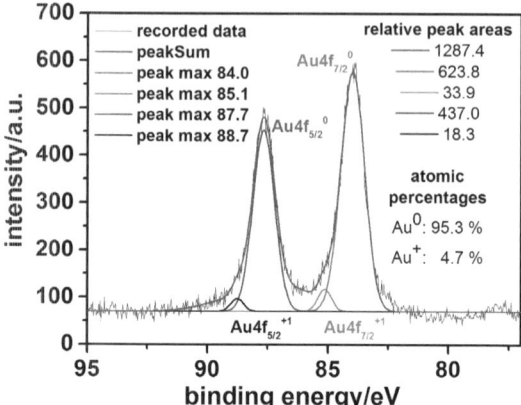

Figure 4.20. High-resolution X-ray photoelectron spectrum of the gold 4f core level. Recorded data were deconvoluted into two peak pairs. Relative peak areas were determined and converted into atomic percentages of gold oxidation states. Adapted with permission from Barchanski et al., copyright 2015 by the American Chemical Society.[428]

The most striking peak pair was determined at 84 eV and 87.7 eV and may be assigned to Au $4f_{7/2}$ and Au $4f_{5/2}$, which corresponds to elemental gold (Au^0). The ratio of the peak areas are nearly equal to the expected 4:3 ratio which is related to the spin multiplicity of each spin-orbit state.[461;462] A second peak pair was identified at 85.1 eV and 88.7 eV, which conforms to the presence of the Au$^+$ gold state. From the relative peak areas, atomic percentages of the gold states were calculated to be 4.7 % for Au$^+$ and 93.7 % for Au^0.

Several studies on the oxidation state analysis of laser-generated AuNPs with XPS are found in the literature to date. The data are summarized in

Table 4.9 with regard to the solvent that was used and the applied laser parameters.

The identification of a partial oxidation of the gold surface in this thesis agrees with the results obtained by Sylvestre et al. who found ratios of 88.7 % Au^0, 6.6 % Au$^+$ and 4.7 % Au^{3+} (**Table 4.9**).[325] Moreover, they correlate with results that were found by Merk et al. for AuNPs that were fabricated in sodium chloride solution (**Table 4.9**, 94.2 % Au^0, 4.3 % Au$^+$ and 1.5 % Au^{3+}).[295] However, in contrast to these findings no secondary gold oxide state (Au^{3+}) was determined for the ps-PLAL fabricated AuNPs in this thesis.

Interestingly, Giusti et al. and Giorgetti et al. demonstrated an increased production of Au(III) as a fragmentation byproduct when irradiating an AuNP colloidal solution with UV or visible laser wavelength, while for an irradiation with 1064 nm pulses this effect was completely absent.[341;342]

Table 4.9. The oxidation states (and ratios) that were found for laser-generated AuNPs in the literature, presented with regard to the used solvent and the laser parameters. Ref. = Reference, Sol. = solvent, OS = oxidation states, Fl. = fluence, LP = laser power, RR = repetition rate, PE = pulse energy, PL = pulse length, WL = wavelength, MQ = MilliQ; **1** = Sylvestre et al. 2004; **2** = Merk et al. 2014; **3** = Fong et al. 2013; **4** = Muto et al. 2007; **5** = Barchanski et al. 2014.

ID	1	2				3[7]					4[8]	5
Ref.	[325]	[295]				[444]					[77]	[376]
Sol.	MQ	NaCl	NaI	NaBr	NaF	MQ	CT AB	CT AC	Na Cl	Na Br	MQ	MQ
OS/ %	Au^0 88.7 Au^+ 6.6 Au^{3+} 4.7	Au^0 94.2 Au^+ 4.3 Au^{3+} 1.5	Au^0 83.9 Au^+ 7.3 Au^{3+} 4	Au^0 91.8 Au^+ 5.1 Au^{3+} 3.1	Au^0 90.8 Au^+ 6.3 Au^{3+} 2.9	Au^0	10^{-5} M Au^0 10^{-3} M Au^0 Au^{3+}	10^{-5} M Au^0 Au^{3+} 10^{-3} M Au^0 Au^{3+}	Au^0	Au^0	Au^0 Au^+ Au^{3+}	Au^0 95.3 Au^+ 4.7
Fl.	600 J cm^{-1}	N/A				2.2×10^5 mJ cm^{-2} $pulse^{-1}$					N/A	0.5 J cm^{-2}
LP/ W	N/A	4.5				N/A					N/A	0.5
RR/ Hz	1,000	100				10					10	5,000
PE/ mJ	1	45				7.5					80	0.1
PL	fs	ns				ns					ns	ps
WL/ nm	800	1064				1064					1064	1030

This finding indicates that the missing Au^{3+} oxide state might be due to the application of a near-infrared wavelength (1030 nm) for ps-PLAL, instead of 800 nm wavelength as was adopted for fs-PLAL in the study from Sylvestre et al. In contrast, Muto et al. used 1064 nm wavelength and postulated the finding of both Au^+ and Au^{3+} states in addition to the one of Au^0, which invalidated the NIR wavelength-assumption (**Table 4.9**).[77] Unfortunately, they did not provide ratios for each oxidation state. Recently, Fong et al. published an XPS-study on laser-generated AuNPs in cationic surfactant media.[463] They also applied a 1064 ns-pulsed laser, but operated with a lower ablation fluence than Muto et al. and reported the formation of solely Au^0 in water without surfactants (**Table 4.9**).[463] If surfactants such as CTAB or CTAC were used, the oxidations states were found to be highly dependent of surfactant concentration. Using a concentration of

---

[7] Atomic percentages of oxidation states were not numbered. A hardly noticeable amount of Au^{3+} was found for 10^{-5} M CTAC, whereas in 10^{-3} M CTAC and CTAB the peaks were clearly visible and of similar height.
[8] Atomic percentages of oxidation states were not numbered. The intensity of Au^{3+} was much lower than for Au^+.

10^{-5} M, the Au^{3+} state was only found for CTAC surfactant and was very close to the detection limit. Whereas, using a concentration of 10^{-3} M that corresponded to the critical micelle concentration (CMC), a clearly visible peak pair of Au^{3+} with similar height was determined for both surfactants (**Table 4.9**). Furthermore, Merk et al. calculated the deviations between AuNPs in different electrolytes with equal ionic strengths. In detail, they found a total of 5.8 % of the gold surface to be oxidized in sodium chloride, while 11.3 %, 8.2 % and 9.1 % of the gold surface were oxidized in sodium iodide, sodium bromide and sodium fluoride, respectively (**Table 4.9**).[295] Conversely, Fong et al. reported no significant surface oxidation of AuNPs in the presence of sodium chloride or sodium bromide (**Table 4.9**).[463] It may be speculated, that the contradictory results are attributed to the different laser parameters that were used for PLAL fabrication of the AuNPs. In addition to the pulse length, also the pulse energy, the repetition rate, the fluence and the laser wavelength varied widely among the studies (**Table 4.9**). However, a close correlation between laser parameters and formation of oxidation states on laser-generated AuNPs has not been found yet and a systematic study on this topic is strongly required.

In any case, the AuNPs that were fabricated with ps-PLAL in this thesis could easily be functionalized with biomolecules *in situ* and the designed AuNP bioconjugates were successfully applied for *in vitro* immunolabeling (see **Chapter 4.1.1.5**).

In conclusion, the fabrication of AuNPs with ps-PLAL using a fluence of 0.5 J cm^{-2} yielded partially oxidized particles ($\sim$ 5 % of atoms) with an Au^+ configuration and a negative zeta potential of up to -30 mV. No other gold configurations were detected, excluding the potential formation of toxic Au^{3+} organogold compounds.[464] However, it must be considered that the NP charge discussed in this chapter was related solely to the particle surface charge related to the partial surface oxidation. Whereas, the particle net charge that interacts with a cell could further be determined with adsorbed molecule species from the incubation medium or stabilizing ions which cover the particle surface by electrostatic forces, possibly yielding a differently charged particle-molecule complex. Nevertheless, these aspects will be discussed in the following sections that cover the consideration areas II and III.

4.2.2.   Optimization of conjugation parameters

The parameters that affect AuNP *in situ* conjugation with biomolecules can be divided into environmental effects of the ablation medium (2nd consideration area) and the characteristics of the applied ligands (3rd consideration area). Both areas will be discussed in the following subchapters and a guideline for parameter optimization will be presented.

4.2.2.1.   The surrounding medium

Because nanobioconjugates are designed for biomedical applications, the solvent they are dispersed in must be highly biocompatible. This demand excludes all types of organic solvents and also restricts the adoption of *Milli-Q* water for osmotic pressure reasons.

DNA derivates are not stable for a long period of time in the slightly acidic *Milli-Q* water (pH ~ 5.8), because they require a more alkaline medium such as Tris-EDTA (TE) buffer (pH 8) to fold into their functional three-dimensional structure. However, a higher pH ($\geq$ 9) should be avoided for double-stranded DNA because it could melt the hydrogen bonds and induce denaturation. DNA denaturation is also favored in media with low salt concentrations, while high salt concentrations stabilize the helical structure.[465;466]

Conversely, proteins are highly sensitive to high salt concentrations because precipitation may occur as a result of protein-salt interactions and the formation of hydrophobic patches on the protein surface.[467] Precipitation may also occur at a pH value that is near the isoelectric point (pI) of the protein at which its net primary charge is zero. This means that every protein has an optimal pH for its biological functionality and even slight changes might affect the activity.

Thus in summary, the specific needs of pH value and salt concentration must be individually set for each biomolecule, depending on its molecule class and characteristics. Furthermore, the optimal ablation medium should provide good electrostability of AuNPs, should enable a high binding efficiency of ligands to AuNPs and needs be highly biocompatible, such as cell culture or buffer media for instance.

In an initial study, the suitability of cell culture media (CCM) and various buffer media for the PLAL-fabrication of AuNPs was screened. Most CCM are not suitable for the monitoring of AuNP-PLAL fabrication, due to the presence of a phenol red pH indicator, which absorbs in the same spectral range as AuNPs (**Figure SI 14**) and hinders the analysis of concentration and agglomeration index. Thus, phenol red-free Roswell Park Memorial Institute medium (RPMI) 1640 of a known composition was chosen for the experiment and was applied once with and once without serum proteins. Furthermore, 10 mM phosphate-buffered saline (PBS), 10 mM HEPES (4-(2-hydroxy-ethyl)-1-piperazine-ethane-sulfonic acid), 10 mM TE (Tris-EDTA) and 10 mM Tris buffer were selected as standard buffers and also adopted with and without serum protein addition. The electrostatic stability and agglomeration indices of AuNPs were analyzed and the results have been summarized in **Figure 4.21**.

The colloids fabricated in PBS, TE, HEPES and CCM featured a dark-purple or grey coloration (**Figure 4.21a**), which arose from plasmon resonance frequencies of AuNP agglomerates because the high-concentrated salts shield the stabilizing surface charge of nanoparticles (see **Chapter 2.8**).

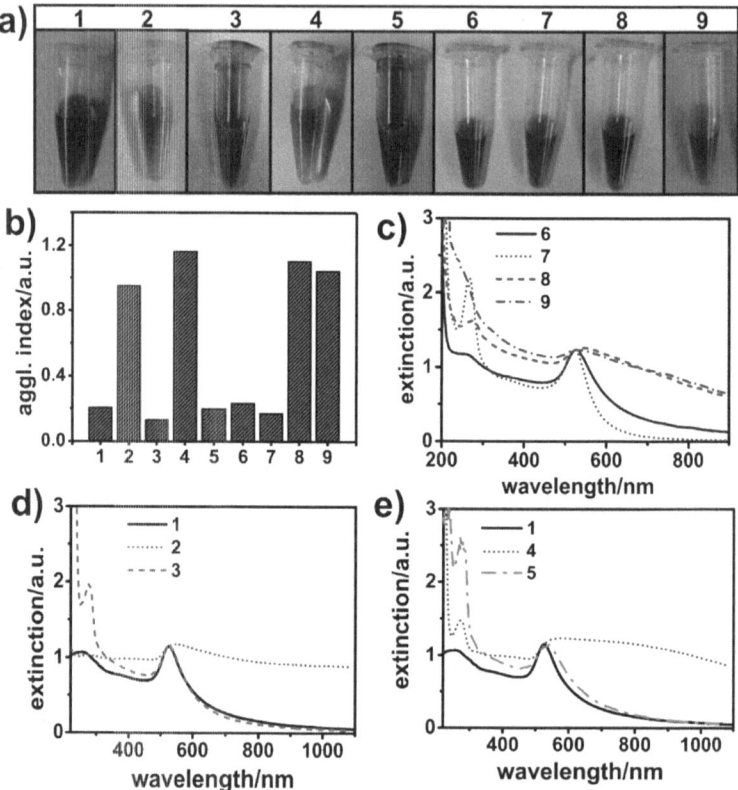

Figure 4.21. Electrostatic stability results of AuNPs in cell culture and buffer media.
a) Photographies of colloidal solutions, with **1** = AuNP (ddH₂O), **2** = AuNP (PBS), **3** = AuNP (PBS + BSA), **4** = AuNP (CCM), **5** = AuNP (CCM + BSA), **6** = AuNP (TRIS), **7** = AuNP (TRIS + ssDNA), **8** = AuNP (TE), **9** = AuNP (HEPES). **b)** Agglomeration index of colloidal solutions. **c)-e)** Normalized extinction spectra of AuNPs ablated in buffer and cell culture media (**1** = black solid line, **2** = red dotted line, **3** = green dashed line, **4** = purple dotted line, **5** = orange dash-dotted line, **6** = blue solid line, **7** = grey dotted line, **8** = pink dashed line, **9** = brown dash-dotted line).

This resulted in an intense NIR absorption (**Figure 4.21c–e**) and agglomeration index ≥ 0.9 (**Figure 4.21b**) compared to an agglomeration index of 0.2 for ligand-free, red-colored AuNPs (**Figure 4.21b**). Interestingly, if the saline media were supplemented with BSA or thiolated ssDNA, no agglomeration occurred (**Figure 4.21b**) and red-colored colloids (**Figure 4.21a**) were obtained, because the biomolecules coordinated and stabilized the AuNPs for up to several weeks. The biomolecules were further found to quench the particle size distribution, which was indicated by a lower absorption in the NIR and a lower agglomeration index (**Figure 4.21b–d**) than for ligand-free AuNPs.

However, in CCM solution the stabilization trend was accompanied by a broadening of the SPR peak (**Figure 4.21b** and **Figure 4.21e**). This broadening may be due to interactions and/or attachment between AuNP conjugates and other ingredients of the CCM, which consist of salts, numerous amino acids, fetal calf serum (FCS) and other nutrients (**Table SI 3**). The adsorption of these ingredients onto AuNPs and AuNP bioconjugates immediately after addition, in known to cause the formation of a *corona* around the particles[370] [367;368] which interact with the cells and may hinder the functionality of conjugated biomolecules.

Thus, as an overall trend, the fabrication of AuNPs in a standard buffer and CCM is not feasible due to the high ionic strengths of the media. However, long-term stable AuNP bioconjugates may develop in these solutions because the particle coordination with biomolecules prevents agglomeration with charge shielding effects.

The potential to generate stable AuNPs in buffer media can be described as a function of buffer concentration. Thus, a dilution series of a buffer or CCM may deliver a threshold concentration ($conc_{th}$) that is acceptable for the fabrication of stable AuNPs (**Figure 4.21, Figure SI 15**).

Depending on the definition of the maximum acceptable agglomeration index, the PBS $conc_{th}$ is derived from the diagram trend to be e.g. 1.6 mM for an agglomeration index of 0.3. This concentration provides sufficient ions for the electrostatic stabilization of AuNPs without the initiation of charge-shielding effects. Interestingly, for PBS concentrations that are below 1 mM, the agglomeration index was found to be even lower than for ligand-free AuNPs (**Figure 4.22**), which indicates an increased electrostatic stabilization of particles by the solution ions or potentially a quenching effect of particle size distribution.

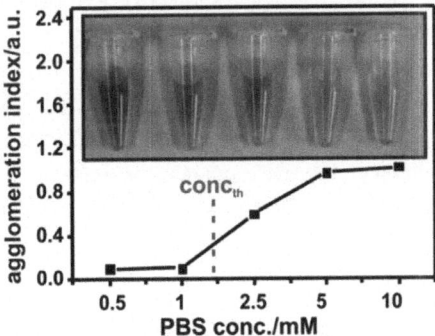

Figure 4.22. Agglomeration index of AuNPs fabricated in PBS as function of PBS concentration. Corresponding photographs of the colloids are presented in the inset. The threshold concentration ($conc_{th}$) for a maximum agglomeration index of 0.3 is marked by the red dashed line.

These results are in agreement with the data from Rehbock et al.,[318] who found an enhancement of electrostability and distinct size-quenching effects for the ablation in media with low ionic strength (1-50 μM). Thus, if ligand-free or undersaturated AuNP bioconjugates are generated in buffer media, the determination of $conc_{th}$ should be performed for each buffer separately, because the ionic strengths differ widely among distinct buffers.

If higher salt concentrations (> 10 mM) of the final medium need to be achieved, the biomolecule addition to the buffer media may not be sufficient to yield non-agglomerated AuNP bioconjugates. In this case, the salt-transfer method may be applied to slowly adapt AuNP bioconjugates that were fabricated in *Milli-Q* water to the ionic strength of the desired buffer with the gradual addition of high concentrated salts (> 1 M) in small volumes. In this regard, the UV-vis spectra of saturated (5 μM ssDNA) and undersaturated (0.5 μM ssDNA) AuNP-ssDNA bioconjugates during gradual salt transfer are presented on **Figure 4.23**.

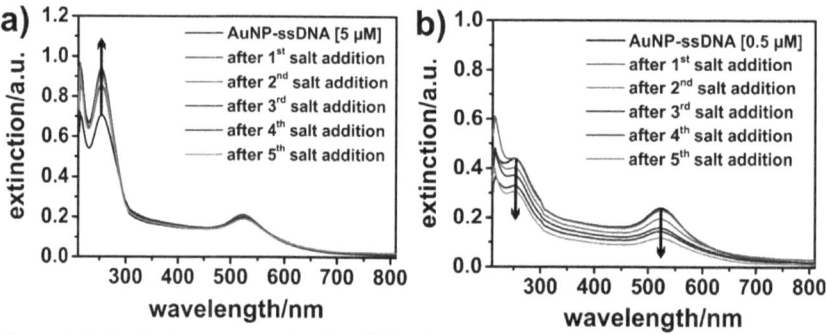

Figure 4.23. **Extinction spectra of gold-ssDNA bioconjugates during salt transfer. a)** Extinction spectra of surface-saturated conjugates, fabricated with 5 μM ssDNA concentration. The peak maximum at 260 nm was increased with salt addition. **b)** Extinction spectra of undersaturated conjugates, fabricated with 0.5 μM ssDNA concentration. The peak maximum at 260 nm and 520 nm dropped with salt addition. Final concentration after 5th salt addition was 150 mM.

It could clearly been seen, that the spectra of saturated gold-ssDNA bioconjugates were not modified by salt addition in the wavelength range from 300 to 800 nm, which implies that the highly stabilized AuNPs were not affected by charge shielding effects up to a salt concentration of 150 mM (**Figure 4.23a**).

Interestingly, the ssDNA peak at 260 nm wavelength was significantly increased during salting, which was most likely due to improved unfolding of nucleotides in the alkalinized medium. In contrast, the spectra of undersaturated nanobioconjugates dropped significantly in intensity with continued salt addition, indicating a defined degree of particle precipitation (**Figure 4.23b**). Thus, if the biomolecule concentration is reduced to an undersaturated concentration (with an AuNP:biomolecule ratio of ≤ 1:0.5), the stabilizing effect of the biomolecules vanishes. In consequence, AuNP bioconjugates should be completely covered with biomolecules to avoid particle losses from precipitation. Then, by using the salt transfer method, the nanobioconjugates can be adapted to salt concentrations of at least 150 mM.

In addition to the variance in ionic strengths, the numerous biological buffers further differ in the pH ranges they work in. This is an important issue for bioconjugation, especially for proteins and antibodies with a specific isoelectric point (pI). The pI is defined as the pH value at which a molecule carries no net charge. Thus, assuming a charged antibody, it carries a positive charge at a pH below their pI, due to the gain of protons and a negative charge at a pH above their pI due to the loss of protons.
The chemical situation on the surface of PLAL-generated AuNPs is determined by its partial oxidation and the correlated electron-accepting properties on the one hand and by the gold material, which allows for the establishment of Au-SR bonds on the other hand.

If metastable atomic and ionic species of gold are ejected from the target during laser ablation in a water-based liquid, the oxygen-containing species in the laser-generated plasma partially oxidizes the surface of the generated AuNPs to Au^+ and Au^{3+}. Furthermore, by chemical reactions between the oxygen-containing species and the AuNPs, a partially hydroxylated surface is formed via Au-OH compounds. With increasing pH, these compounds can lose protons and form Au-O$^-$ groups, relative to the pK value of the hydroxylated surface. Sylvestre et al. conducted this critical point approximately at a pH value of 5, where the Au-OH and Au-O$^-$ compounds are found in equilibrium[325;324]

Thus, if the pH is changed to more acidic conditions (< 5, more protons in solution), the equilibrium is shifted to the Au-OH site, while under more basic conditions (> 5, more hydroxyl in solution), the deprotonated site (Au-O$^-$ + H$^+$) is preferred (**Figure 4.24**).

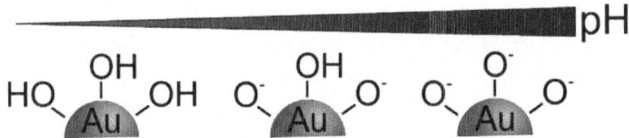

Figure 4.24. **Surface chemistry on PLAL-generated AuNP surface.** The deprotonation of hydroxyl-group is presented as function of increasing pH value. Adapted with permission from L. Gamrad; copyright 2012 by Lisa Gamrad, Master's thesis.[469]

In this regard, the negative charge on the particle surface increases and therefore their zeta potential also increases.[197] The pI values that are reported in the literature for AuNPs in additive-free water vary in a range from 2 to 2.5.[468;324;390]
Thus, for optimal binding of an antibody to the partially oxidized AuNP surface, the pH must be adjusted slightly above the pI of the antibody to result a negative net charge.

However, the pIs of antibodies differ widely, depending on antibody class and species and calculations may not be as simple as they are for proteins.

Therefore, to determine the optimal pH range, a titration of the AuNP-antibody solution must be performed. The titration results for two related antibodies that target the same molecule in different species are presented in **Figure 4.25** and have been evaluated using optical characterization and optical density (OD) measurements at the SPR wavelength.

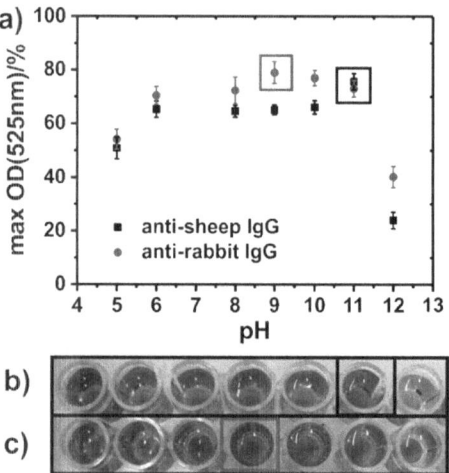

Figure 4.25. Results of AuNP-antibody bioconjugate pH titration. a) The OD results of two titrated AuNP-antibody bioconjugate solutions (anti-sheep IgG = black squares, anti-rabbit-IgG = red dots) in the pH range from 5 to 12, presented in relation to the untreated solution prior titration (OD = 100 %). **b)** Photography of the corresponding colloidal solutions in a well plate, illustrated for AuNP-anti-sheep-IgG. **c)** Photography of the corresponding colloidal solutions in a well plate, illustrated for AuNP-anti-rabbit-IgG. Highest OD values and corresponding wells are boxed.

Compared to the initial OD (100 %) of the untreated AuNP-antibody solution, values between 25 and 80 % were detected for the adjusted colloids (**Figure 4.25a**), due to the modified conditions of the surrounding medium (according to the Drude model) and due to a potential agglomeration of NPs resulting from charge shielding effects. The triplicate testing depicted a maximum OD at the characteristic pH values of 9 for anti-rabbit-IgG and 11 for anti-sheep-IgG. Those solutions actually featured the most reddish colorations which indicates the highest NP stability (**Figure 4.25b–c**) and which is correlated with their biomolecule surface coverage. Conventionally the pI of IgG is found to be between pH 6 and 8.5.[470] For the anti-rabbit IgG, an optimal pH value of 9 was determined by titration, which was slightly above the given pI range and matched the conjugation theory perfectly. The determined optimal pH value of 11 for anti-sheep IgG appears to be relatively high in comparison. However, no data concerning pIs for anti-sheep IgGs was found in the literature. Furthermore, the pI is known to differ widely between species and the results were reproducible. Thus, because the differences were obvious even among closely related antibodies, it is recommended to perform the titration for each antibody prior to conjugation. This is important, because the stability of established nanobioconjugates is determined by the antibody PI.

In summary, buffer media containing high salt concentrations are not suitable for the fabrication of ligand-free or undersaturated AuNP bioconjugates due to charge shielding effects. These particles must be generated in a diluted buffer media below the threshold concentration (compare electrolytes with low ionic strength[318]) or in *Milli-Q* water with slow transfer into saline media. Conversely, if highly attractive molecules (e.g. BSA or thiolized biomolecules) in an oversaturated concentration are added to the concentrated buffer media, the fabrication of highly stabilized nanobioconjugates is also feasible. Furthermore, the pH value for the ablation of pH-sensitive biomolecules should be evaluated using a titration experiment with optical characterization. Appropriate buffer media or pH-adjusted *Milli-Q* water should then be chosen for nanobioconjugate fabrication with regard to the determined results.

4.2.2.2.   Binding stability and functional group

 The main prospect on nanobioconjugates is the binding stability between the particle core and the attached biomolecules because it defines both colloidal stability and nanobioconjugate functionality to a high degree. The nanobioconjugates must be stable enough to resist ionic strength and pH variations on their way through intracellular compartments without decomposition. In addition, ligands such as antibodies should be strongly connected with the particle to exhibit their targeting functionality. However, it should be considered that for certain other ligands, it might be necessary to separate from the particles at the area of interest in order to be functional (e.g. for gene silencing issues with siRNA, see **Chapter 2.5**).

If atoms or molecules strike a solid surface, an adhesion process is initiated which could be either of a physical nature (physisorption) or of a chemical nature (chemisorption). Chemisorption is a type of adsorption that involves a strong interaction between the two components and results in the creation of a new type of chemical bond. It is characterized by chemical specificity. Conventional examples for chemisorption are self-assembled monolayers (SAM), where reactive reagents such as thiols (RS-H) interact with metal surfaces such as gold (Au). In this case, strong Au-SR bonds are formed. Typical binding energy is 1-10 eV, which often involves high activation energies.[471;54] While during physisorption the adsorbents are attached to the surface by weak chemical attraction (mono- and multilayer) and may detach by leaving the solid surface untreated (intact), the adsorbents in chemisorption can change the surface (usually monolayer) and desorption is less prominent.

Chemical bonds are caused either by electrostatic forces between the negatively charged electrons and the positively charged protons inside the nuclei of atoms or they result from dipole attraction. The bond strength is divided into two classes, which include *strong bonds* such as covalent or ionic bonds and *weak bonds* such as hydrogen or van der Waals bonds (**Figure 4.26**).

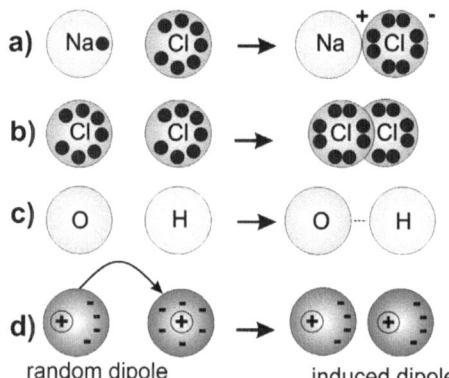

Figure 4.26. Schematic illustration of four atomic bond types. a) Ionic bonds. **b)** Covalent bonds. **c)** Hydrogen bonds. **d)** Van der Waals bonds. Blue circles = electrons.

To achieve an electrostatic conjugation, two solutions with components of opposite charge are simply mixed. However, the stability of the conjugates is often poor, since the ionic interactions are strongly influenced by factors such as ionic strength, concentration and the pH of the solution. Moreover, the orientation of ligands is not predictable with ionic conjugation and may result in functionality reduction or loss when there are orientation-sensitive molecules with active centers such as antibodies (see **Chapter 2.10**). In this context, Mutisya et al. studied the *in situ* and *ex situ* conjugation of unmodified antibodies to AuNPs and found both nanobioconjugate formulations to be biologically active in ELISA testing.[358] However, no *in vitro* test was performed in which the harsh effects of pH or salts could have influenced a separation of the electrostatically bound biomolecules from the particles.

To analyze the effect of the binding stability of biomolecules on the AuNP surface, the titration experiment from **Chapter 4.2.2.1** was repeated with the same antibodies that were functionalized with a hetero-bifunctional OPSS-PEG-NHS (orthopyridyldisulfide-(poly)ethylene glycol-N-Hydroxysuccinimide) linker, and which attach to the AuNPs via the disulfide-containing OPPS group that is located at the distal end of the linker. The results are presented in **Figure 4.27** and are compared to the antibodies without (w/o) linker.

The general trend was found to be the same for both nanobioconjugates, while the absolute values were significantly higher for the OPSS-containing antibodies (**Figure 4.27a**). The optimal pH value of 11 was determined to have nearly the same OD as the untreated colloid, which indicates no agglomeration and only shows small changes in the liquid environment. Moreover, the color of the colloidal solutions was more intense and more reddish for nanobioconjugates with disulfide-containing antibodies than for AuNPs that were conjugated to antibodies without OPSS (**Figure 4.27b–c**).

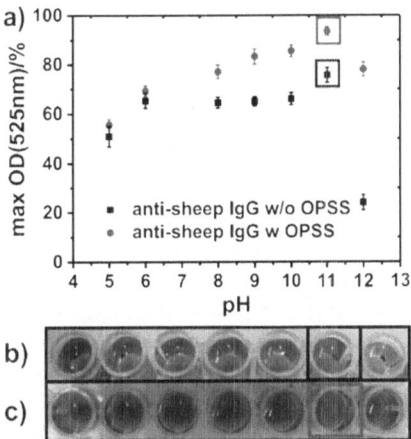

Figure 4.27. Results of pH titration study for covalently and electrostatically conjugated AuNP-antibody bioconjugates. a) The OD results of two titrated AuNP-anti-sheep-IgG solutions, one coupled by hetero-bifunctional OPSS-PEG-NHS linker (w OPSS = red dots) and one without the linker (w/o OPSS = black squares), presented in the pH range from 5 to 12 and in relation to the untreated solution (OD = 100 %). **b)** Photographies of the corresponding colloidal solutions in a well plate are shown for nanobioconjugates without thiol function. **c)** Photographies of the corresponding colloidal solutions in a well plate are shown for nanobioconjugates with thiol function. Highest OD values and corresponding wells are boxed.

However, the most important difference was found with application in a *golden blot* assay, which resulted in significantly intensified labeling of the IgG analyte by covalently attached antibodies than for those without a linker (see **Chapter 4.2.2.4**).

Thus, the covalent attachment of an antibody to AuNPs using a sulphur-containing, hetero-bifunctional linker was found to be more stable and functional than an electrostatic attachment, which is most likely due to a strong gold-disulfide bonding (**eq 2.1**).

As aforementioned, the oxidative addition of sulphuric functions to a bare gold surface is enabled by the mechanisms that are summarized in **eq 2.1–eq 2.2**.

In addition, the surface chemistry (see **Chapters 2.8** and **4.2.2.1**) allows for the pH-dependent reactions of PLAL-generated AuNPs with sulphur-containing molecules which are summarized in **eq 4.1–eq 4.2**.[469]

$$Au\text{-}OH + HS\text{-}R \rightarrow Au\text{-}SR + H_2O \ (pH < 5.8)$$

eq 4.1

$$Au\text{-}O^- + HS\text{-}R \rightarrow Au\text{-}SR + OH^- \ (pH > 5.8)$$

eq 4.2

Thus, to estimate whether the sulphuric function generally binds to the oxidized gold atoms (Au^+/Au^{3+}) or to the neutral gold atoms (Au^0), the data from **Chapter 4.1.1** on the AuNP bioconjugate fabrication with ssDNA and BSA using ps-PLAL can be used for a brief thought process. One gold nanoparticle with a diameter of 9 nm features a total surface area of 254 nm^2. Considering a gold atom to have an approximated diameter of 0.27 nm, the total amount of surface atoms can be calculated to be 4456. Muto et al. proposed that 3.3–6.6 % of the surface atoms of a single AuNP are oxidized. This would be 147–294 atoms for the 9 nm particle. The *in situ* bioconjugation with ps-PLAL was calculated to result in 163 ssDNA and 309 BSA molecules attached to the gold surface (**Table 4.1**). These values correspond perfectly with the determined amount of oxidized surface atoms, especially if the discussed multilayer formation of BSA molecules is considered. On the other hand, thiols and disulfides are hydrophobic moieties, which suggest that they are more attracted towards the Au^0 neutral gold atoms.

In order to examine a preference for a thiol or disulfide function for PLAL-AuNP coordination, four different biomolecules were applied for a focused study. The naturally occurring amino acid L-cysteine and the pharmaceutical drug acetylcysteine (*N*-acetyl-L-cysteine, NALC) were chosen as representatives of thiol-containing molecules, while the oxidized form of L-cysteine (L-cystine) and DL-α-lipoic acid (DL-alpha) were adopted as molecules that have a disulfide function (**Figure SI 16**). The ligands were freshly dissolved to a 1 µM concentration in *Milli-Q* water and subsequently used for ps-PLAL. The expected attachment behavior of the biomolecules to AuNPs is illustrated in **Figure 4.28**. Thereby, the attachment by the sulphuric function is assumed to be superior to the electrostatic coordination by NH_2 and COOH electron-donor moieties.

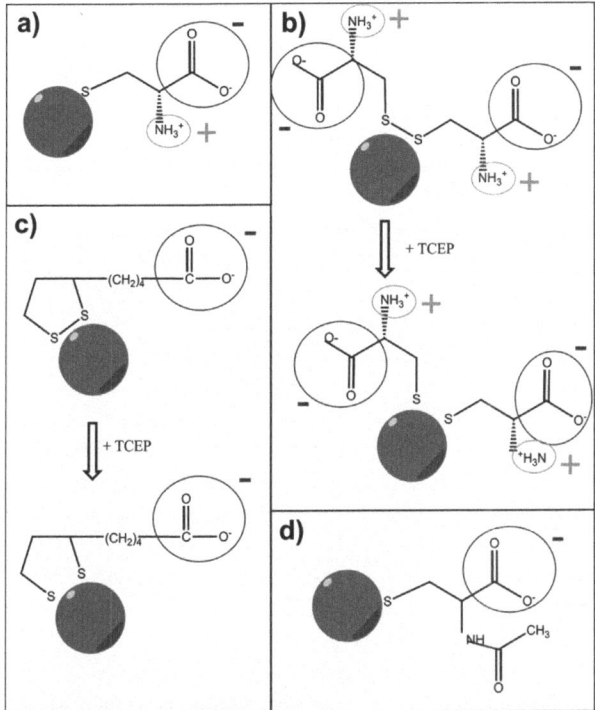

Figure 4.28. Chemical structures of the nanoparticle-attached dissociated ligands prior to and after TCEP treatment. a) Attachment behavior of L-cysteine. **b)** Attachment behavior of L-cystine. **c)** Attachment behavior of DL-α-lipoic acid. **d)** Attachment behavior of N-acetyl-L-cysteine.

The laser ablation in L-cysteine, NALC and DL-alpha solutions yielded stable colloids, with zeta potentials of -19 mV, -28 mV and -25 mV, while in L-cystine solution, a purple coloration with a zeta potential of -4 mV indicated low stability and NP agglomeration (**Figure 4.29**).

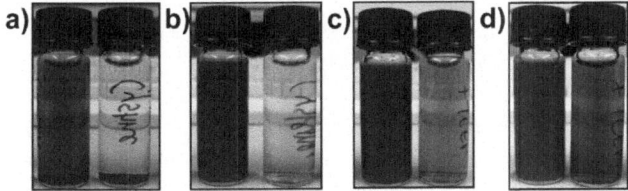

Figure 4.29. Photographs of PLAL-fabricated AuNP bioconjugates equipped with thiolated and disulfide-containing ligands. a) AuNP-L-cystine colloid. **b)** AuNP-L-cysteine colloid. **c)** AuNP-DL-alpha colloid. **d)** AuNP-NALC colloid. The untreated sample directly after ablation (left vessel) is presented in comparison to the sample after 24h treatment with TCEP (right vessel).

This finding may be explained by the chemical structure of the coordinating ligands (**Figure SI 16**). Both NALC and DL-alpha feature a carboxyl function, which may dissociate in water yielding a negatively charged carboxylate ion. These carboxylate ions induce electrostatic repulsion between the ligand-covered AuNP bioconjugates, and thus enhance their stability and yield a high zeta potential. In contrast, L-cysteine and L-cystine molecules both include a primary amine function within their structure in addition to the carboxyl function. This primary amine may be protonated forming a quaternary ammonium ion, which in turn may attract other dissociated carboxylate ions from the solvent while forming a multilayer complex on the AuNP surface. This effect is reflected by the reduced zeta potential.

However, this effect is much more distinct for L-cystine than for L-cysteine, as a single L-cystine molecule features two amine and carboxyl groups at once. Moreover, due to the twofold molecular size and steric dimension of L-cystine, fewer molecules are able to cover the AuNP surface. This is associated with an increased concentration of free L-cystine molecules in the solution which may be attracted by the conjugated molecules (and even by the electron-accepting AuNPs themselves), thereby enhancing the multilayer formation and it may even initiate a bridge agglomeration between the AuNPs.

The colloids were split after ablation and one portion was treated with a reducing agent Tris(2-carboxyethyl)phosphine hydrochloride (TCEP) to break the disulfide bonds (**Figure 4.28b–c**). The characteristic is that TCEP is not only odorless compared to conventional DTT, but also an irreversible reducing agent, yielding two free thiol groups.

Interestingly, 24 hours after TCEP addition the colloidal coloration was significantly reduced or it became transparent in the ranking of: NALC < DL-alpha < L-cysteine < L-cystine, although no obvious NP precipitation was observed.

These findings are in agreement with results from Wang et al., who determined a color vanishing of AuNP-cysteine solution after a 6-hour reaction time. They declared a spontaneous fragmentation/dissolution effect of AuNPs by electron transfer between L-cysteine molecules and the AuNPs, accompanied by oxidation of L-cysteine to L-cystine. The complete loss of color indicated that the NPs completely disintegrated to molecular-sized species.[472;473] The effect was not found for L-alanine, indicating that the mercapto (-SH) group is necessary and implying that TCEP catalyze the reaction in this study at least for the disulfide-bearing molecules.

Moreover, Wang et al. determined the electron-transfer to be a function of local biomolecule concentration around the AuNPs, since the decoloration was only found for biomolecules bearing an additional carboxyl function in addition to the mercapto group that allowed for an electrostatic coordination of the biomolecules around the AuNP surface. For cysteamine on the other hand, which is missing a carboxyl function and thus accumulates far away from the AuNP surface, no decoloration/fragmentation was determined. [472;473]

Transferring these findings to this study, high local biomolecule concentrations around the particles were enabled, since all biomolecules bear carboxyl functions, while L-cysteine and L-cystine further featured an additional amine function. Thus, the distance between the free L-cystine/L-cysteine and AuNPs was distinctly shortened, which is beneficial when injecting electrons into the AuNPs. L-cystine is thereby more efficient due to the lower conjugation efficiency and thus higher concentration of free molecules as discussed previously. In comparison, DL-alpha and NALC transfer fewer electrons into the AuNPs because they feature a single carboxyl function only. Moreover, the effect is more distinct for DL-alpha than for NALC, most likely due to the separation of the disulfide function, yielding two free mercapto groups which in turn could be oxidized back to the disulfide during electron transfer.

In summary it was determined that biomolecules that bear electron-donor moieties may coordinate the electron-accepting PLAL-AuNPs, but that the covalent attachment of ligands with a thiol or disulfide function yields much more stable and functional AuNP bioconjugates. A significant preference for covalent attachment of either thiol- or disulfide-containing ligands was not determined. However, the molecule structure should be considered carefully prior to conjugation, since electron-transfer related spontaneous fragmentation/dissolution of AuNPs by electron-donor-containing molecules could occur, especially if reducing agents are applied.

4.2.2.3.   Ligand amount and ligand net charge

Petersen et al. 2011 [XIII], Cooperation LZH-FLI-MHH

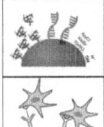
From an economic point of view, the reduction of costs for nanobioconjugate fabrication is correlated with a reduction of applied ligand concentration. In fact, in an optimal scenario nanobioconjugate functionality is given by a single attached ligand and in actuality by 1 to 10 ligands for statistical reasons. However unfortunately, such a low amount of ligands may not be sufficient to stabilize the nanobioconjugates (especially during salt transfer, **Chapter 4.2.2.1**) and the isolated ligands may also tend to wrap around the empty area on the nanoparticle surface (see **Chapter 4.2.2.4**). On the other hand, the ligand concentration should not be too high because it might hinder the particle formation during PLAL or induce multilayer development on the particle surface. Thus, there are two threshold values that are defined as minimum concentration ($conc_{min}$) and maximum concentration ($conc_{max}$), and the adopted ligand concentration should be within a concentration win-

dow that is defined by those thresholds to achieve optimal nanobioconjugate formation and functionality (**Figure 4.30**).

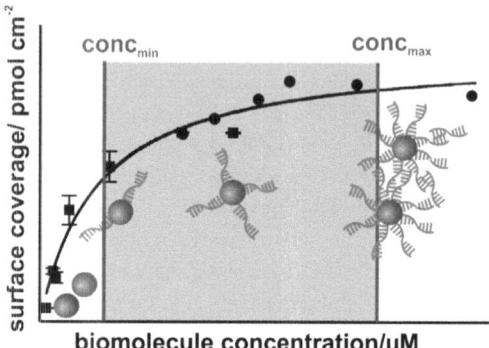

Figure 4.30. Surface coverage trend as a function of biomolecule concentration. With two threshold values termed $conc_{min}$ (minimal ligand concentration required for particle stabilization) and $conc_{max}$ (maximal ligand concentration which may be adopted without forming multilayers on the particle surface) an optimal concentration window for the fabrication of stable, monolayered nanobioconjugates is defined (blue area).

$Conc_{min}$ can be determined by adopting a variation of the titration assay presented in **Chapter 4.2.2.2.** Therefore, AuNPs are mixed with thiolated ligands in varied concentration before highly-concentrated NaCl is added to test the stability of nanobioconjugates. An example with thiolated antibodies as model ligands is presented in **Figure 4.31**.

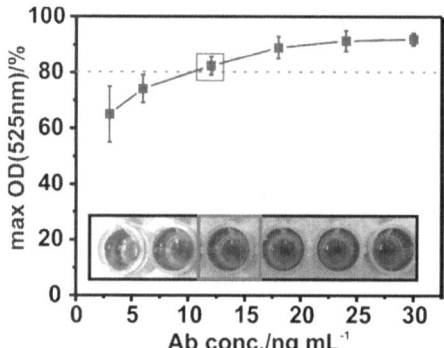

Figure 4.31. Determination of $conc_{min}$ by titration of AuNP-antibody bioconjugates. The OD results of titrated AuNP-antibody solution are presented in the concentration range of antibody (Ab) from 3 to 30 ng mL⁻¹. An untreated solution was referred to an OD of 100 %. Photographies of the corresponding colloidal solutions in a well plate are presented in the inset of the graph. Highest OD value and corresponding well is boxed in blue color. $Conc_{min}$ is defined as minimum concentration yielding an OD value > 80 % (green dashed line).

At a concentration that is below conc$_{min}$, the colloids turn violet and the OD drops below 80 % of the maximum OD. This ligand concentration is the minimum that can be adopted for nanobioconjugate fabrication. Applied concentrations above conc$_{min}$ will yield red-colored colloids with plateaued OD values between 80 and 100 % of maximum OD, which are optimal for biological applications.

If the ligand concentration must be kept below conc$_{min}$, e.g. in the case of highly valuable/expensive pharmaceutical agents, then the adoption of a secondary, dummy ligand may be required in order to reach sufficient electrosteric stabilization of the AuNPs. For this intent, thiolated poly(ethylene)glycol is highly suitable because it does not feature any reactive function and acts in a neutral/inert manner towards any matter. It can be purchased in different chain lengths and because to the thiol function it binds strongly with gold surfaces. The bivalent functionalization of AuNPs with two or more ligands will be extensively presented and discussed in **Chapter 4.2.2.5**.

However, insufficient biomolecule concentrations below conc$_{min}$ will not have drastic effects on the nanobioconjugates except for reduced stabilization and effectivity. On the other hand, if exceeding biomolecule concentrations above conc$_{max}$ are applied, the intrinsic characteristics of the AuNP bioconjugates may change.

For instance, comparing the extinction spectra of ligand-free AuNPs and AuNP-penetratin bioconjugates with penetratin concentrations that range from 1 to 20 μM, a significant NIR contribution has been found for nanobioconjugates fabricated with 2.5 μM or higher penetratin concentration (**Figure 4.32**).

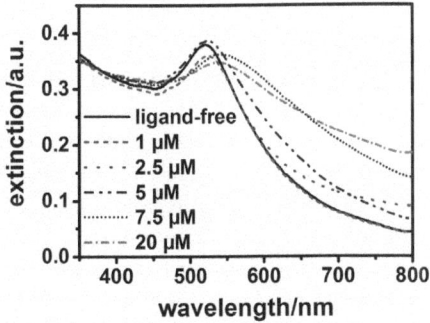

Figure 4.32. Extinction spectra of ligand-free AuNPs and AuNP-penetratin bioconjugates. Colloids were generated by PLAL in *Milli-Q* water (ligand-free) or penetratin solutions (nanobioconjugates) with concentrations from 1 to 20 μM penetratin. Spectra are normalized to 380 nm wavelength.

In addition, starting from a 5 μM penetratin concentration and more distinctly for 7.5 and 20 μM penetratin concentrations, the SPR maximum is shifted by 5 to 30 nm to longer wavelengths for nanobioconjugates, accompanied by a significant peak broadening

(**Figure 4.32**). These spectra modifications indicate an agglomeration/aggregation effect on AuNPs as a function of ligand concentration, which may influence the functionality of AuNP bioconjugates.

The TEM analysis of AuNP-penetratin bioconjugates presented a significant impact of penetratin concentration on gold nanoparticle size and agglomeration (**Figure 4.33**).

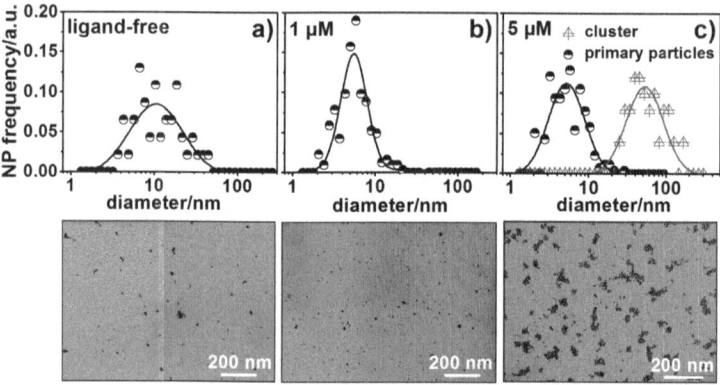

Figure 4.33. Influence of penetratin concentration on nanoparticle size and particle agglomeration. Transmission electron micrographs (bottom) and presented with the corresponding histograms (top) for ligand-free and penetratin-conjugated AuNPs that were generated by PLAL. **a)** Ligand-free AuNPs generated in *MilliQ* water. **b)** AuNP-Pen bioconjugates generated in 1 µM penetratin solution. **c)** AuNP-Pen bioconjugates generated in 5 µM penetratin solution. Reprinted with permission from Petersen et al., copyright 2011 by the American Chemical Society.[197]

While ligand-free AuNPs featured a broad size distribution with a mean primary particle size of 15 nm, the *in situ* conjugation with a 1 µM penetratin resulted in a significant size quenching effect with smaller size distribution and a mean primary particle size of 7 nm. In contrast, the *in situ* conjugation with a 5 µM penetratin resulted in size quenching of the primary particles as well, but further lead to an agglomeration of those primary particles to large clusters with mean sizes of 70 nm (**Figure 4.33**).

This clustering is very likely the result of reduced electrostatic repulsions between the net-charge negative AuNPs after dense covering with the net-charge positive penetratin peptide. The zeta potential was found to be reduced from -28 mV for ligand-free AuNPs to -12 mV for AuNP-Pen conjugates with a 5 µM penetratin concentration. This reduced the interparticle distance and allowed for agglomeration by van der Waals interactions.[197] These results are in agreement with findings from Gamrad et al. on laser-generated AuNP-CPP bioconjugates.[335] They determined that depending on the CPP net charge and concentration, there are two regimes above and below the pI of the bioconjugates which allow for the fabrication of stable nanobioconjugates, because of charge compensa-

tion between the net-charge positive CPPs and the net-charge negative AuNPs. Conversely, a low colloidal stability is found close to the pI. In detail they declared, that a CPP with
high positive net charge ($\geq$ +8) will yield stable AuNP-CPP bioconjugates only at very
low ligand dose (negative zeta potential) and very high ligand dose (positive zeta potential); whereas, a CPP with lower positive net charge (~ +3) will broaden the regime of
stable AuNP conjugates at low ligand doses, while higher ligand amounts will require and
additional steric stabilization.[335]

In the same manner, the penetratin peptide with a net charge between +4 and +5 yielded
stable AuNP-Pen bioconjugates only in low ligand doses, while higher concentrations led
to particle clustering in this study. Moreover, it was shown by Petersen et al. that for
AuNP-Pen bioconjugates with a penetratin concentration of 7.5 μM the colloidal stability
and thereby the zeta potential dropped to -9 mV.[197] Thus, penetratin concentrations
above 7.5 μM would most likely yield completely instable AuNP-Pen bioconjugates with a
zeta potential around zero at.

The cellular uptake of the size-quenched primary nanobioconjugates and the nanobioconjugate clusters was analyzed with confocal microscopy and transmission electron microscopy and results are presented in **Figure 4.34** and **Figure SI 18**.

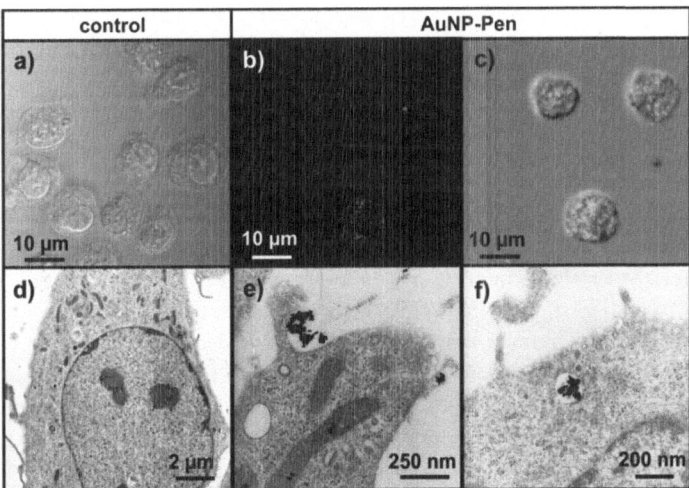

Figure 4.34. Cellular uptake of AuNP-penetratin bioconjugates (AuNP-Pen). a)–c) Representative
laser scanning confocal micrographs that depict immortalized bovine endothelial cells (GM7373) in differential contrast (DIC) mode prior to nanoparticle treatment **(a)** and after co-incubation with AuNP-Pen
(b, c) The backscatter of AuNPs after excitation at 543 nm is imaged in false-red color and presented as
single channel **(b)** and in overlay with the corresponding DIC image **(c)**. **d)–f)** Representative transmission electron micrographs that depict M3E3/C3 pluripotent cells prior to nanoparticle treatment **(d)** and
after co-incubation with AuNP-Pen **(e–f)**. Nanoparticle uptake was found to occur via micropinocytosis.
Adapted in part with permission from Petersen et al., copyright 2011 by the American Chemical
Society.[197]

A large number of AuNP-Pen conjugates was found to be associated with GM7373 endothelial cells by CLSM, which was indicated by strong scattering signals of the gold nanoparticles that were co-localized with the incubated cells (**Figure 4.34b–c**) and that were
absent for untreated cells (**Figure 4.34a**). However, the penetratin peptide is a common
representative of the CPP class which stimulates a receptor-mediated NP uptake into somatic cells. Thus, further TEM analysis of ultrathin cell sections verified the micropinocytotic internalization of AuNP-Pen conjugates (**Figure 4.34e**) into endosomes
(**Figure 4.34f**) on M3E3/C3 pluripotent cells, while no particles were detected on untreated cell sections (**Figure 4.34d**). Thereby, neither qualitative, nor quantitative penetration differences between the primary nanobioconjugates and the nanobioconjugate clusters were observed (**Figure SI 18**).[474]

The micropinocytotic uptake mechanism allows that singular particles as well as particle
clusters are internalized into the cells, so that the cluster size did not have a negative effect on the nanobioconjugate function. Moreover, if a photothermal application was focused then the cluster uptake would be preferred compared to the internalization of singular particles, because the clusters are featuring a higher NIR absorption. Conversely, for
other applications, such as the therapeutic treatment of the lysosomal storage disease, the
clustering would reduce the active surface area of nanoparticles and thereby possibly the
effectivity of the therapy.

Thus, if the focus of nanobioconjugate design is not set on the maximization of ligand
load for applications such as the transport of high amount of cargo into the cells, then
extremely high ligand concentrations should be avoided (especially if an oppositely
net-charged ligand is applied). This applies to cost effectiveness and avoids the potential
ligand-induced clustering of AuNPs which may influence their biological functionality.

In summary, there are two critical ligand concentrations to be considered for nanobioconjugate fabrication. First, the minimum ligand concentration ($conc_{min}$) must be exceeded.
This is necessary to gain stable conjugates without the tendency of precipitation in saline
media or the risk that the biomolecules will wrap around the particle surface while be-
coming functionless. Second, exceeding of the maximum ligand concentration ($conc_{max}$)
should also be avoided. This will prevent unwanted multilayer or particle cluster formation and allow for a cost-effective nanobioconjugate production (if maximized ligand
transport is not aimed). As a consequence, there is a defined concentration window confined by $conc_{min}$ and $conc_{max}$. This window is highly dependent on the nanoparticles' intrinsic parameters and the biomolecule characteristics and needs to be determined for
each ligand, especially in an automated fabrication process with fixed production conditions.

4.2.2.4. Biomolecule length, dimension and binding orientation

Barchanski et al. 2012 [VIII], LZH

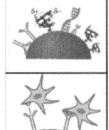

Having considered the environmental conditions of an appropriate ablation medium, an optimal ligand concentration between $conc_{min}$ and $conc_{max}$ and the requirement of a sulphuric function for covalent bonding, the reachable surface coverage may be further dependent on some biomolecule-related characteristics such as its length, dimension and binding orientation.

In addition to carbohydrates and lipids, there are two other main macromolecule classes that may be distinguished and that are frequently used in biomedical research. *Proteins* are biomolecules which have many different functions within a cell, such as structure stabilization, transportation, catalysis or immune defense. Conversely, *nucleotides* are mainly required for the storage of genetic information and for signal transduction.

The spatial resolution of proteins is complex with at least three or even four different structure levels (**Figure 4.35a**).[466;465]

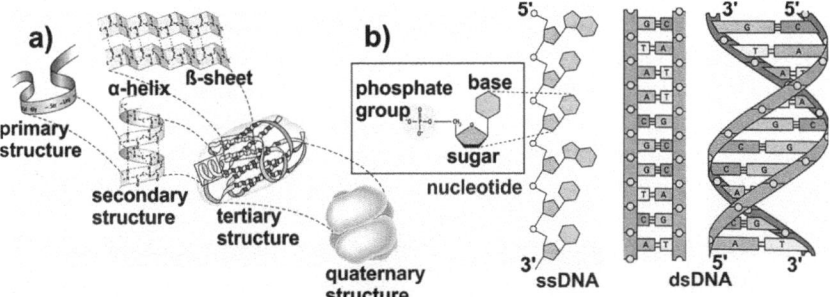

Figure 4.35. **Fundamental configuration of proteins and DNA. a)** The four structure levels of proteins, which are the primary, the secondary, the tertiary and the quaternary structure. [466] The secondary structure may either appear as alpha helix or beta sheet. Reprinted from N. A. Campbell et al., copyright 1997 by Spektrum Akademischer Verlag.[466] **b)** Structure of deoxyribonucleic acid (DNA). Nucleotides are composed of nitrogen-containing base, a pentose sugar and a phosphate group. Single-stranded oligonucleotides (ssO), also termed *ssDNA* are strands of nucleotides which are connected between sugar and phosphate group. And double-stranded DNA (dsDNA) is composed of two antiparallel ssDNA strands which are held together by hydrogen bonds between two paired bases (hybridization). Reprinted in parts from N. A. Campbell et al., copyright 1997 by Spektrum Akademischer Verlag & from B. Alberts et al., copyright 2002 by Garland Science.[466;465]

The basic amino acid sequence is called a *primary structure*. If the sequence folds up into distinct helical, sheet or coiled motives by hydrogen bonds, this is termed a *secondary structure* (**Figure 4.35a**). Further folding includes disulfide-bonds as well as ionic and van der Waals forces and results in a spatial, *tertiary structure* (**Figure 4.35a**). Finally, proteins sometimes need to assemble into complexes, which are named *quatary structures*, in order to be functional (**Figure 4.35a**). For example, immunoglobulins (antibodies) are quaternary-structured with two light and two heavy protein-chains (via disulfide-bonding). If the structure contains less than 50 amino acids and thus a low molecular weight then the molecule is termed *peptide* instead of protein. Peptides generally feature a fibrous (linear) or cylindrical structure. However, the number of amino acids is generally fixed for each peptide/protein, depending on its function and must not be modified. Two protein classes can be distinguished; proteins in which the tertiary or quaternary structure is globular and resembles a sphere and fibrilic-like proteins, which feature a more elongated format.[466;465] Fortunately, most spatial protein structures are summarized in the international protein data base (PDB). Thus, if the precise location of the sulphuric function on the biomolecule is known, then the footprint of the biomolecule (the area it occupies on the nanoparticle surface) can be determined according to the protein dimensions that are provided by the PDB.

Conversely to proteins, the nucleic acids generally exhibit a linear structure, which is required for DNA hybridization. Thereby, three components; a nitrogen-containing base, a pentose sugar and a phosphate group form the main unit that is termed a *nucleotide* (**Figure 4.35a**).[466;465] In total, there are four different nucleotides distinguished, namely adenine (A), guanine (G), cytosine (C) and thymine (T). A connection between the sugar and phosphate groups of the nucleotides results in a single-strand with a 3'- and a 5'-terminus that are important for DNA replication (**Figure 4.35b**). These strands are commonly termed as *single-stranded DNA* (ssDNA). Furthermore, A and T bases as well as C and G bases may pair by hydrogen bonds (hybridization), yielding two connected antiparallel single-strands called *double-stranded DNA* (dsDNA) which is the primary structure of the genetic code (**Figure 4.35b**). The primary structure can twist into a secondary double helix (**Figure 4.35b**) and further it may develop coiled structures by using histones, yielding finally the chromatin which is the main material of chromosomes. The strand length of ssDNA can be varied in a broad range, depending on the required genetic information. Short sequences of nucleotides are usually termed *single-stranded oligonucleotides* (ssO) and they can easily coil from their linearly primary structure into secondary hairpin or loop structures.[466;465] Moreover, there is a specific species of nucleic acids termed *ribonucleic acid* (RNA) which includes a ribose sugar instead of pentose and which is required for information transport and protein synthesis.[466;465] Interestingly, small RNA molecules can be applied to interfere with the conventional intracellular RNA and to stop their coding of specific genes. These *small interfering RNA* (siRNA) are commonly used in biomedical research to down-regulate specific genes.

The linearity of ssDNA minimizes their footprint on the nanoparticle surface if the sulphuric function is attached at one of the strand-ends. However, the aforementioned coiling effects could increase the spatial dimension and reduce the number of attachable ligands on the particle surface.

Thus, the length and dimension of a biomolecule are mainly related on the applied biomolecule class, while the binding orientation results from the intramolecular location of the sulphuric function. To discuss these factors in detail, a linearly random ssO of varied length and with a varied insertion position of a thiol function will be examined. Second, a large globular protein will be directly compared to a more linearly and small peptide. Finally, the importance of correct binding orientation will be analyzed with the example of a functional antibody.

Single-stranded oligonucleotides (ssO)

To investigate the influence of ssO length and binding orientation on PLAL-AuNPs, five ssO designs were applied, which are summarized in **Table 4.10**. The standard molecule (ssO18-3') featured an 18-mer nucleotide sequence, with a thiol function (**HS**) at 3' terminus. Two molecules included an additional 10-mer oligothymidine spacer (**T10**) prior to **HS** and differed in insertion position of **T10-HS** at either ssO 5' terminus (ssO28-**T10-5'**) or ssO 3' terminus (ssO28-**T10-3'**). While the last two molecules included a 20-mer oligothymidine spacer (**T20**) prior **HS** and also differed in insertion position of **T20-HS** at 5' terminus (ssO38-**T20-5'**) or 3' terminus (ssO38-**T20-3'**).

Table 4.10. Overview of adopted gold-ssO species. Adapted with permission from Barchanski et al., copyright 2015 by the American Chemical Society.[420]

Conjugate ID	Length	Sequence
ssO18-3'	18mer	5'-GGCGACTGTGCAAGCAGA-3'-$(CH_2)_3$-**SH**
ssO28-**T10-5'**	28mer	**HS**-$(CH_2)_6$-(T_{10})-5'-GGCGACTGTGCAAGCAGA-3'
ssO38-**T20-5'**	38mer	**HS**-$(CH_2)_6$-(T_{20})-5'-GGCGACTGTGCAAGCAGA-3'
ssO28-**T10-3'**	28mer	5'-GGCGACTGTGCAAGCAGA-3'-(T_{10})-$(CH_2)_3$-**SH**
ssO38-**T20-3'**	38mer	5'-GGCGACTGTGCAAGCAGA-3'-(T_{20})-$(CH_2)_3$-**SH**

Gold ablation in ssO solutions resulted in red-colored, stable colloids with high zeta potential values of -30 mV to -40 mV (**Figure SI 19**), which is most likely due to the negatively net-charged phosphate backbone of ssO. Ablation in the ssO38-**T20** sequences with prolonged phosphate backbone actually yielded noticeably higher zeta potential values than ablation in ssO28-**T10** sequences (**Figure SI 19**).

All colloids exhibited the SPR of AuNPs, which was slightly blue-shifted by 1 to 2 nm for the nanobioconjugates compared to ligand-free AuNPs (**Figure SI 20**), which indicates a size-quenching effect. In fact, the calculated Feret diameter of AuNP bioconjugates was found to be 2 to 3 nm smaller than for ligand-free AuNPs (**Table SI 4**). However, detailed electron microscopy analysis was not performed. In addition to the SPR, all nanobioconjugates featured an additional maximum at 260 nm, resulting from the attached nucleotides, which inclined proportionally with increasing ssO concentration (**Figure 4.36**).

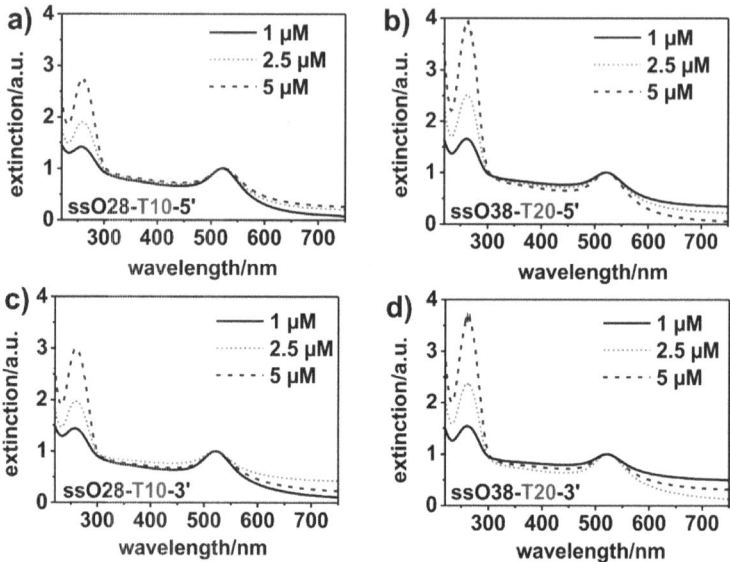

Figure 4.36. Normalized extinction spectra of the four gold-ssO-spacer nanobioconjugates. a) Extinction spectra of ssO28-**T10-5'** nanobioconjugates. b) Extinction spectra of ssO38-**T20-5'** nanobioconjugates. c) Extinction spectra of ssO28-**T10-3'** nanobioconjugates. d) Extinction spectra of ssO-**T20-3'** nanobioconjugates. Varied ssO concentrations of 1 µM (black solid line), 2.5 µM (orange dotted line) and 5 µM (purple dashed line) are presented. Reprinted with permission from Barchanski et al., copyright 2015 by the American Chemical Society.[420]

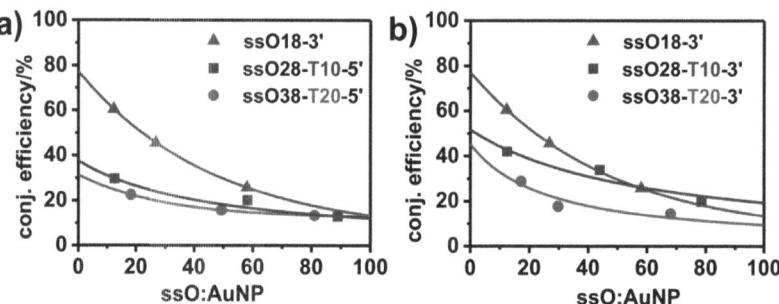

Figure 4.37. Mean conjugation efficiencies of PLAL-generated gold-ssO-spacer nanobioconjugates. a) Differences between 18-mer model nanobioconjugates without spacer (ssO18-**3'** = green triangles) and **5'**-spacer-nanobioconjugates (ssO28-**T10-5'** = red squares, ssO38-**T20-5'** = blue dots). b) Differences between 18-mer nanobioconjugates without spacer (ssO18-**3'** = green triangles) and **3'**-spacer-nanobioconjugates (ssO28-**T10-3'** = red squares, ssO38-T20-3' = blue dots). Results are exponentially fitted and plotted against the ssO concentration, which was depicted as ratio of ssO molecules towards the number of PLAL-generated AuNPs in solution (ssO:AuNP). Adapted with permission from Barchanski et al., copyright 2015 by the American Chemical Society.[420]

The conjugation efficiencies of ligands and surface coverage values on AuNPs were plotted against the ssO concentration, which was recalculated as ratio of ssO molecules towards the number of PLAL-generated AuNPs. The results are summarized in **Figure 4.37** and **Figure SI 21**.

The ssO18-**3'** spacer-less nanobioconjugates featured a conjugation efficiency of 62 % for an ssO:AuNP ratio of 16:1, which dropped exponentially to 28 % for an ssO:AuNP ratio of 59:1. This was most likely due to limited binding places on the AuNP surface (**Figure 4.37**). In comparison, the conjugation efficiencies of the spacer-containing nanobioconjugates were decreased by 30 to 60 % while featuring a comparable, exponential development.

In principle, the same, but inversed, logarithmic trend was found for the ligand surface coverage on AuNPs (**Figure SI 21**). This trend corresponds to data from Steel et al., who calculated that surface coverages on a gold substrate decrease significantly for ssO strands that are longer than 24 bases.[475]
It is important to note that the quantitative comparability of the data is limited because the ssO:AuNP ratios that resulted from the ablation-related, fluctuating yield and the standard deviations that resulted from the variance of three *in situ* conjugation runs per ssO concentration and spacer design, differed significantly. However, qualitatively, the shorter ssO28-**T10** nanobioconjugates seemed to exceed the conjugation efficiencies of the longer ssO38-**T20** nanobioconjugates by 5 to 30 % and exhibited higher mean coverage values by 15 to 45 % (**Figure 4.37** and **Figure SI 21**). In addition, for nanobioconjugates with a thiol function at the 3'-end (ssO28-**T10**-**3'** and ssO38-**T20**-**3'**) the efficiencies exceeded those of nanobioconjugates with a thiol function at the 5'-end (ssO28-**T10**-**5'** and ssO38-**T20**-**5'**) by 7 to 33 %, while higher mean coverage values of 5 to 45 % were reached (**Figure 4.37** and **Figure SI 21**). Thus, both ssO length and binding orientation have an impact on conjugation efficiency and the data recommend a short ssO with a thiol function at the **3'** sequence ending.

The varying results of nanobioconjugates with different binding orientations can be explained by the production-related failure of ssO molecules with thiol modifications at the **3'** end. During synthesis, not only the full-length products but also the failure-related, preterm-capped strands will contain the thiol group, because synthesis direction is conventionally completed from the **3'** towards the **5'** strand end. On the contrary, for synthesis of an ssO with a **5'**-thiol modification, only the full-length product will contain the modification, because the thiol function will be added in the last step. Thus, because the enthalpy of the conjugation process is determined exclusively by the thiol function on the ssO, the ssO-**3'** nanobioconjugates will yield increased conjugation efficiencies and surface coverage values.[420] On the other hand, only full-length ssO (**5'** modification) feature an enhanced functionality due to higher hybridization ability and thus, the question of strand end thiolization should be considered carefully.

In order to further investigate the impact of ssO length and concentration, the steric attachment behavior was identified using a cross-related analysis of the hydrodynamic diameter of nanobioconjugates as well as the biomolecule footprint and deflection angle of the biomolecules, which have been summarized in **Table 4.11**.

For the short ssO18-**3'** nanobioconjugates, a reduction of the biomolecule footprint from 75 to 26 nm^2 and of the deflection angle from 84 to 59° was determined for ssO concentrations ranging from 1 to 5 µM. Furthermore, the hydrodynamic size of AuNPs was found to be nearly constant about 60 nm for 1 µM and 2.5 µM ssO concentrations, but then it increased to 69 nm for the 5 µM ssO concentration. The same parameter trend that was a function of ssO concentration was also observed for all spacer-containing nanobioconjugates.

Table 4.11. Calculated parameters of AuNP-ssO bioconjugates, including footprint of ssO, radius of footprint approximation on the AuNP surface (R) and the deflection angle of ssO. Adapted with permission from Barchanski et al., copyright 2015 by the American Chemical Society.[420]

Conjugate ID	ssO Concentration/µM	Footprint/nm^2	R/nm	Deflection Angle/°
ssO18-3'	1	75	5	84
	2.5	44	4	66
	5	26	3	59
ssO28-T10-3'	1	110	6	101
	2.5	61	4	76
	5	31	3	58
ssO38-T20-3'	1	151	7	122
	2.5	104	6	104
	5	46	4	60
ssO28-T10-5'	1	129	6	121
	2.5	84	5	98
	5	33	3	64
ssO38-T20-5'	1	167	7	138
	2.5	102	6	111
	5	32	3	63

Furthermore, all data on the biomolecule footprint, deflection angle and hydrodynamic diameter of AuNPs rose significantly in conjunction of ssO length (**Table 4.11**). For instance, considering the ssO-**3'** conjugates, the ssO footprint rose from 75 to 110 to 151 nm^2, while the deflection angle enhanced from 84 to 101 to 122°. In addition, the hydrodynamic diameter increased from 59 to 74 to 84 nm for ssO18-**3'**, ssO28-**T10-3'** and ssO38-**T20-3'**, respectively.

From those data, three scenarios can be developed for nanobioconjugate formation as they relate to ssO length and concentration. These scenarios are illustrated on **Figure 4.38**.

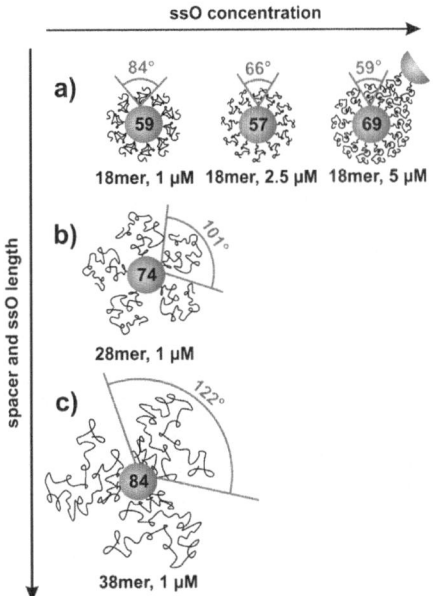

Figure 4.38. Schematic illustration of the steric configuration of nanobioconjugates related to ssO length and concentration. a) *scenario (I):* stretched conformation of short ssO18 with increasing surface coverage and hydrodynamic diameter and decreasing deflection angle as function of ssO concentration. **b)** *scenario (II):* increased flexibility of prolonged ssO28 enables unspecific interactions between the ssO chain and the AuNP surface. **c)** *scenario (III):* maximal flexibility of long ssO38 causes significant wrapping of the polymer-like, coiled ssO chains around the AuNP surface. Deflection angles are illustrated in blue color. Number on AuNPs = hydrodynamic diameter of nanobioconjugates. Reprinted with permission from Barchanski et al., copyright 2015 by the American Chemical Society.[420]

Scenario (I): Short, 18mer ssO are aligned in a mostly *stretched* configuration pointing perpendicular to the gold nanoparticle surface. For low ssO concentrations, the entire AuNP surface area is not occupied by the ssO, which enables an unspecific attraction between the amino-containing bases or the DNA phosphate backbone.[476-478] These interactions result in a wrapped, worm-like ssO-structure with a high deflection angle (**Figure 4.38a**). For increased ssO concentration, more AuNP surface area is occupied by the ssO molecules and biomolecule wrapping is omitted. This confirmation facilitates densely packed ssO ligands on the AuNP, which results in decreased deflection angles and high surface coverage values (**Figure 4.38a**). Similar conformation was also observed by Parak et al. for different surface coverages and ssO lengths.[479] As the ssO concentration further increased, the formation of a biomolecule bilayer seems very likely to explain the enhanced hydrodynamic diameter of AuNPs, due to decreased electrostatic repulsion of ssO molecules within the concentrated sample (**Figure 4.38a**). This interpretation is in agreement with results from Zhang et al., who determined that electrostatic and complementary base-pairing interactions of ssO take place in high-salt solutions only.[480]

Scenario (II): As the ssO length increases with spacer insertion, the flexibility of the elongated ssO is also enhanced, enabling more unspecific interactions among the ssO nucleotides and the phosphate backbone (resulting in hairpin coiling structures) and between those moieties and the AuNP surface (resulting in a stronger wrapping effect) (**Figure 4.38b**). Thus, the conjugated ssO number on the AuNP surface is significantly decreased, while the deflection angle of ssO strongly increases. This assumption is in line with results from Steel et al., who found a less orderly arrangement of long ssO chains on a planar gold surface, compared to shorter ssO chains.[475] These findings reflect an increasing polymeric behavior because the effect of a single thiol group on the ssO attachment becomes less significant than the adsorptive biomolecule-gold interactions.

Scenario (III): The flexibility increase of a 38mer spacer-prolonged ssO sequence enables the ssO to wrap completely around the gold nanoparticle surface resulting in a polymer-like coiled formation (**Figure 4.38c**). The surface coverage is reduced and the deflection angle is significantly increased.

These data correlate with reports from Shlyakhtenko et al. who used atomic force microscopy to determine that ssDNA that is immobilized on a surface exists in a globular, freely jointed and highly flexible worm-like chain conformation.[481] However, there are also reports that assume that because of steric hindrance from nearby molecules, ssDNA may change its conformation from a random coil to a more extended form.[475]

In summary, both the increased charge and the flexibility of an increased ssO chain length causes enhanced coiling and wrapping effects of the ligands around the AuNP surface, which significantly limits the ssO surface coverage. When using nucleotides that had a thiol function at their **3'**-end, the resulting nanobioconjugates had high conjugation efficiencies and surface coverage values, because all full-length or capped failure nucleotide product will contain the modification. Conversely, when using nucleotides with a thiol function at their **5'**-end, the resulting nanobioconjugates had low conjugation efficiencies and surface coverage values, because only the full-length product will be modified.

Thus, a short ssO chain length or the pre-saturation of the AuNP surface with a dummy ligand (see **Chapter 4.2.2.5**) is strongly recommended and the strand end thiolization should be considered carefully.

Proteins

To study the effects of molecular size and three-dimensional structure, a huge globular protein (anti-IgG antibody) and a small cylindrical peptide (penetratin) were chosen and their conjugation behavior onto AuNPs was compared to each other. Penetratin (Pen) is a conventionally used cell-penetrating peptide of 2,391.8 g mol⁻¹. According to the PDB, it features a small dimension with a 0.8 nm diameter at the binding site (**Figure 4.39**).

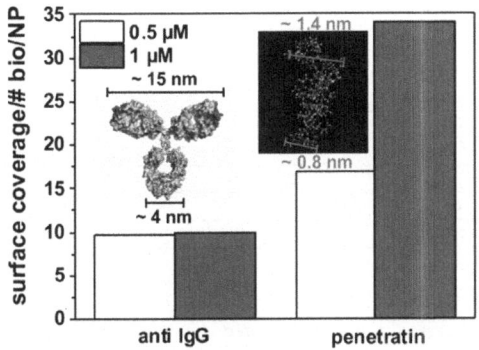

Figure 4.39. Surface coverage values of anti-IgG antibody and penetratin on AuNPs. Data are presented for 0.5 µM concentration (**white bars**) and for 1 µM concentration (**green bars**). Corresponding molecular images of biomolecules are presented in the insets. The anti-IgG antibody molecular image was adapted from Klein et al., copyright 2010 by J. S. Klein and P. J. Bjorkman, PLoS Pathogens.[482] and the penetratin molecular image was adapted from the RCSB Protein Data Bank (PDB ID: 1KZ0).[483]

On the other hand, the anti-IgG antibody exhibits 150,000 g mol⁻¹ and a diameter of ~ 4 nm at the binding site (**Figure 4.39**). Both biomolecules were applied for *in situ* bioconjugation in concentrations of 0.5 µM and 1 µM, respectively and the fabricated nanobioconjugates were thoroughly characterized.

Ablation in penetratin solution resulted in AuNP-Pen nanobioconjugates with a number-weighted modal diameter of 6.7 nm as determined by TEM (**Figure 4.39**). Assuming a biomolecule footprint of ~ 2 nm², a maximum conjugable number of 76 Pen molecules per NP was calculated (**Table 4.12**). However, attached numbers of 17 Pen and 34 Pen were determined for the 0.5 and 1 µM concentration (**Figure 4.39**).

These values seem relatively low upon the first view; however the conjugation efficiency should be considered which is 99 and 98 % for 0.5 and 1 µM, respectively (**Table 4.12**). Nearly the entire amount of added pen has attached to the NPs, indicating that the 0.5 and 1 µM concentrations were too little to cover the entire surface areas relative to the given NP number and that the calculated maximum of 76 Pen per NP could be reached by using a higher Pen concentration.

Table 4.12. Parameters and results for the *in situ* bioconjugation of an anti-IgG antibody and penetratin to AuNPs, covering general information on the biomolecules, calculations for theoretically attachable biomolecule numbers and determined biomolecule numbers per NP.

Biomolecule	Anti IgG		Penetratin	
3D Format	Y-shape		cylindrical	
Footprint /nm²	~ 16		~ 2	
NP Modal Diameter /nm	9		6.7	
NP Surface Area /nm²	254		141	
NP Concentration /μg mL⁻¹	100		52	
Max. # bio/NP (Calculated)	~ 16		~ 76	
Bio Conc. /μM	0.5	1	0.5	1
Ratio Bio/NP	18:1	36:1	17:1	35:1
Conj. Efficiency /%	54	30	99	98
# Bio/NP (Determined)	9.7	10	17	34

On the other hand, ablation in an anti-IgG antibody solution yielded AuNP-anti-IgG nanobioconjugates with a number-weighted modal diameter of 9 nm as determined by TEM (**Figure SI 22**). The footprint was assumed to be 16 nm², which would result in a maximum conjugable number of 16 anti-IgG antibodies per NP by calculation

However, in practice, only ~ 10 anti-IgG antibodies per NP were determined for the 0.5 and 1 μM concentrations, respectively (**Figure 4.39**), which is significantly below the expected value of 16. In these cases, conjugation efficiencies of 54 and 30 % were determined, which indicate that the 0.5 and 1 μM anti-IgG antibody concentrations already exceeded the required biomolecule amount to cover the complete surface area relative to the given NP number. Concerning the calculated number of 16 anti-IgG antibodies per NP it should be taken into consideration, that only the modal NP diameter was used for calculation, and that the given dimensions of anti-IgG antibodies are simply approximated data. Thus, marginal changes will yield different values. For instance assuming a diameter of 3 nm and a footprint of 9 nm², a maximum conjugable number of 28 anti-IgG antibodies per NP is calculated, while a diameter of 5 nm and a footprint of 25 nm² will yield exactly 10 anti-IgG molecules per NP. Thus, perfectly accurate data cannot be calculated in advance, however the approximation is quite good and allows for an assessment of applicable biomolecule concentration and surface coverage on the NPs. Regarding the biomolecule dimension, a clear size trend was observed, yielding lower surface coverage values for larger-sized proteins than for smaller peptides.

In addition, for proteins the binding orientation towards NP may be of high importance. In general, not many proteins and peptides feature activity *per se* and they are generally used for structural issues *in vivo*. However, there are at least two protein classes that bear active sites for targeting function; enzymes in metabolism and antibodies in immune defense. If their active sites are not accessible for specific ligands, there is no functionality provided and the system will not working properly. Thus, the correct binding orientation of these proteins on the AuNPs is the goal. In these cases, adsorptive attachment is not recommended, but binding should be controlled by direct insertion of a sulphuric function on the opposite part of the active site.

This effect is demonstrated on a highly specific immunoblot assay, which compares PLAL-generated AuNP-antibody bioconjugates in which anti-IgG antibodies were thiolized in a controlled manner using a heterobifunctional hydrazide-PEG-dithiol linker (AuNP-IgG-SH) with AuNP-antibody bioconjugates where the anti-IgG antibodies were used unmodified (AuNP-IgG) for attachment during *in situ* bioconjugation.

For optimal comparability of both nanobioconjugates, identical laser parameters yielding the same NP concentration were applied. Moreover, the same antibody concentration that is slightly above $conc_{min}$ was chosen for each fabrication route and the nanobioconjugates were not purified by centrifugation to avoid detachment of weakly bound (physisorbed) ligands. For this reason, it was not possible to calculate the effective NP surface coverage with antibodies. However, a theoretically calculated maximum ligand load of 6 antibodies per NP could be reached. For the *golden blot* assay, an IgG from a rabbit was used as a specific analyte on the blotting membrane, while an unspecific TAT peptide was used as negative control. The results of the immunoblotting assay are presented in **Figure 4.40**.

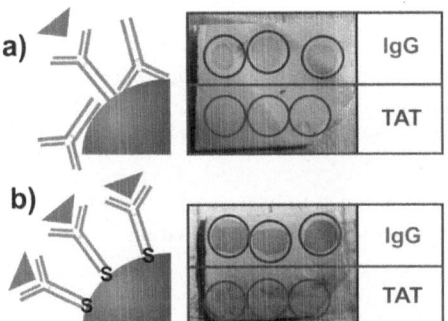

Figure 4.40. Functionality proof of directed and randomly oriented AuNP-antibody nanobioconjugates using a *golden blot* assay. a) Results for AuNP bioconjugates with randomly oriented anti-IgG antibodies. **b)** Results for AuNP bioconjugates with anti-IgG antibodies attached in directed orientation by an OPPS-PEG-SH linker. IgG was immobilized on the membrane within the red circles and unspecific TAT peptide within the purple circles, respectively. Antibody = green color, specific ligand = blue color.

Both AuNP bioconjugate probes specifically targeted the IgG analyte and not the TAT control peptide, depicting the functionality of both conjugates. However, the labeling intensity of AuNP-IgG conjugates was significantly lower (**Figure 4.40a**) than for AuNP-IgG-SH conjugates (**Figure 4.40b**), despite equal treatment.

This result most likely indicates: *i)* a low surface coverage of AuNPs with unmodified IgG, *ii)* an incorrectly-oriented attachment of IgG to the AuNP surface, *iii)* a mixture of both scenarios. Nevertheless, if a controlled orientation can be provided, this should be the method of choice to obtain functional AuNP-IgG bioconjugates.

It can be summarized, that the molecular size of globular ligands has an impact on conjugation efficiency and surface coverage that is similar to the length of linear ligands. In detail, the larger the molecular size, the lower the number of attachable ligands to AuNPs due to a larger molecular footprint. However, binding orientation and thus the insert position of the sulphuric function has a tremendous effect on ligand functionality, at least if ligands with active centers as enzymes or antibodies are applied.

4.2.2.5.   *Bivalent functionalization and surface saturation*

Barchanski et al. 2011 [XII], *LZH*

 For certain applications, it might be necessary to have two or more different ligands attached to the nanoparticle surface, yielding a bivalent nanobioconjugate. For instance a cell-penetrating peptide may be required to provide cellular internalization of the NP, while a secondary nucleic derivate or pharmaceutical agent will trigger an intracellular effect. Moreover, a non-functional dummy ligand might be applied for *i)* NP surface pre-saturation in order to control the surface coverage of a functional secondary ligand or *ii)* NP surface or post-saturation in order to increase the nanobioconjugate stability and to enable its functionality even in highly saline media. Furthermore for dummy pre-saturation treatment, the expenses of fabrication will be reduced because the ratio of NPs to functional ligands can be kept small (1:1).

Considering the *in situ* conjugation technique, three approaches for AuNP functionalization with two different ligands (A and B) can be traced:
- **Approach 1** covers the *in situ* conjugation with ligand A and the *ex situ* conjugation with ligand B.
- **Approach 2** examines *in situ* conjugation with ligand B and the *ex situ* conjugation with ligand A.
- A simultaneous *in situ* co-conjugation with both ligands A and B is handled in **Approach 3**.

A detailed comparison of those scenarios is found in the Bachelor thesis from C. Sehring.[414] Interestingly, the feasibility to obtain bivalent AuNP conjugates was ensured by all three approaches. This result indicates that **Approach 3** is the optimal method due to the ease of accomplishment in a one-step process. However, whether and how precise the number of attached ligands A and B can be controlled during the bivalent conjugation process is unknown to date.

Thus, to study those issues, two experimental setups were analyzed, covering a dummy pre- and post-saturation with **Approaches 1 and 2** and a simultaneous co-conjugation with two functional ligands using **Approach 3**.

The simultaneous *in situ* co-conjugation of AuNPs with **Approach 3** was performed using mixtures of the cell-penetrating peptide penetratin and an aptamer sequence termed *miniStrep*, targeting the streptavidin protein. Both ligands were bearing a sulphuric function for covalent AuNP binding and were applied in varied concentrations according to **Table 4.13**, yielding the bivalent AuNP conjugates **bi-con I–bi-con VI**.

Table 4.13. Overview of fabricated bivalent AuNP conjugates (**bi-con I–VI**). Specified characteristics are the applied concentrations of penetratin (pen$_{conc}$) and aptamer (apt$_{conc}$) as well as determined conjugation efficiencies (CE$_{pen}$, CE$_{apt}$), surface coverages (SC$_{pen}$, SC$_{apt}$) in both pmol cm^{-1} and number of biomolecules per nanoparticle unit and total number of biomolecules per nanoparticle (# Bio NP^{-1}).

ID bi-con	pen$_{conc}$ μM	apt$_{conc}$ μM	CE$_{pen}$ %	CE$_{apt}$ %	SC$_{pen}$ pmol cm^{-1}	SC$_{apt}$ pmol cm^{-1}	SC$_{pen}$ #pen NP^{-1}	SC$_{apt}$ #apt NP^{-1}	# Bio NP^{-1}
I	0.25	1.5	100	78	10.3	54	16	100	116
II	0.25	4.5	100	52	12.4	91.5	16	155	171
III	1	1.5	99	47	33	32	66	56	122
IV	1	4.5	99	37	30	61	66	106	172
V	2.5	1.5	56	81	63.7	72.2	91	105	196
VI	2.5	4.5	54	64	57.5	149.4	88	190	278

For the fabricated bivalent AuNP conjugates, a constant average NP diameter of 10 to 12 nm (**Figure 4.41a**) and a constant AuNP yield of 70 μg mL^{-1} were determined, independent of the applied ligand concentrations. This allows for ideal comparability of the fabricated colloids. In contrast to mono-conjugation,[197;269] no distinct size quenching effect was found, which was likely due to electrosterical binding hindrance between the positive net-charged peptide and the negative net-charged aptamer.

However, the obtained gold spheres were perfectly shaped and non-agglomerated, which can be seen as an indication for significant ligand coverage and electrosteric stabilization. High zeta potential values between -32 mV to -42 mV were determined for all bivalent nanobioconjugates.[63] This is highly interesting because it was worked out earlier that the stability of AuNP-penetratin mono-conjugates is strongly dependent on the CPP net charge and concentration, yielding stable AuNP bioconjugates with a zeta potential of -28 mV at low penetratin concentration (≤ 1μM) and less stable AuNP bioconjugates with a zeta potential of only -12 mV at a penetratin concentration of 5 μM (see **Chapter 4.2.2.3**). This charge balancing effect between negatively charged AuNPs and positively charged CPPs was studied by Gamrad et al. and they declared that for CPPs with a net charge ~ +3, such as penetratin, there is a broad regime of low ligand doses that will yield stable conjugates.[335] However, if higher ligand doses are applied the resulting bioconjugates will be less stable and will require additional ligands for steric stabilization.[335] In this study, the additional stabilization is given with the net-charge negative *miniStrep* aptamer, yielding stable conjugates with the aforementioned high zeta potentials.

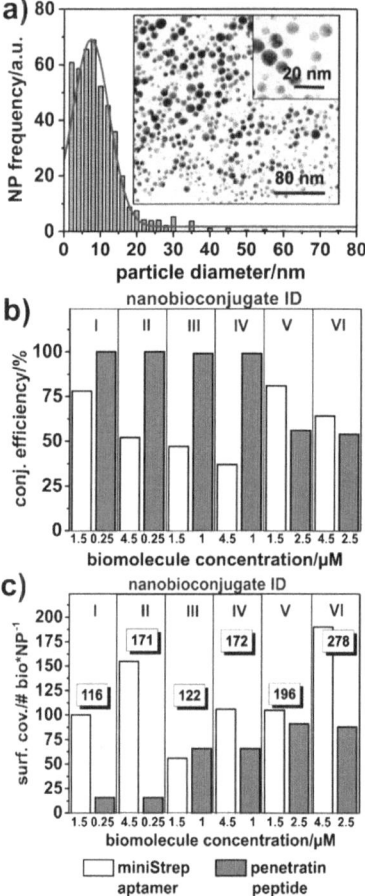

Figure 4.41. Characteristics of PLAL-fabricated bivalent AuNP conjugates I–VI. a) Representative particle size distribution with Gaussian fitting function and transmission electron micrograph (inset) for **bi-con IV. b)** Conjugation efficiencies of miniStrep aptamer (white bars) and penetratin peptide (green bars) for **bi-cons I–VI. c)** Surface coverage values of miniStrep aptamer (**white bars**) and penetratin peptide (**green bars**) as number of biomolecules per nanoparticle expression for **bi-cons I–VI.** Adapted in parts with permission from Barchanski et al., copyright 2011 by the Japan Laser Processing Society.[63]

A high number of biomolecules was detected on the bivalent AuNP nanobioconjugates. In detail, penetratin concentrations of 0.25 μM, 1 μM and 2.5 μM yielded mean conjugation efficiencies of 100 % (**bi-con I** and **bi-con II**), 99 % (**bi-con III** and **bi-con IV**) and 55 % (**bi-con V** and **bi-con IV**) and mean surface coverage values of 16, 66 and 90 molecules per AuNP, respectively (**Table 4.13** and **Figure 4.41b**). The miniStrep aptamer was applied in concentrations of 1.5 μM and 4.5 μM and yielded highly differing, but significantly lower conjugation efficiencies between 47 and 78 % for 1.5 μM (**bi-con I** and **bi-con III** and **bi-con V**) and between 37 and 64 % for 4.5 μM (**bi-con II** and **bi-con IV** and **bi-con VI**). Surface coverage values ranged from 56 to 105 molecules per AuNP for 1.5 μM concentration and from 106 to 190 molecules per AuNP for 4.5 μM concentration, respectively (**Table 4.13** and **Figure 4.41c**).

For the **bi-cons I–IV**, nearly 100 % conjugation efficiency of the penetratin molecule was achieved, while significantly lower efficiencies of the miniStrep aptamer were determined. For this reason, a higher affinity of penetratin towards the AuNP surface can be assumed. However, this hypothesis was excluded, because gold-affine functions (thiol, disulfide) were implemented. Furthermore, it can be assumed that penetratin features a higher mobility than the miniStrep aptamer. Actually, both ligands vary significantly in molecular size, resulting in different diffusion coefficients of 8.8×10^{-11} m^2 s^{-1} for penetratin and 5.7×10^{-12} m^2 s^{-1} for the aptamer while indicating a higher mobility and thus faster particle coordination of penetratin molecules. In addition, it seems very likely that the saturation concentration needed to cover the entire AuNP surface is not reached by using 1 μM of penetratin, allowing up to 66 penetratin molecules to be attached to the AuNP surface simultaneously. However, using a 2.5 μM penetratin concentration, a total number of 90 molecules per AuNP was determined while the conjugation efficiency dropped significantly; nearly in half, which indicates a high excess of molecules in the solution (oversaturation). Moreover, the formation of multilayers should be considered, because it was shown in **Chapter 4.2.2.3** on the example of penetratin, that peptide multilayers can form above a threshold concentration of 2 μM.

The overall low efficiency values that were determined for the miniStrep aptamer indicate that the saturation concentration to cover the AuNP surface was exceeded for all of the bi-conjugates that were analyzed. Considering the surface coverage, values of 100 aptamer and 16 penetratin molecules per AuNP were determined for **bi-con I**, yielding a total number of 116 molecules per AuNP (**Table 4.13** and **Figure 4.41c**). This value is assumed to be the maximum monolayer load due to an excess of the saturation concentration. For **bi-con III**, a comparable total amount of 122 molecules per AuNP was distributed onto 66 penetratin molecules and 56 aptamer molecules (due to the different supply concentrations) (**Table 4.13** and **Figure 4.41c**), which supports the maximum monolayer assumption (**Figure 4.42**).

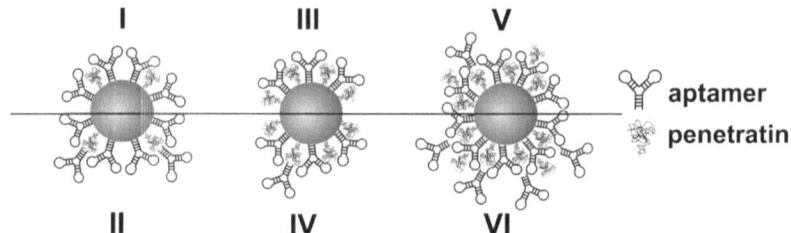

Figure 4.42. Illustration of the ligand distribution on bivalent gold-penetratin-aptamer conjugates I-VI in correlation to the ligand concentration. The illustration is depicting an optimal scenario in which the secondary biomolecule layers are perfectly oriented and which probably is not reached in effect.

However, for **bi-con II** and **bi-con IV**, higher coverage values were reached. For **bi-con II** an amount of 155 aptamer and 16 penetratin molecules yielded a total of 171 molecules per AuNP. A total of 172 molecules per AuNP were distributed onto 106 aptamer and 66 penetratin molecules (due to the different supply concentrations) for **bi-con IV** (**Table 4.13** and **Figure 4.41c**). Thus, if there is an excess molecule supply and due to the fact, that reciprocal attraction exists between the oppositely charged penetratin and aptamer molecules, the additional attachment of singular aptamer molecules on the penetratin molecules seems very likely (**Figure 4.42**).

Interestingly, for **bi-con V** and **bi-con VI**, the conjugation efficiencies of both the min-iStrep aptamer and penetratin were below 100 % and much more similar, while total surface coverage values of 196 and 278 molecules per AuNP were determined, which are significantly higher than for **bi-cons I–IV** (**Table 4.13** and **Figure 4.41c**). In this case, a mixed multilayer formation can be assumed, because excess molecule supply and intermolecular attraction was significantly enhanced (**Figure 4.42**).

In summary, the successful fabrication of bivalent AuNP conjugates was enabled by *in situ* co-conjugation with two different ligands (**Approach 3**). Thereby, several parameters that affected the attached number of each biomolecule were identified:

- Small biomolecules feature a high mobility and coordinate the NPs more rapidly than large biomolecules, which results in a higher number of attached ligands even at a supply concentration than is lower than for the larger-sized molecules.
- For low biomolecule supply concentrations that are below, equal or slightly above the saturation concentration, a mixed monolayer is formed on the AuNPs.
- For biomolecule supply concentrations that are above the saturation concentration, additional biomolecules may attach to the monolayer, especially if they are attracted by opposite (net) charges.

- If extremely high supply concentrations are applied and even more if the biomolecules bear opposite (net) charges, mixed multilayers may form on the AuNP surface.

Thus, by paying attention to the applied biomolecule type, size, charge and supply concentration, the molecule distribution on the AuNP surface may be estimated and even controlled to some extent. This highlights the single-step **Approach 3** as the optimal method for the quick fabrication of bivalent AuNP conjugates. However, if a very precise ligand amount/distribution is required or if the biomolecule handling is complex (e.g. requiring salt transfer), then the **Approaches 1** and **2** may be superior.

In another experimental setup, the biological functionality of laser-generated bivalent AuNP conjugates was successfully verified. In this setup, AuNPs were functionalized with a distinct cell penetrating peptide for nuclear entry and a DNA derivate for DNA hybridization. For stability and functionality issues with the DNA derivate, it was necessary to perform a salt transfer, which could not be run in a batch with the peptide due to the risk of protein precipitation. Thus, in this case, the bivalent functionalization **Approach 1** was adopted for nanobioconjugate fabrication and the colloids were applied for a cellular penetration study of bovine spermatozoa. In summary, a cellular and even nuclear penetration into acrosome-reacted spermatozoa was successfully verified and was discussed in detail in **Chapter 4.2.2.6.** However, intracellular DNA hybridization of bivalent conjugates has not reached thus far, due to the disability of nanobioconjugates to move freely inside the highly condensed chromatin and to reach the DNA sequence of interest *in vitro*.

From an economic point of view, the reduction of costs for nanobioconjugate fabrication is correlated with a reduction of the applied ligand number. In an optimal scenario, nanobioconjugate functionality would already be given by a single attached ligand and in reality, by 1-10 ligands for statistical reasons. Unfortunately, 1-10 ligands are not sufficient to stabilize the nanobioconjugate and if long chain-forming ligands are used, they may tend to wrap around the empty space on the nanoparticle surface (see **Chapter 4.2.2.4**).
However, using bivalent functionalization **Approaches 1** and **2**, the controlled surface saturation and the blocking of empty surface area/potential binding places may be achieved if a non-functional dummy) ligand without any function is applied. For this intent, the 5 kDa thiol-functionalized methoxyl (poly)ethylene glycol (mPEG-SH) is an appropriate dummy molecule because it features an extended configuration for excellent particle stabilization and provides good monolayer coverage within a few hours. It can either be implemented for pre-saturation when the conjugation is carried out *in situ* while the functional ligand is attached with *ex situ* conjugation to the remaining free surface (**Approach 1**) or for post-passivation when the functional ligand is conjugated *in situ*

and the dummy saturates the remaining free surface area with *ex situ* conjugation (**Approach 2**).

To study both approaches and to analyze the effectivity of the methods, a study was conducted using a concentration series of mPEG-SH (1 nM–2.5 μM) as a dummy ligand and a Cy5-labeled ssDNA in an oversaturated concentration (5 μM) for **Approach 1** and in an undersaturated concentration (0.5 μM) for **Approach 2**. The laser ablation time was fixed to yield a mean particle concentration of 75 μg mL⁻¹ in each colloid. A modal particle size of 11.4 nm was determined for **Approach 1** and 10.8 nm was calculated for **Approach 2**. This comparability is most likely due to the similar molecular size of the applied molecules. Nanobioconjugates were purified after the *in situ* conjugation step as well as after the *ex situ* conjugation step to avoid the detection of unbound molecules. This purification step was enabled, because both ligands were equipped with a thiol function for covalent AuNP binding that excludes the undesired removal of electrostatically bound ligands with centrifugal force. After centrifugation, the number of attached ssDNA-Cy5 ligands was determined with fluorescence analysis and found to be approximately 15 for the *in situ* conjugation and approximately 45 for the *ex situ* conjugation without mPEG-SH addition.

In analyzing the results for **Approach 1**, a clear trend was determined (**Figure 4.43a**).

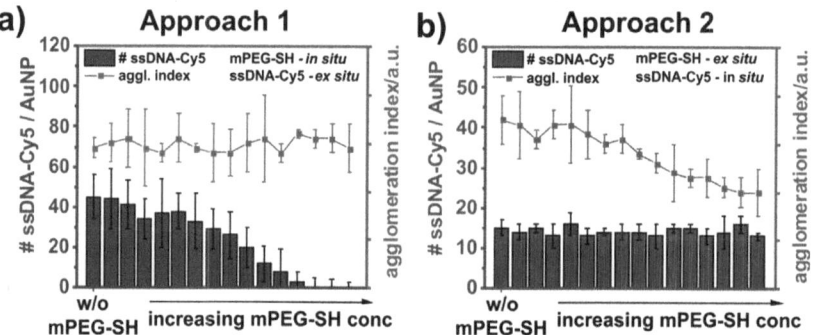

Figure 4.43. Surface coverage values and agglomeration index of pre-saturation and post-passivation approaches. a) Surface coverage (red bars) and agglomeration index (blue squares) of ssDNA-Cy5-AuNP-mPEG-SH bi-conjugates with *ex situ* conjugated ssDNA-Cy5 as function of increasing mPEG-SH concentration, which was used as a pre-saturation *in situ* ligand. **b)** Surface coverage (red bars) and agglomeration index (blue squares) ssDNA-Cy5-AuNP-mPEG-SH bi-conjugates with *in situ* conjugated ssDNA-Cy5 as function of increasing mPEG-SH concentration, which was used as a post-passivation *ex situ* ligand.

In detail, the number of ssDNA molecules on the AuNP surface was significantly reduced as function of increasing mPEG-SH concentration, because the increased number of dummy molecules was occupying more of the potential binding places for the nucleotides. A multilayer formation was excluded because the dummy did not feature an opposite net charge compared to the ssDNA. Interestingly, the pre-saturation effect was nearly linear (**Figure 4.43a**), with only some small fluctuations in the low dummy ligand concentration range (1–10 nM), which was possibly due to the sterical wrapping and therefore area-blocking effect of the long-chained ligand (see **Chapter 4.2.2.4**). However, for higher concentrations, a precise calculation of the functional ligand number can be enabled, making the pre-saturation technique very attractive for reducing the cost of expensive ligands.

For **Approach 2** on the other hand, nearly the same surface coverage of ssDNA was found to be independent on the applied mPEG-SH concentration (**Figure 4.43b**). This result excludes the option of ligand substitution (exchange) where the primary conjugated ligand is replaced by the secondary added ligand. To analyze the stability of the bivalent conjugates as function of mPEG-SH concentration, 500 µL of PBS was added and the agglomeration index was calculated after 30 min of incubation. Interestingly, the agglomeration index was found to decrease as function of mPEG-SH concentration, which proves the enhancement of colloidal stability, most likely by additional steric stabilization of the undersaturated AuNP-ssDNA conjugates with the mPEG-SH (**Figure 4.43b**). Conversely, for **Approach 1**, there was no change in agglomeration index, because the over-saturated amount of ssDNA enabled a good colloidal stability for all samples (**Figure 4.43a**). Thus, post-passivation appears to be an interesting option in order to increase the stability of AuNP bioconjugates.

Unfortunately, the fabricated mPEG-SH-AuNP-ssDNA-Cy5 bivalent conjugates could not be adopted for a biological functionality test, because a random ssDNA sequence without a biological function was used. However, all AuNP-antibody conjugates that were fabricated and discussed in the framework of this thesis were mPEG-SH post-passivated in order to enhance the stability of the AuNP conjugates. Their functionality was successfully proven in **Chapter 4.1.2.1**, **Chapter 4.1.1.5**, and **Chapter 4.2.2.4**.

In summary, the fabrication of bivalent AuNP conjugates was enabled with three different **Approaches**, including the *in situ* conjugation with ligand A and the *ex situ* conjugation with ligand B, the *in situ* conjugation with ligand B and the *ex situ* conjugation with ligand A and the *in situ* co-conjugation with both ligands at once. Among those **Approaches**, the co-conjugation was determined to be the most effective method, due to the achievement of bivalent functionalization in a single-step process. Moreover, the ligand distribution on the AuNP surface was found to be a function of ligand size (molecular weight and dimension), their net charge and their supply concentrations. The bivalent functionalization method may also be applied to pre-saturate the AuNP surface with a functionless, dummy ligand in order to increase the cost effectiveness of the expensive, functional ligands. Alternatively the bivalent functionalization method may be applied to post-passivate the leftover free surface after conjugation with a functional ligand in order to increase the colloid stability in high saline media.

4.2.2.6.   Amphiphilic ligand nature

Barchanski et al. 2015 [111], Cooperation LZH-FLI-UDE

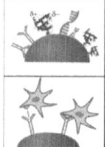 When looking at reproductive biology, the vital labeling of genetically interesting DNA-sequences is required for sorting issues to prevent the inheritance of defective alleles or to select beneficial traits in livestock. While the genetic analysis of oocytes before *in vitro* fertilization (IVF) is an established technique,[111] the performance of genetic tests on spermatozoa is still challenging. In the early 90s, Johnson et al. and Levinson et al. used the fluorescence-*in-situ*-hybridization (FISH) to validate the separation of X and Y chromosome-bearing spermatozoa using flow cytometry for the prevention of X-linked diseases.[484;485] For this technique fluorophore-labeled DNA probes were applied, which hybridize specifically to complementary target sequences on permeabilized sperm cells after DNA denaturation. Unfortunately, a vital labeling has not been demonstrated to date, since oligonucleotide probes cannot penetrate the sperm membrane or hybridize to non-denatured DNA. Thus, bivalent nanobioconjugates functionalized with a DNA derivate for hybridization issues and a peptide for cell or nucleus penetration may enable this complicate endeavor.

Biological membranes of somatic cells feature a unique architecture. The basic structure is a lipid bilayer with embedded proteins. Membranes are selectively permeable for small, diffusive molecules and they are able to internalize larger molecules and particles using transport proteins or by means of endocytotic processes. The bilayer is composed mainly of phospholipids, which are amphiphilic molecules with a hydrophilic head and two hydrophobic (lipophilic) tails that arrange in water by self-organization into a two-layered sheet (bilayer) with the tails pointing towards the sheet center. The membrane is structured in tightly packed and ordered microdomains that contain various glycosphingolipids, cholesterol and proteins, which are termed *lipid rafts* and which float freely on the bilayer and increase its fluidity (fluid mosaic model). Specific molecules such as CPPs are required to surmount these complex obstacles and it should be emphasized that both the charge of a nanobioconjugate (which is determined by the ligand coating)[180] and the hydrophobicity of a ligand[189] significantly influence the interaction with and the potential crossing of biological membranes.

Different from somatic cells, spermatozoa feature a very specialized membrane structure with reduced metabolism (lacking conventional active membrane transport mechanisms such as endocytosis) and advanced cell segmentation. Spermatozoa contain only a minimum of cytoplasm. Beneath the usual, all-enveloping plasma membrane (PM), the anterior part of the nucleus is entirely covered by the acrosome, a hat-shaped vesicle containing various substances such as enzymes that are required for fertilization. These are bordered by an outer acrosomal membrane (OAS) and an inner acrosomal membrane (IAM). The posterior part of the plasma membrane is reinforced on the inside and stiffened by a protein-rich electron-dense layer called the *post-acrosomal sheath* (PAS). The border between the anterior and the posterior part of the plasma membrane is marked by the equatorial segment comprising of the abruptly narrowed caudal portion of the acrosomal cap. The double-membrane nuclear envelope (NE) is the only membrane that has pores at its distal end (**Figure 4.44**).

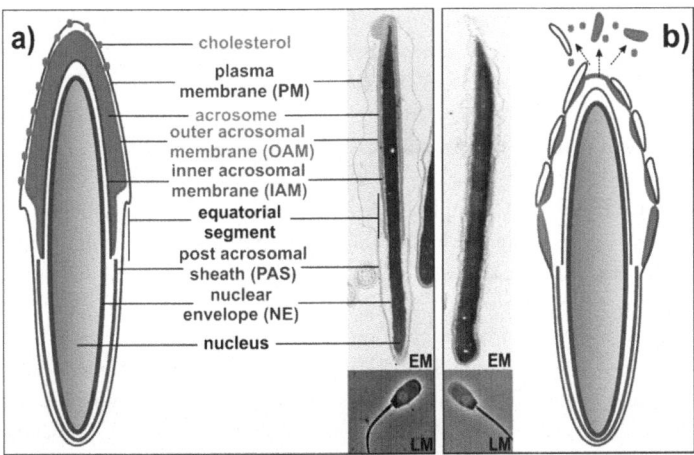

Figure 4.44. Schematic comparison between acrosome-intact and acrosome-reacted spermatozoa. a) Illustration and equivalent transmission electron micrograph (EM) of acrosome-intact sperm ultrastructure depicting a median sagittal section of the sperm head. **b)** Illustration and EM of spermatozoon ultrastructure after acrosome-reaction. The morphological modifications are mainly the release of cholesterol from the plasma membrane (PM) and the ejection of hydrolytic enzymes from the acrosome, accompanied by membrane component migration. Prompt differentiation between acrosome-intact and acrosome-reacted spermatozoa in fresh ejaculate is depicted on light micrographs (LM) in the inset. Adapted with permission from Barchanski et al., copyright 2015 by the American Scientific Publishers.[198]

The lipid composition of the PM is extremely complex,[486] and contains a high amount of cholesterol and represents a very effective biological barrier.[487;488] The PM further undergoes extreme alterations during sperm maturation and fertilization, which imply the release of cholesterol from the PM and the subsequent ejection of hydrolytic enzymes from the acrosome, which is supported by changes in membrane fluidity with the migration of membrane proteins and lipids. This process is commonly termed an *acrosome reaction* and highlights a particular challenge with respect to nanoparticle internalization.

Although the acrosome reaction is a step during sperm maturation, which normally occurs when spermatozoa are already associated with the oocyte, there is always a subpopulation of spermatozoa (5 to 10 %) even in fresh ejaculates where a premature spontaneous acrosome reaction has already taken place. In TEM sections and on light micrographs, such spermatozoa are clearly distinguishable by their lack of an acrosomal cap, which usually covers the anterior part of the nucleus (**Figure 4.44**).

The acrosome was found to be an insurmountable obstacle for AuNP bioconjugates and no particle internalization has been found thus for acrosome-intact spermatozoa. Internalization behavior has only been documented on acrosome-reacted sperm cells.

The AuNP bioconjugates for the spermatozoa penetration study were fabricated with the bi-functionalization **Approach 1** (see **Chapter 4.2.2.5**). A net-charge negative DNA derivate termed *locked nucleic acid* (LNA) was first attached for triplex hybridization with non-denatured DNA sequences. To allow for a broad spectrum of different parameter effects, the additional benefit of three subsequently conjugated, net-charge positive CPPs with varied chemical composition was screened, which yielded bivalent AuNP bioconjugates.

The Deca-Arginine (**10R**) was applied as a representative of the polycationic, highly hydrophilic CPP class, containing a high number of positively-charged amino acids such as arginine and lysine and featuring a pI of 13.2 and a zeta potential of +6 mV (**Table SI 6**). In addition, the Transactivator of Transcription (**TAT**) featured a mixed cationic-neutral composition with a small hydrophobic content and was characterized by a pI of 12.9 and a zeta potential of +4 mV (**Table SI 6**). Finally, the Simian Virus 40 Large T Antigen Nuclear localization signal (**NLS**) was used as a representative of the amphiphilic CPP class with alternating patterns of (poly)cationic, neutral and (poly)anionic domains and increased hydrophobicity, yielding a pI of 9.6 and a zeta potential of -10 mV (**Table SI 6**). In addition to the CPP-AuNP-LNA bivalent conjugates, also ligand-free AuNPs, monovalent AuNP-LNA and AuNP-NLS bioconjugates were tested as a negative control (**Figure 4.45, Table 4.14**).

Table 4.14. Overview of the samples that were used for the spermatozoa penetration study.

Sample ID	Type of Conjugation	Use	Ligand Composition
AuNP	none / ligand-free	negative control	/
AuNP-LNA	monovalent	negative control	hydrophilic
AuNP-NLS	monovalent	negative control	amphiphilic
10R-AuNP-LNA	bivalent	sperm penetration	cationic, hydrophilic
TAT-AuNP-LNA	bivalent	sperm penetration	cationic-neutral, hydrophilic
NLS-AuNP-LNA	bivalent	sperm penetration	amphiphilic

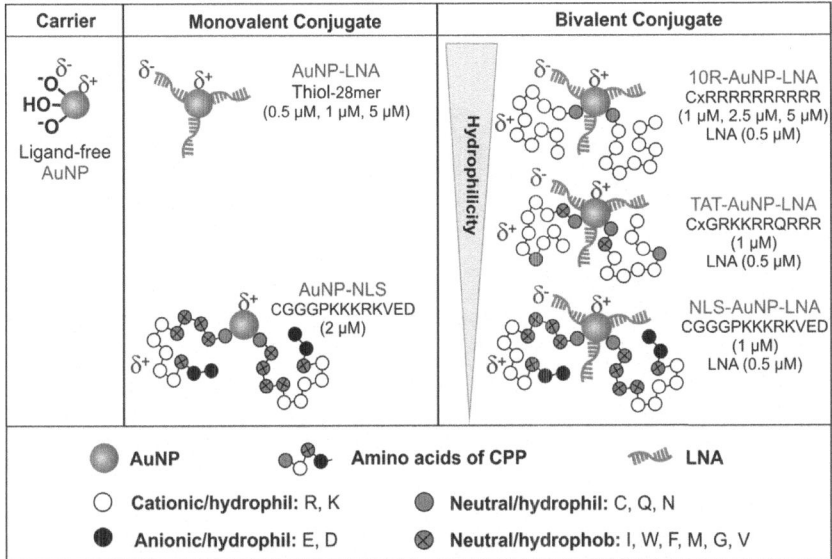

Carrier	Monovalent Conjugate	Bivalent Conjugate
Ligand-free AuNP	AuNP-LNA Thiol-28mer (0.5 µM, 1 µM, 5 µM)	10R-AuNP-LNA CxRRRRRRRRRR (1 µM, 2.5 µM, 5 µM) LNA (0.5 µM)
		TAT-AuNP-LNA CxGRKKRRQRRR (1 µM) LNA (0.5 µM)
	AuNP-NLS CGGGPKKKRKVED (2 µM)	NLS-AuNP-LNA CGGGPKKKRKVED (1 µM) LNA (0.5 µM)

● AuNP   ●○✕● Amino acids of CPP   〰〰 LNA

○ Cationic/hydrophil: R, K   ● Neutral/hydrophil: C, Q, N

● Anionic/hydrophil: E, D   ⊗ Neutral/hydrophob: I, W, F, M, G, V

Figure 4.45. Schematic overview of the applied sample designs for the spermatozoa penetration study. Samples are classified in ligand-free AuNPs, monovalent AuNP conjugates and bivalent AuNP conjugates. Bivalent AuNP conjugates are functionalized with negative LNA strands and net-positive CPPs as ligands. The CPP sequence is varied, consisting of cationic and neutral, as well as twisted anionic-cationic amino acids featuring reduced hydrophilic properties. Monovalent AuNP conjugates are equipped with one representative of either LNA or CPP class and ligand-free AuNPs feature a positive surface charge. Adapted with permission from Barchanski et al., copyright 2015 by the American Scientific Publishers.[198]

A nanoparticle concentration of 10 µg per mL was used, which can be calculated to a number dose of 1.1×10^5 nanoparticles per sperm cell and a surface dose of 4.4×10^6 nm^2 nanoparticle surface per sperm cell. The penetration behavior of the fabricated probes (**Table 4.14**) was evaluated after co-incubation with fresh bovine spermatozoa using transmission electron microscopy. Detailed experimental procedures are found in **Chapter 3.3.4**.

Only a few ligand-free, primary AuNPs were found sporadically attached to the plasma membrane (**Figure 4.46a**), which is likely due to aggregation behavior of the purely electrostatically stabilized particles in the saline spermatozoa buffer and their subsequent precipitation (**Figure 4.46b–c** and **Figure SI 23a**).

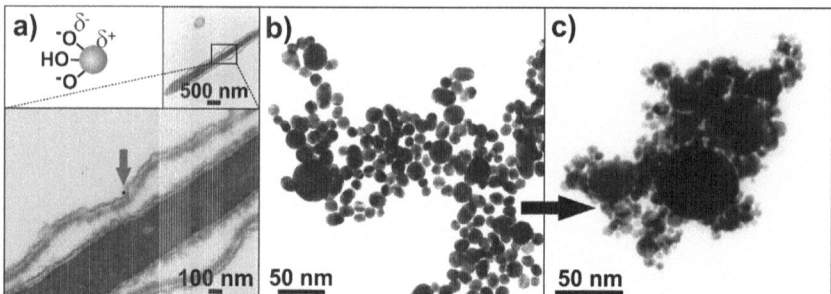

Figure 4.46. Transmission electron micrographs of bovine spermatozoa after co-incubation with ligand-free AuNPs and particle conformation prior and after salt transfer. a) Transmission electron micrograph of a single ligand-free AuNP (**blue arrow**) attached to the outer cell membrane of spermatozoa. **b)** Transmission electron micrographs of ligand-free AuNPs before transfer into salt-containing media. **c)** Transmission electron micrographs of ligand-free AuNPs after transfer into salt-containing media. Adapted with permission from Barchanski et al., copyright 2015 by the American Scientific Publishers.[198]

Conversely, for AuNP-LNA monovalent conjugates that were protected against agglomeration by steric ligand coordination (**Figure SI 24** and **Figure SI 23a**), high numbers of primary nanoparticles were found all over the PM of acrosome-intact spermatozoa (**Figure 4.47a**) and in the post-equatorial region between the PAS and the NE of acrosome-reacted spermatozoa. This result was independent of the applied LNA concentration (**Figure 4.47b**).

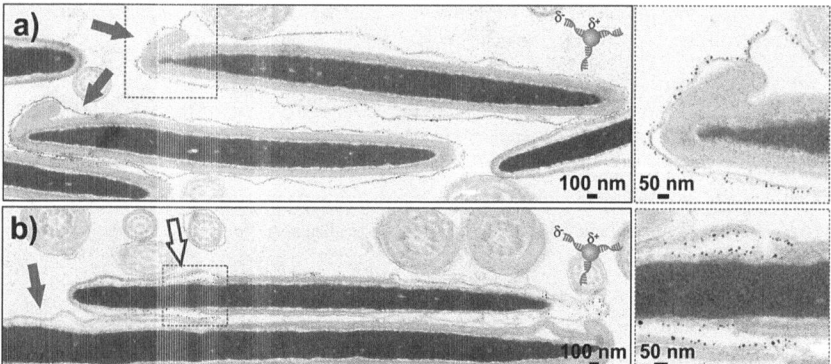

Figure 4.47. Transmission electron micrographs of bovine spermatozoa after co-incubation with AuNP-LNA monovalent bioconjugates. a) Singular nanobioconjugates are attached to the outer cell membrane of acrosome-intact spermatozoa. **b)** Accumulated nanobioconjugates are detected between the PAS and the NE of acrosome-reacted sperm cells. Red dashed boxes are presented in magnification on the right. Green arrows = acrosome-intact spermatozoa, black-framed arrow = acrosome-reacted spermatozoa. Adapted with permission from Barchanski et al., copyright 2015 by the American Scientific Publishers.[198]

The same scenario was observed for all CPP-conjugated and thus stabilized (**Figure SI 23a**), bivalent AuNP bioconjugates with 1 µM CPP concentration (**Figure 4.48a–c**).

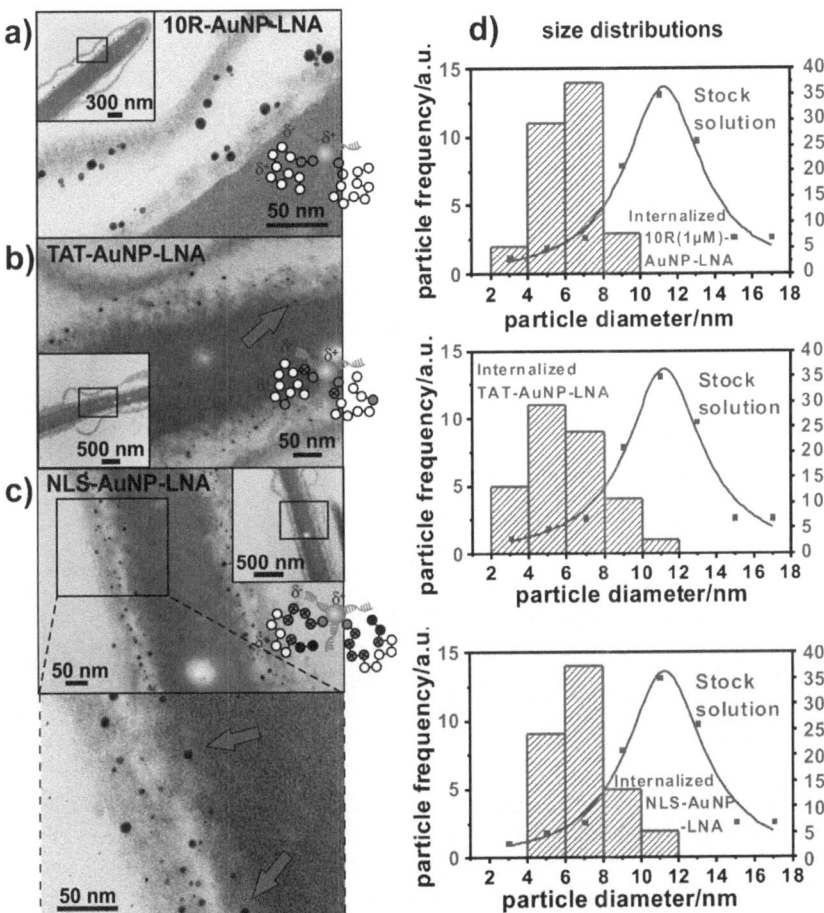

Figure 4.48. Transmission electron micrographs of bovine spermatozoa after co-incubation with CPP-AuNP-LNA bivalent conjugates. a) Micrograph of 10R-AuNP-LNA bivalent conjugates. **b)** Micrograph of TAT-AuNP-LNA bivalent conjugates. **c)** Micrograph of NLS-AuNP-LNA conjugates. For 10R-AuNP-LNA conjugates, AuNPs are detected close to the NE, while singular particles were located sporadically within sperm nucleus for the TAT-AuNP-LNA conjugates (red arrow). Efficient internalization of AuNPs into spermatozoa nuclei was visualized for NLS-AuNP-LNA conjugates (red arrows). The size distributions of the bivalent AuNP conjugates that penetrated the spermatozoa are presented in comparison to the AuNP-LNA stock solution to which the CPPs were conjugated to by *ex situ* method. Reprinted with permission from Barchanski et al., copyright 2015 by the American Scientific Publishers.[198]

Interestingly, if higher concentrations of CPP were applied for conjugation, a ligand-induced agglomeration occurred and only particle clusters were detected on the electron micrographs (**Figure SI 25**). This was due to the charge compensation effect that was discussed previously (see **Chapter 4.2.2.3**).

However, an interesting and statistically ensured difference in the penetration depth of the different bivalent probes was noticed. The trend was obvious in relation to both the applied CPP and the nanoparticle sizes. Isolated particles of 10R-AuNP-LNA bioconjugates were detected in high amounts between the PAS and the NE, mostly attached to one of these cell barriers (**Figure 4.48a**).

For bivalent TAT-AuNP-LNA bioconjugates more NPs were found attached to the NE and they were also found in different penetration depths within the equatorial region of the NE, while singular particles were detected inside the border area of the nucleus (**Figure 4.48b**). The most significant results were obtained for bivalent NLS-AuNP-LNA bioconjugates, where a high number of particles were detected between the PAS and the NE regions and a considerable amount had entered the nucleus superficially (**Figure 4.48c**). However, no particles were observed in the nuclear center within an incubation time of two hours.

Regarding NP size, the diameters of internalized bivalent conjugates ranged from 2 nm to 14 nm with a maximum that ranged between 4 and 8 nm (**Figure 4.48d**), while larger sized particles were attached to the PM exclusively.

Finally, analyzing monovalent AuNP-NLS conjugates, no particles were found to be associated with the spermatozoa membrane, neither with intact, nor with acrosome-reacted sperm.

The specific hybridization of LNA with complementary DNA sequences was not achieved due to the inferior nuclear internalization of bivalent AuNP bioconjugates and the disability of AuNPs to move freely within the condensed chromatin. Thus, additional or different ligands still need to be analyzed.

As particle internalization was only found on acrosome-reacted spermatozoa, it is speculated that biochemical membrane modifications during acrosome reaction facilitate/enable the nanoparticle entry. This process is characterized by an efflux of decapacitation factors and membrane molecules as cholesterol and is accompanied by a local reduction of the negative membrane charge and calcium influx. Furthermore, the membrane fluidity is enhanced by migration and lateral re-organization of the membrane proteins and lipids, which aggregate in lipid rafts at the apical PM and create lateral membrane heterogeneity.[489-492;487;488;493] Thus, it seems very likely that the NP attachment to the PM of acrosome-intact spermatozoa without internalization is due to the presence of cholesterol and the correlated compact membrane status.[494] Pawar et al. hypothesized that when a significant amount of cholesterol is effluxed from the PM during acrosome reaction, then the membrane fluidity/permeability increases and facilitates NP internalization.[494] Moreover, Welsher and Yang recently found a correlation between small fluidity hotspots and local NP dynamics using a 3D dynamics heat mapping technique.[495]

Comparing the internalization of the three different bivalent AuNP bioconjugates, the NLS-AuNP-LNA bioconjugates were mainly found at all penetration depths throughout the post-equatorial region of PAS and NE, with distinct particle internalization into the nucleus. Thus, since the hydrophilic, polycationic 10R-AuNP-LNA bioconjugates and cationic-neutral TAT-AuNP-LNA bioconjugates did not enter the nucleus, it can be assumed that an amphiphilic CPP such as NLS that feature a twisted anionic-neutral-cationic composition with significant hydrophobic content and in conjunction with a positive net-charge is required to trigger nuclear penetration, most likely with the formation of transitory structures.[496] This assumption is also in line with conclusions from Verma et al., who described an optimal cellular penetration of AuNPs with alternating anionic and hydrophobic end groups.[189]

The internalized NPs featured diameters that ranged from 2 to 10 nm, while the particle size distribution in the incubation solution ranged between 2 and 22 nm. Thus, a clear correlation between cellular uptake and NP size with the preference of mainly small particles < 10 nm is demonstrated, which correlates with the literature, to explain the size-selective NP uptake and intracellular distribution.[82;178;204]

To date, only a few publications have reported results on nanoparticle interaction with spermatozoa.[138;497;494;135;498] For instance, Wiwanitkit et al. claimed that there was a spontaneous translocation of ligand-free AuNPs into human sperm heads and tails[135], and Moretti et al. found AuNPs in the sperm nuclei.[138] Makhluf et al. discussed the uptake of polyvinyl alcohol-conjugated iron nanoparticles into bovine spermatozoa and their accumulation on intracellular organelles such as acrosome and mitochondria.[498;499] All of those study designs differed significantly from each other and did not aim to control nanoparticle internalization into the spermatozoa nucleus.

In summary, the successive cellular penetration of bi-conjugated AuNP bioconjugates into advanced cells such as acrosome-reacted spermatozoa was found to be highly dependent on the chemical composition of attached penetration ligands. The internalization results and depths of the 6 nanobioconjugate probes that were examined are depicted in **Table 4.15** and **Figure 4.49**.

Table 4.15. Summarized results for the penetration behavior of ligand-free AuNPs as well as of monovalent and bivalent AuNP conjugates on acrosome-intact and acrosome-reacted spermatozoa. Adapted with permission from Barchanski et al., copyright 2015 by the American Scientific Publishers.[198]

Sample	Acrosome-Intact Spermatozoa	Acrosome-Reacted Spermatozoa
Ligand-Free AuNP	sporadic attachment to PM	no particles
Monovalent AuNP-LNA	all-over attachment to PM	accumulation between PAS and NE mostly attached to one of those barriers
Monovalent AuNP-NLS	no particles	no particles
Bivalent 10R-AuNP-LNA	attachment to PM	accumulation between PAS and NE mostly attached to one of those barriers
Bivalent TAT-AuNP-LNA	attachment to PM	accumulation between PAS and NE mostly attached to NE and singular particles detected inside the border zone of N
Bivalent NLS-AuNP-LNA	attachment to PM	intense accumulation between PAS and NE; many particles attached to NE and several entered the N superficially

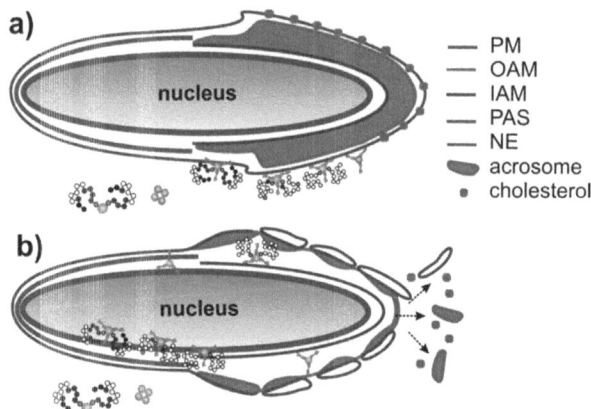

Figure 4.49. Illustration of sperm membrane association and cellular penetration of the six adopted AuNP bioconjugate probes. a) On the example of an acrosome-intact spermatozoon; **b)** On the example of an acrosome-reacted spermatozoon. Adapted with permission from Barchanski et al., copyright 2015 by the American Scientific Publishers.[198]

While small (< 10 nm) monovalent AuNP-LNA bioconjugates and the polycationic, bivalent 10R-AuNP-LNA bioconjugates featured an accumulation between the PAS and the NE, singular cationic-neutral, bivalent TAT-AuNP-LNA bioconjugates were detected inside the border zone of the nucleus and solely amphiphilic NLS-AuNP-LNA bioconjugates were even found inside the nucleus. These bioconjugates had a twisted anionic-neutral-cationic composition and a high hydrophobic content and were able to enter the nucleus superficially. Thus, a significant correlation between the chemical composition of nanobioconjugate ligand shell and their interaction with biological membranes was clearly demonstrated.

4.2.3. Summary and discussion

Depending on a defined application, various aspects of the nanoparticles' characteristics, the nanoenvironment, the biomolecules and the aimed properties of the resulting nanobioconjugates must be considered before the PLAL fabrication is started. Most aspects that were examined within this Chapter and their impacts on each other have been identified and are summarized in **Figure SI 26**.
The main recommendation is to make a detailed plan of attention prior to nanobioconjugate fabrication by considering the structure-function relationship.

 The nanoparticles' intrinsic characteristics such as the particle size and charge are of main importance for NP uptake, cytotoxicity and imaging properties and must be adjusted carefully.

Primary particle size
The primary particle size must be adjusted according to the aimed cell penetration mechanism (diffusive crossing versus receptor-mediated endocytosis) and the adopted visualization technique (light microscopy versus electron microscopy).
PLAL-fabricated AuNPs in ultrapure water are generally not monodisperse and feature a broad PSD. The PSD can be narrowed with *in situ* photofragmentation method; however ps-pulses and NIR-wavelength have rarely been used for this attempt. In this study, an AuNP primary size reduction from 34 to approximately 6 nm diameter with accompanied PSD narrowing was successfully enabled and most likely due to a highly efficient second harmonic generation and energy transfer of absorbed laser light during the photofragmentation.
Alternatively to photofragmentation, the separation of individual particle size classes can be aimed. This was so far achieved with fine adjustment of the laser process parameters. However, this is a complex method because a very broad parameter screening series for each desired size class is required. In this study, the separation of size classes was accomplished with *ex situ* centrifugal processing of the fabricated colloids. Using this approach, an AuNP primary size reduction from 42 nm to a diameter of approximately 7 nm was reached. Moreover, the required centrifugal force and time can be estimated by calculation with Svedberg equation.

However, in contrary to *in situ* conjugation, monodispersity will not be reached with these approaches. Furthermore, it should be considered that the photofragmentation method cannot be performed with nanobioconjugates.

Nanoparticle charge

Concerning biological applications, it is known, that the charge of nanoparticles strongly influences their cellular uptake behavior and their toxicity. Thus, detailed knowledge about the amount of charged surface atoms is essential in order to gain specific biological functionality. There was no knowledge about that topic for ps-PLAL generated AuNPs to date. In this study, the fabrication of AuNPs with ps-PLAL using a fluence of 0.5 J cm^{-2} yielded partially oxidized particles ($\sim$ 5 % of atoms) with an Au$^+$ configuration and a negative zeta potential of up to -30 mV. No other gold configurations were detected.

Compared to other studies, the extent of nanoparticle surface oxidation seems to be varying with laser parameters such as the pulse length, the pulse energy, the repetition rate, the fluence and the laser wavelength. However, a close correlation between the laser parameters and the formation of oxidation states on laser-generated AuNPs has not been found yet and a systematic study on this topic is strongly required.

 When the work on this thesis started, there was not much information about the interactions of laser-generated nanoparticles with their close surroundings; especially during particle formation and *in situ* bioconjugation. However, it was assumed that these surface interactions are mainly determining the colloidal stability and the feasibility for nanobioconjugate formation; thus they are of high importance and should be considered carefully.

The surrounding medium

The ablation solvent should provide optimal conditions for the electrostability of nanoparticles (depending on their surface oxidation) and biomolecules (above pI of proteins; slightly alkaline for single-stranded DNA) and it should allow for the dilution into biological media such as cell culture media or buffer media, because the applicability of *Milli-Q* water is limited due to osmotic pressure.

However, it was found that the fabrication of AuNPs in a standard buffer such as PBS and cell culture media is not feasible due to the high ionic strengths of the media, which lead to charge shielding effects. Moreover, the adsorption of serum proteins and the formation of a protein corona, which can change the AuNP functionality, have to be considered. Thus, to generate the nanoparticles in these media, the threshold concentration has to be determined, which defines the ionic strength of the medium that allows AuNP formation without charge shielding and which differs widely among distinct buffers. Alternatively, the possibility to transfer the water-generated nanoparticles into the buffer must be considered. For this intent, a salt transfer method was established that allows the nanobioconjugates to adapt to salt concentrations of at least 150 mM. However, the AuNP bioconjugates should be completely covered with biomolecules to avoid particle losses from precipitation with this method.

For the optimal binding of proteins such as e.g. antibodies to the partially oxidized AuNP surface, the pH must be adjusted slightly above the pI of the antibody to result in a negative net charge. To determine the optimal pH range, it is recommended to perform a titration test with optical characterization with the AuNP-antibody solution.

Binding stability

The main prospect on nanobioconjugates is the binding stability between the particle and the attached biomolecules because it defines both colloidal stability and nanobioconjugate functionality to a high degree. The nanobioconjugates must be stable enough to resist ionic strength and pH variations on their way through intracellular compartments without decomposition. In addition, ligands such as antibodies should be strongly connected with the particle to accomplish their distinct function.

In this study, it was determined that biomolecules that bear electron-donor moieties such as COOH or NH$_2$ may coordinate the electron-accepting PLAL-AuNPs, but that the covalent attachment of ligands with a thiol or disulfide function yields much more stable and functional AuNP bioconjugates, which can resist (ultra)centrifugal forces and high ionic strengths.

Functional group

The conjugation with a specific, material-affine function such as thiol/disulfide in the case of gold enables a controlled conjugation and should also be adopted whenever a covalent bonding without the requirement of ligand separation is needed.

A significant preference for covalent attachment of either thiol-containing or disulfide-containing ligands to the gold surface was not determined. However it was found, that the molecule structure should be considered carefully prior to conjugation, since electron-transfer-related spontaneous fragmentation/dissolution of AuNPs by electron-donor-containing moieties could occur, especially if reducing agents are applied.

To estimate whether the sulphuric function generally binds to the oxidized gold atoms (Au$^+$/Au^{3+}) or to the neutral gold atoms (Au^0), a thought process was performed, which supported strongly the theory that covalent bonding is generally occurring with the oxidized surface atoms of AuNPs.

Ligand amount

From an economic point of view, the reduction of costs for nanobioconjugate fabrication is correlated with a reduction of applied ligand concentration. However, if a too low amount of ligands are used they may not be sufficient to stabilize the nanobioconjugates from precipitation in saline media and the isolated ligands may also tend to wrap around the empty area on the nanoparticle surface. On the other hand, the ligand concentration should also not be too high, because it might hinder the particle formation during PLAL or induce multilayer formation on the particle surface. Thus, two threshold values were defined, termed as minimum ligand concentration and maximum ligand concentration, which can be determined by titration. It is recommended, that the adopted ligand concentration for nanobioconjugate formulation should be within a concentration window that is defined by those thresholds to achieve optimal nanobioconjugate formation and functionality.

Ligand net charge

The net charge of ligands can have a tremendous effect on the colloidal stability, especially if oppositely charged nanoparticles and ligands are applied. It was found, that net-charge positive peptides can initiate a particle aggregation as function of ligand concentration. This aggregation is very likely the result of charge compensation between the net-charge negative AuNPs after dense covering with the net-charge positive peptide. This reduces the interparticle distance and allows for agglomeration by van der Waals interactions.

It has to be considered, that a peptide with high positive net charge ($\geq$ +8) will yield stable AuNP bioconjugates only at very low ligand dose (negative zeta potential) and very high ligand dose (positive zeta potential); whereas, a peptide with lower positive net charge ($\sim$ +3) will broaden the regime of stable AuNP conjugates at low ligand doses, while higher ligand amounts will require and additional steric stabilization. [335]

 The characteristics of a ligand such as dimension, orientation and composition are strongly determining the properties of the resulting nanobiohybrids. If the nanobioconjugates need to accomplish several functions such as specific cell targeting, cell penetration and drug delivery, a co-conjugation of antibodies, CPPs and drugs on a single NP may be necessary. This bivalent or multivalent conjugation complicates the whole conjugation scenario because the ligand characteristics may supplement or erase each other.

Biomolecule length and dimension

The length and dimension of a biomolecule will strongly impact the surface coverage of the fabricated nanobioconjugates and are mainly related on the applied biomolecule class (nucleotide or protein).

It was found in this study that the charge and flexibility of a prolonged, linear nucleotide chain causes enhanced coiling and wrapping effects of the ligands around the AuNP surface, which significantly limits the nucleotide surface coverage. Thus, a short nucleotide chain length or the pre-saturation of the AuNP surface with a dummy ligand is strongly recommended.

Moreover, the molecular size of globular proteins has an impact on conjugation efficiency and surface coverage that is similar to the length of linear nucleotides. In detail, the larger the molecular size, the lower the number of attachable ligands to AuNPs due to a larger molecular footstep.

Binding orientation
Especially for targeting and catalytic function, the binding orientation of biomolecules such as antibodies and enzymes onto the NPs is not trivial because their active centers need to be accessible. In this context it was determined, that correctly oriented antibodies allowed the nanobioconjugates to have a significantly higher targeting functionality than the nanobioconjugates with randomly oriented antibodies.
It was further demonstrated, that the binding orientation results from the intramolecular location of the sulphuric function; especially for linear ligands. When using nucleotides that had a thiol function at their 3'-end, the resulting nanobioconjugates had high conjugation efficiencies and surface coverage values, because all full-length or capped failure nucleotide product will contain the modification. Conversely, when using nucleotides with a thiol function at their 5'-end, the resulting nanobioconjugates had low conjugation efficiencies and surface coverage values, because only the full-length product will be modified.

Bivalent functionality
For certain applications, it might be necessary to have two or more different ligands attached to the nanoparticle surface, yielding a bivalent nanobioconjugate. For instance a cell-penetrating peptide may be required to provide cellular internalization of the NP, while a secondary nucleic derivate or pharmaceutical agent will trigger an intracellular effect.
In this study, the fabrication of bivalent AuNP conjugates was enabled with three different approaches, including *i)* the *in situ* conjugation with ligand A and the *ex situ* conjugation with ligand B, *ii)* the *in situ* conjugation with ligand B and the *ex situ* conjugation with ligand A and *iii)* the *in situ* co-conjugation with both ligands at once. Among those approaches, the co-conjugation was determined to be the most effective method, due to the achievement of bivalent functionalization in a single-step process. Moreover, the ligand distribution on the AuNP surface was found to be a function of ligand size (molecular weight and dimension), their net charge and their supply concentrations. However, if a very precise ligand amount/distribution is required or if the biomolecule handling is complex (e.g. requiring salt transfer), then the other approaches may be superior.

Surface saturation
A non-functional dummy ligand might be applied for *i)* NP surface pre-saturation in order to control the surface coverage of a functional secondary ligand or *ii)* NP surface or

post-saturation in order to increase the nanobioconjugate stability and to enable its functionality even in highly saline media. Furthermore for dummy pre-saturation treatment, the expenses of fabrication will be reduced because the ratio of NPs to functional ligands can be kept small (1:1).

In this study, the bivalent functionalization method was applied to pre-saturate the AuNP surface with a functionless, dummy ligand in order to increase the cost effectiveness of the expensive, functional ligands. Alternatively it was shown, that the bivalent functionalization method can also be applied to post-passivate the leftover free surface after conjugation with a functional ligand in order to increase the colloid stability in high saline media.

Amphiphilic ligand nature

The amphiphilic nature of the biomolecule may influence the biological functionality of the nanobioconjugates, especially if cell penetration through the highly fluidic lipid bilayer is aimed.

In the study, the successive cellular penetration of bivalent AuNP bioconjugates into advanced cells such as acrosome-reacted spermatozoa was found to be highly dependent on the chemical composition of attached penetration ligands. While polycationic, bivalent bioconjugates featured an accumulation between the post acrosomal sheath and the nuclear envelope of acrosome-reacted spermatozoa, singular cationic-neutral, bivalent bioconjugates were detected inside the border zone of the nucleus and solely amphiphilic bioconjugates were even found inside of the nucleus. These bioconjugates had a twisted anionic-neutral-cationic composition and a high hydrophobic content, which enabled the multiple membrane crossings most likely with the formation of transitory structures.

If those points are considered carefully, the PLAL-fabrication of customized and functional nanobioconjugates for nearly every type of biological application is feasible, which allows for highly specific and perfectly biocompatible action due to the water-based fabrication method.

4.3.   Transfer of the *in situ* bioconjugation method to other materials

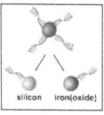

The *in situ* bioconjugation of gold nanoparticles has become an established technique to equip AuNPs with biological functions during the PLAL process.[39;269;420] With the adoption of thiol or disulfide containing biomolecules an oriented and covalent binding can be obtained, resulting in highly stable nanobioconjugates with biological activity. Moreover, the *in situ* functionalization of AuNPs during PLAL is a single-step process, while conventional *ex situ* conjugation requires several processing steps.

However, in addition to gold there are other biomedically relevant materials such as iron that must be bioconjugated to achieve biological compatibility and functionality. Because these nanomaterials are conventionally fabricated on an organic solvent basis and because they require a multi-step purification and conjugation procedure, the possibility of water-based PLAL fabrication with bioconjugation in a single step is very interesting. Therefore, the facility and efficiency of transferring the *in situ* bioconjugation method onto other materials than gold has to be studied. In addition, the analysis of the formed bond types and the functionality of the novel fabricated nanobioconjugates are of high interest and will be discussed on the examples of silicon and magnetic, iron-based NPs in the following subchapters.

4.3.1.   Silicon and silicon-based nanoparticles

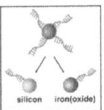 The element silicon (latin, *silex* or *silicis* = flint, Si) is a grey-colored, very brittle, tetravalent metalloid with the atomic number 14, which was discovered in the late 18th or early 19th century. Its specific characteristics are summarized in **Table SI 7**.

All naturally occurring types of silicon are non-toxic. Only the inhalation of fine particles of silica is known to cause silicosis disease.[48-50] Silicon features the electron configuration [Ne] $3s^2\ 3p^2$ and can donate or share its four outer electrons which allows for different chemical bonds, although it is relatively inert in crystallized form and most acids do not have an effect on it. It is usually not oxidized when dissolved in water because a protective surface layer of silicon dioxide (SiO_2) is formed rapidly (**eq 4.3**). However, ortho silicic acid (H_4SiO_4) is mainly present, since silicon dioxide dissolves slowly in water with a solubility of 0.12 g L^{-1} (**eq 4.4–eq 4.5**)

$$Si\ (s) + 2\ H_2O\ (l) \rightarrow SiO_2\ (s) + 2\ H_2\ (g)$$

eq 4.3

$$SiO_2\ (s) + 2\ H_2O\ (l) \leftarrow \rightarrow H_4SiO_4\ (s)$$

eq 4.4

$$H_4SiO_4\ (s) + H_2O\ (l) \leftarrow \rightarrow H_3O^+\ (aq) + H_3SiO_4^-\ (aq)$$

eq 4.5

Recently, nanoscaled silicon has attracted the interest of scientists due to its excellent biocompatibility and biodegradability.[500] In addition it has a high photostability without photobleaching.[501] These features highlight the material as an interesting alternative to cadmium-containing quantum dots (QDs), which are the most commonly used QDs to improve luminescence yields but feature a heavy-metal-related cytotoxicity via Cd^{2+} ion release.[502;503] For instance, silicon nanoparticles (SiNPs) are considered promising as bio-sensors or as luminescent *in vivo* markers for cellular labeling, especially if conjugated to biological moieties such as DNA and proteins.[504-506]

Conventional fabrication methods of SiNPs mainly cover wet chemistry reduction routes,[507-509] electrochemical etching[510;511] and the gaseous phase decomposition of silanes.[512;513] However, recently the fabrication of SiNPs was presented with PLAL as an appealing alternative approach.[514-518]

For the bioconjugation of SiNPs, a variety of chemical methods have already been employed.[519;520] For instance, Erogbogbo et al. have presented a multi-step, chemical-based method that includes particle size reduction and passivation using hydrofluoric acid etching, functionalization of the NP surface with carboxyl groups and bioconjugation via a coupling reaction.[505] Obviously, such techniques are extremely time-consuming and thus emphasize the interest for the application of the one-step *in situ* bioconjugation method during PLAL.

4.3.2.   PLAL-generated SiNPs and SiNP bioconjugates

Intartaglia et al. 2012 [IX]*; Cooperation LZH-IIT*
Bagga and Barchanski et al. 2013 [VI]*, Cooperation LZH-IIT*

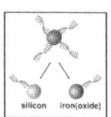 For the transferability study of *in situ* bioconjugation method, SiNPs should be equipped with nucleotides and proteins. As a representative of the nucleotide class an ssDNA with thiol-termination was chosen that was conventionally used for AuNP functionalization. The ssDNA featured various potential binding moieties such as the aromatic nucleotides with primary amines, the phosphate backbone, the thiol modification and an overall negative net-charge, which allowed for the analysis of preferred NP conjugation. Conversely, Protein A from *Staphylococcus aureus* was applied as a representative of the protein class. Protein A has a high affinity for the constant Fc (crystallizable fragment) portion of mammalian IgG with 5 antibody-binding domains[521] and features a good stability in a wide pH and temperature range.[521] The adoption of Protein A as a linker molecule with nanoparticles may enable the selective targeting of antibodies for immunolabeling application and diverse antibody classes may even be targeted because the Fc region is constant for all antibodies. Furthermore, proteins have a favorable affinity for Si surfaces[522] and Protein A in particular is already known to attach to Si-based surfaces via physical adsorption.[523-525]

Both biomolecules were applied to the *in situ* bioconjugation and the fabricated SiNP bioconjugates were purified with ultracentrifugation. The binding efficiencies and binding mechanisms were determined by UV-vis spectrophotometry and micro-probe Raman spectroscopy and the biological functionality of SiNP bioconjugates was examined with immunolabeling and TEM.

The ablation of silicon in *Milli-Q* water with PLAL using the process parameters defined in **Table 3.4** and **Table 3.6** resulted in stable colloidal solutions with a zeta potential of -23 mV and a yellow/brownish coloration (**Figure 4.50a**, inset).

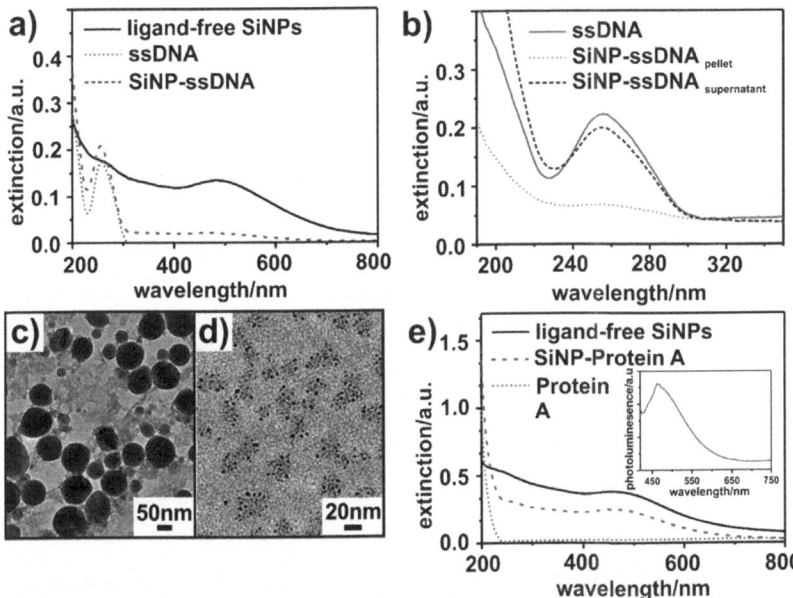

Figure 4.50. Characteristics of PLAL-fabricated SiNP and SiNP bioconjugates. a) Extinction spectra of the ligand-free SiNPs (black solid line), ssDNA solution prior conjugation (green dotted line) and SiNP-ssO bioconjugates (red dashed line) produced by *in situ* bioconjugation. Photography of SiNP colloidal solution is presented in the inset. **b)** Extinction spectra of ssDNA (green solid line), of SiNP-ssDNA pellet (orange dotted line) and SiNP-ssDNA supernatant (purple dashed line) after purification by ultracentrifugation. **c)** Transmission electron micrograph of ligand-free SiNPs. **d)** Transmission electron micrograph of SiNP-ssDNA bioconjugates. **e)** Extinction spectra of ligand-free SiNPs (black solid line), Protein A solution prior conjugation (**green dotted line**) and SiNP-Protein-A bioconjugates (red dashed line) produced by *in situ* bioconjugation. Photoluminescence spectrum of SiNP-Protein A bioconjugates after excitation with 400 nm is presented in the inset of **e)**. Images **a)-d)** were adapted with permission from Intartaglia et al., copyright 2012 by the Royal Society of Chemistry.[526] Image **e)** was adapted with permission from Bagga et al., copyright 2013 by IOP Publishing.[527]

The characteristic SiNP extinction spectra featured a low offset in the NIR regime, a characteristic maximum of approximately 485 nm, a shoulder at 270 nm and a UV contribution that ranges from 190 to 250 nm (**Figure 4.50a** and **Figure 4.50e**). However, if ablation was performed in a biomolecule solution a noticeable blue shift of the peak maximum to 460 nm was observed.[526] The shifting on the absorption edge could be ascribed to changes in nanoparticle size (quantum confinement effect) due to biomolecule coordination.

Interestingly, if SiNPs were generated in a biomolecule solution, the particle yield was significantly reduced by 40 % for proteins and by 80 % for ssDNA (**Figure 4.50a** and **Figure 4.50e**), which is a contrary observation compared to AuNPs where the yield was enhanced by biomolecule presence (**Chapter 4.1.1.1.**). It can be speculated that these findings were attributed to the vertical ablation setup, where the biomolecule sedimentation onto the target hinders the formation and release of NPs into the solvent. The biomolecules did not seem to be degraded to a significant extent during *in situ* conjugation, because the characteristic extinction intensities of the biomolecules at 260 nm (ssO) and 190–220 nm (Protein A) were not reduced after ablation (**Figure 4.50a** and **Figure 4.50e**). However, a detailed degradation study similar to that for AuNPs (**Chapter 4.1.1.4**) has not been performed yet.

Examining high-resolution TEM data (**Figure 4.50c–d**), isolated SiNPs with a pseudo-spherical morphology and the crystalline structure of bulk silicon were observed after ablation in *Milli-Q* water. A mean NP size of 60 nm was determined, while bioconjugated SiNPs exhibited a narrowed particle size with a mean of 3.5 nm for SiNP-ssDNA bioconjugates[526] and 8 nm for SiNP-Protein A bioconjugates,[527] which confirms the claimed size quenching effect due to biomolecule coordination. Moreover, in accordance with the findings on the diffusion coefficient that is explained in **Chapter 4.1.1.3**, the smaller and thus more mobile ssDNA molecule is able to diffuse faster and it can quench the size of SiNPs to a greater extent than Protein A, resulting in a smaller nanoparticle size.

After purification with ultracentrifugation, the UV peak intensities of SiNP bioconjugates and untreated biomolecule solutions were compared and the characteristic biomolecule peaks for ssDNA (**Figure 4.50b**) and for Protein A[527] were clearly recovered in the pellets. Conjugation efficiencies and surface coverage values were calculated from UV-vis spectra of pellets and supernatants and the results are summarized in **Table 4.16**.

Table 4.16. Calculated results on conjugation efficiency and surface coverage of SiNP bioconjugates.

SiNP Bioconjugates	Conjugation Efficiency/%	# Biomolecules per SiNP
SiNP-ssDNA [1 μM]	20	1.5
SiNP-Protein A [1.68 μM]	26	5

The conjugation efficiency was between 20 to 26 % and thus very similar for both biomolecules, while the surface coverage was found to be threefold higher for Protein A than for ssDNA. These values resulted from an equal recovery of NP concentration in the pellet, but had higher particle number of SiNP-ssDNA bioconjugates compared to SiNP-Protein A bioconjugates in cause of the lower mean NP size.

To analyze whether bioconjugation changes the optical characteristics of SiNPs, photo-luminescence measurement was performed on SiNP bioconjugates. Using either 350 nm (SiNP-ssDNA) or 400 nm (SiNP-Protein A) UV excitation, blue-green emission peaks at 450 nm[526] and at 475 nm (**Figure 4.50e**, inset) were clearly observed, which appeared to be similar to the luminescence response of ligand-free SiNPs.[514]

An elemental analysis of SiNP-ssDNA bioconjugates was performed with HAADF-STEM-EDX on areas with agglomerated SiNP clusters and on areas without NPs on the TEM grid (**Figure 4.51a**).

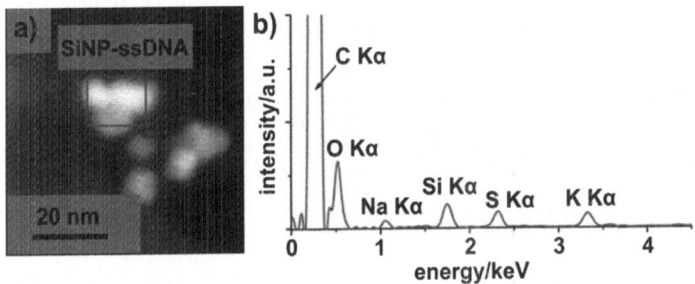

Figure 4.51. Results of the HAADF-STEM analysis of SiNP-ssDNA bioconjugates. a) HAADF-STEM image displaying the measurement area (red-framed box) for STEM-EDX on a SiNP agglomerate. **b)** STEM-EDX elemental characterization results. Adapted with permission from Intartaglia et al., copyright 2012 by the Royal Society of Chemistry [526]

Both signals of silicon-Kα and sulfur-Kα were found on the NP areas (**Figure 4.51b**) while no sulfur signal was detected on the regions without SiNPs (**Figure SI 27**). This indicates a close association between the biomolecules and the SiNPs. The C-Kα peak was derived from the support C-film on the TEM grid, while the O-Kα, Na-Kα and K-Kα peaks are most likely residues from the *Milli-Q* water and could also be detected in pure *Milli-Q* water with inductively coupled plasma atomic emission spectroscopy (ICP-OES) (**Table SI 8**).

Micro-probe Raman scattering was performed to analyze the structure and chemical bonding of SiNP-ssDNA bioconjugates and the results are presented in **Figure 4.52**.

Because the drop-casted samples show a coffee ring structure with an alteration of the molecular concentration from the periphery to the center (**Figure 4.52a**, inset), meas-urements at different areas were performed. The presence of SiNPs was verified as having an asymmetric, sharp peak around 520 cm^{-1} (**Figure 4.52a**), which is in agreement with data from the available literature.[514;528;529] The analysis of the broad band at approximate-ly 480 cm^{-1} indicates the presence of SiO$_x$ (with x = 1-3) which is most likely in the form of a very tiny shell around the NPs, while the spectra in the range of 1200-3200 cm^{-1} indi-cate vibrational bands of inter/intra DNA molecules attached to the SiNP-SiO$_x$ shell.

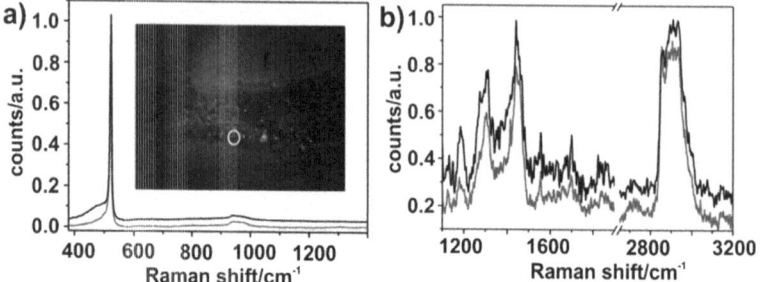

Figure 4.52. Results of Raman analysis of SiNP-ssDNA bioconjugates. a) Raman spectra in low frequency range from 380–1400 cm^{-1}. **b)** Raman spectra in high frequency range from 1100–3200 cm^{-1}. The optical image of the coffee ring and the measurement areas are illustrated in the inset of **a)**. The measurement area marked by a red-framed box is corresponding to the red spectrum, while the measurement area marked by a yellow circle is corresponding to the black spectrum. Adapted with permission from Intartaglia et al., copyright 2012 by the Royal Society of Chemistry [526]

In detail, three bands at 1182 cm^{-1}, 1260 cm^{-1} and 1305 cm^{-1} were identified, which are related to the combinations of thymine (T) and cytosine (C) nucleic acid bases, the combination of adenine (A) and C and to the combination of A and C, respectively (**Figure 4.52b**).[530] Furthermore, two bands at 1450 and 2900 cm^{-1} can be associated with C-H bending and stretching vibrations of the biomolecule, respectively (**Figure 4.52b**).[530] Interestingly, a shoulder peak at approximately 1277 cm^{-1} could be related to the stretching vibration of Si-C/Si-CH in the configuration of H$_x$C$_{4-x}$-SiO$_x$, with x = 1–3 (**Figure 4.52b**),[531;532] which indicates an unspecific interaction between the SiNPs and the ssDNA nucleotides. According to a publication from Knoop et al. it can be speculated that the reaction between SiNPs and ssDNA occurs on the defect site of the SiO$_x$.[533]

Regarding the binding of Protein A to SiNPs, it is already known that physical adsorption is found to occur on the Si surface with multiple binding sites per particle.[523] Moreover, the significant value of conjugation is also supported by the fact that the binding process during laser ablation occurs slightly above the theoretical pI of Protein A (pI = 5.1), which favors the stability and steric orientation of the biomolecule in solution.[534]

Finally, the biological functionality of *in situ* conjugated SiNPs was demonstrated on Protein A-capped nanoparticles with immunolabeling. First, human fibroblasts were labeled with a primary antibody that targets the cytoskeleton protein vinculin. Then, the antibody-labeled cells were incubated with SiNP-Protein A bioconjugates to initiate Protein A attachment to the Fc protein of the antibody. After removal of unbound biomolecules by extensive washing, fluorescence signals were detected using confocal microscopy (**Figure 4.53b–c**).

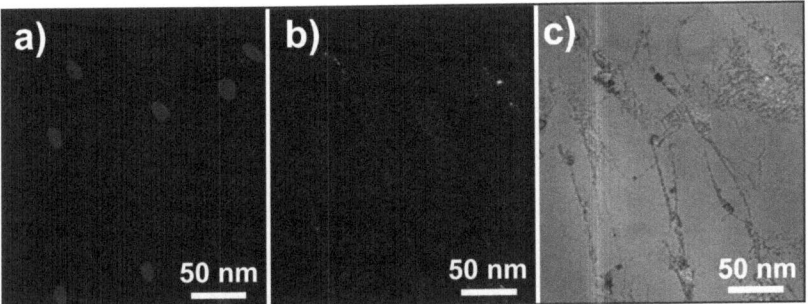

Figure 4.53. Confocal imaging of the membrane-skeleton protein vinculin labeled with SiNP-Protein A bioconjugates. a) Nuclei staining (blue color) with Hoechst dye. **b)** Signal from SiNP-Protein A bioconjugates (green color) indicating vinculin distribution. **c)** Overlap of image **a)** and image **b)** in transmission mode. Adapted with permission from Bagga et al., copyright 2013 by IOP Publishing.[527]

Nuclear staining with Hoechst dye is presented in **Figure 4.53a.** Because vinculin is distributed all over the cytoskeleton, the entire structure of human fibroblasts was clearly labeled with SiNP-Protein A bioconjugates attached to anti-vinculin primary antibodies (**Figure 4.53b–c**). A false-positive signal of cellular autofluorescence was excluded by using a negative control (**Figure SI 28**).

To complement the imaging analysis, transmission electron micrographs were recorded on sections of the fibroblasts, which were previously analyzed with confocal microscopy (**Figure 4.54**).

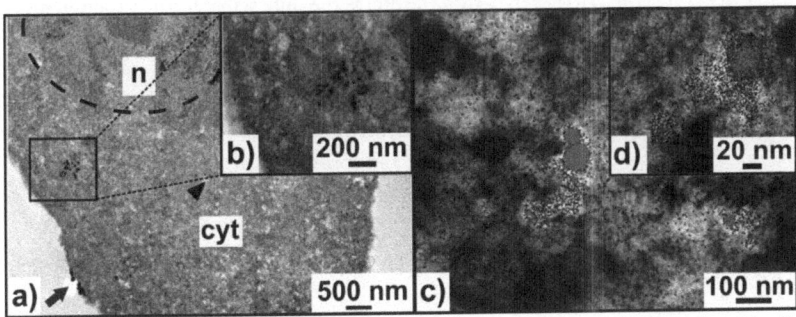

Figure 4.54. Transmission electron micrographs of fibroblasts incubated with Protein A-conjugated SiNPs. a) Parasagittal section through a fibroblast. Cluster of small electron-dense nanoparticles (boxed) and small singular NPs (black arrowhead) are visible inside the cell cytoplasm while some nanoparticle cluster were detected attached to the cell membrane (blue arrow). **b)** Enlarged view of the NP cluster inside the cell cytoplasm. **c)–d)** Superimposed EDXS mapping on transmission electron micrographs at increasing magnification with the distribution of silicon shown in red color. cyt = cytoplasm; n = nucleus. Adapted with permission from Bagga et al., copyright 2013 by IOP Publishing.[527]

Cellular structures such as nucleus and cytoplasm were clearly visible on the micrographs. Furthermore, small electron-dense, spherical particles with diameters that ranged from 5 to 20 nm were detected. They were found in both singular (**Figure 4.54a**, black arrow) and clustered (**Figure 4.54a**, boxed, and **Figure 4.54b**) distributions within the cytoplasm and they were attached to the cell membrane (**Figure 4.54a**, blue arrow). No particles were found inside the nucleus. Using energy-dispersive spectroscopy mapping, the particles were clearly assigned to silicon material (**Figure 4.54c–d**).

In summary, the *in situ* bioconjugation of SiNPs was confirmed as an alternative method to equip silicon nanoparticles with biological function, resulting in a distinct size quenching effect during fabrication and either an adsorptive conjugation of proteins or an unspecific binding of Si-C/Si-CH in the configuration of H_xC_{4-x}-SiO_x, with x = 1–3 to a thin SiO_x-shell around the SiNPs. The SiNP bioconjugates featured biological activity and could be efficiently used as cellular labeling markers.

4.3.3.   Magnetic iron-based nanoparticles

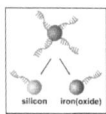 Iron is a soft, silvery-gray colored metal of the first transition series with the atomic number 26 and electron configuration [Ar] $3d^6$ $4s^2$ (latin, *ferrum*, Fe). It is the fourth most common element in the Earth's crust, which is mostly found combined with oxygen as iron oxide minerals, and has been used since ancient times. Its specific characteristics are summarized in **Table SI 9**.

Iron represents an example of allotropy in a metal with at least 4 allotropic forms (alpha, gamma, delta, epsilon). However, α-Iron (ferrite) is the only stable form at room temperature. Pure iron is not stable and reacts with oxygen in the air to form various oxide and hydroxide compounds (mainly in the +II and +III oxidation state, **Table SI 10**). The most common are iron(II,III)oxide (Fe_3O_4, magnetite) and iron(III)oxide (Fe_2O_3, hematite; γ-Fe_2O_3, maghemit). The non-stoichiometric iron(II)oxide (FeO, wustite) also exists, although it is unstable at room temperature and forms only at temperatures > 567 °C. It decomposes during controlled cooling to α-Fe and Fe_3O_4, while a metastable compound may form during splat cooling, which is also stable at room temperature.[535;536]

The anaerobic oxidation of iron at a high temperature can be schematically represented by **eq 4.6–eq 4.8**.

$$Fe\ (s) + H_2O\ (l) \rightarrow FeO\ (s) + H_2\ (g)$$

eq 4.6

$$2\ Fe\ (s) + 3\ H_2O\ (l) \rightarrow Fe_2O_3\ (s) + 3\ H_2\ (g)$$

eq 4.7

$$3 \ Fe \ (s) + 4 \ H_2O \ (l) \rightarrow Fe_3O_4 \ (s) + 4 \ H_2 \ (g)$$

eq 4.8

When both water and oxygen are present (moist air) elementary iron corrodes (oxidizes) and its silvery color changes to a reddish-brown color because hydrated oxides are formed (**eq 4.9**).[536]

$$4 \ Fe \ (s) + 3 \ O_2 \ (g) + 6 \ H_2O \ (l) \rightarrow 4 \ Fe^{3+} \ (aq) + 12 \ OH^- \ (aq) \rightarrow 4 \ Fe(OH)_3 \ (s) \ or$$

$$4 \ \alpha\text{-}Fe^{3+}O(OH) \ (s) + 4 \ H_2O \ (l)$$

eq 4.9

Naturally occurring iron oxide, iron hydroxide, iron carbide and iron penta carbonyl are water insoluble. Usually there is a difference between water-soluble Fe^{2+} compounds and generally water insoluble Fe^{3+} compounds. The latter are only water soluble in strongly acidic solutions, or if they are reduced to Fe^{2+}.

Iron often forms chelation complexes that play an important role in nature, for instance Fe^{2+} as a central atom of the co-factor Häm B in hemoglobin, which is important for the oxygen transport within the human body. It is also used at the active site of many redox enzymes dealing with cellular respiration. Green plants apply iron for energy transformation processes and some bacteria internalize iron particles and convert them to magnetite for application as a magnetic compass.

Magnetic nanoparticles (MNPs) that are based on the metal iron are applied frequently in the biomedical sector, particularly for hyperthermia treatment of cancer,[240] for thermosensitive drug release[537;538] and for magnetic cell separation and sorting issues.[539] Conventional synthesis methods cover top down techniques such as wet grinding of iron powder using a planetary ball mill[540] and bottom up procedures such as chemical vapor deposition[541] or chemical synthesis with precipitation from iron(II)-chloride and iron(III)-chloride solutions to fabricate iron-oxide nanoparticles (IONPs, **eq 4.10–eq 4.11**).

$$Fe^{2+} \ (aq) + 2 \ Fe^{3+} \ (aq) + 8 \ OH^- \ (aq) \rightarrow Fe_3O_4 \ (s) + 4 \ H_2O \ (l)$$

eq 4.10

$$Fe_3O_4 \ (s) + \tfrac{1}{4} \ O_2 \ (g) + 4 \ \tfrac{1}{2} \ H_2O \ (l) \rightarrow 3 \ Fe(OH)_3 \ (s)$$

eq 4.11

Thereby, the nature of developing iron-oxide states depends on the applied salts, the ratio between Fe^{2+} and Fe^{3+} and the pH value of the solution.[542-544]

Recently, magnetic nanoparticles were also produced by the PLAL method.[545;418;546] If iron is ablated with the ns-PLAL technique different phases have been obtained, depending on the applied solvent and adjusted process parameters.[545] In water Amendola et al. determined magnetite NPs with small fractions of hematite, wustite and α-iron.[418] *Ex situ* bioconjugation of ns-PLAL fabricated MNPs with diverse (bio)molecules (N-phosphonomethyl iminodiacetic acid hydrate - PMIDA; Albumin-fluorescein isothio- cyanate; Fluorescein isothiocyanate isomer) was also demonstrated by Amendola et al. and a ligand coverage of 2.3–12.9 nanomoles (bio)molecules per mg of Fe was reached.[418] However, due to the adoption of ns pulse duration and related heat dissipa- tion into the medium ps-laser *in situ* conjugation has not been a topic of investigation yet. Thus, the focus of this study was on the PLAL fabrication of highly magnetic NPs with ps pulse duration, their thorough characterization and *in situ* bioconjugation with proteins and finally their application for cellular manipulation.

4.3.4.   PLAL-generation of MNPs and magnetic nanobioconjugates

Bachelors' thesis M. Meißner 2013, LZH
Bachelors' thesis M. Merkle 2013, LZH

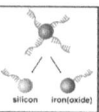 MNPs were fabricated with the ps-PLAL of an iron target in *Milli-Q* water using the process parameters that were defined in **Table 3.5**. With the opti- mal target position for *in situ* bioconjugation (**Chapter 4.1.1.4**, target posi- tion: 1 mm behind the determined focal point in air = -1 mm), the generated MNPs featured an average yield of 112 µg mL^{-1} min^{-1}, a zeta potential of +13.3 mV, a lowest hydrodynamic diameter of 60 nm number mean and a PDI of 0.25 (**Figure SI 29**). The NPs were observed to be black-greyish in color and highly magnetic by attraction with an NdFeB induction disc (**Figure 4.55a**). Their Feret size distribution was determined to range from 5 to 250 nm with a maximum at 52±1.5 nm (**Figure 4.55b**), while the mean hydrodynamic diameter was 63 nm (**Figure SI 29**). On the scanning electron micrograph it can clearly be seen, that the spherical primary MNPs are embedded into a distinct, amorphous matrix, which is most likely due to iron ox- ide/hydroxide formation (**Figure 4.55b**, inset).

Deconvolution of the Mössbauer spectra resulted in a magnetic sextet and two non-magnetic doublets (**Figure 4.55c** and **Table SI 11**). Sextet formation is characteristic for magnetic states, as it is results from a magnetic hyperfine structure. In this case, the iron core receives a magnetic field of 32.8 T (**Table SI 11**) which is typical for α-Fe while it would be higher for other states. Other attributes for α-Fe are a missing quadrupol splitting (QS) and an isomeric shift (IS) of nearly 0 mm s^{-1}.[547] The doublets can be con- strued with two theories. There may be an indication of wustite, which is presented with the overlap of two doublets. In this case, the first doublet with an IS of 1 mm s^{-1} and a QS of 0.7 mm s^{-1} would represent the Fe^{2+} ions and the second spectrum with an IS of 0.4 mm s^{-1} and a QS of 0 mm s^{-1} would result from the interplay of the Fe^{3+} ions.[547] Although FeO is magnetic no sextet develops, which may be an indication of a core-shell structure with α-Fe core and a thin FeO shell that would not result in a magnetic splitting. This unusual metastable phase may develop by the extremely confined conditions within the plasma plume. It is also possible that the Fe^{3+} ions originate from an iron hydroxide (e.g. goethite) most likely within the amorphous matrix found on the SEM images (**Figure 4.55b**).

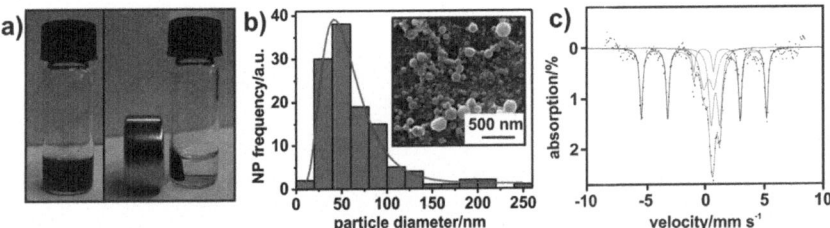

Figure 4.55. Characteristics of PLAL-generated MNPs. a) MNPs in ddH₂O before and after 10 s attraction with a NdFeB induction disc. **b)** Primary particle size distribution of MNPs in ddH₂O with LogNormal fitting and scanning electron micrograph in the inset. **c)** Mössbauer spectrum of MNPs in ddH₂O. Black dots = raw data, blue line = best fitting curve and pink lines = deconvoluted spectra. Adapted with permission from M. Meißner, copyright 2013 by Marita Meißner, Bachelor thesis.[415]

These results correlate with the obtained X-ray diffraction spectrum of MNPs, identifying a main peak of α-Fe at $2\theta = 44.6°$ and of wustite at $2\theta = 36.2, 42.1$ and $61.0°$ (**Figure 4.56a**).

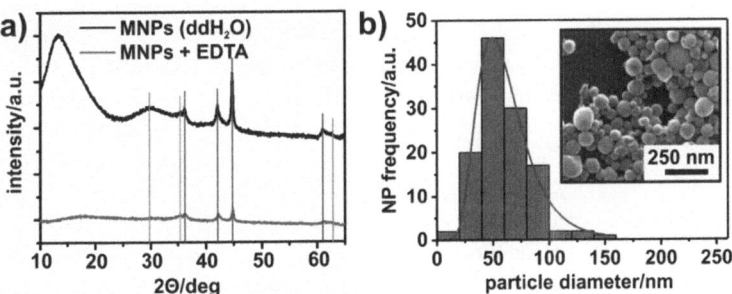

Figure 4.56. Results of EDTA treatment on MNPs. a) X-ray diffraction spectra of MNPs in ddH₂O and after treatment with EDTA. Green lines = magnetite, blue lines = wustite, pink line = α-iron. **b)** Primary particle size distribution of MNPs in ddH₂O after treatment with EDTA with logNormal fitting and corresponding scanning electron micrograph in the inset. Adapted with permission from M. Meißner, copyright 2013 by Marita Meißner, Bachelor thesis.[415]

Interestingly, two other reflections at $2\theta = 30.2$ and $62.5°$ were assigned to magnetite, which has most likely been built by partial decomposition of wustite to α-Fe and magnetite. An intense and broad peak was further found at $2\theta = 12.3°$ which the software was not able to assign a material to. It can be assumed that the peak represents the amorphous iron oxide/hydroxide portion, because it vanished completely after EDTA treatment, which was applied to dissolve the amorphous matrix (MNPs + EDTA, **Figure 4.56a**).

The treatment was performed because the amorphous hydroxides could be an issue for cellular entry or for kidney removal of MNPs from the body, due to increased aggregate size.

The process can be schematically represented by **eq 4.12**.

$$Fe(OH)_2\ (s) + H_2EDTA\ (aq) \leftarrow Fe^{2+}\ (aq) + 2\ OH^-\ (aq) + 2\ H^+\ (aq) + EDTA^{2-} \rightarrow$$
$$Fe(EDTA)\ (aq) + 2\ H_2O\ (l)$$

eq 4.12

Thereby, the EDTA acts as Lewis-Base and Fe acts as a Lewis-Acid, building a chelate complex when the hydroxyl groups of the $Fe(OH)_2$ react with the protons of the EDTA. A short EDTA treatment resulted in singular and separated spheres (**Figure 4.56b**, inset) with a comparable primary particle size distribution of 51 ± 7 nm as the untreated MNPs (**Figure 4.56b**). However, it should be noted that a long treatment also affected the other iron phase(s), which resulted in lower peak intensities on the XRD spectrum (**Figure 4.56a**) and a color-change of the solution from dark-greyish to nearly transparent. The dissolution can also be tracked using UV-vis spectrophotometry over time (**Figure SI 30a–b**) with constant extinction reduction in the NIR while ending up in an aggressive hole formation on the MNPs (**Figure SI 30c–d**).

To test the *in situ* bioconjugation method on MNPs, the particles were fabricated by PLAL in a solution of Alexa594 fluorophore-labeled BSA (BSA-Alexa594), treated with EDTA and purified thoroughly with three centrifugation steps. The FT-IR spectra of ligand-free MNPs, pure BSA-Alexa594 molecules and MNP-BSA-Alexa594 conjugates are presented in **Figure 4.57a**.

The bands between 1400 and 1000 cm^{-1} could not be assigned to a single iron oxide state, which could indicate a mixture of oxide states in the PLAL-generated colloid. However, the distinct FT-IR bands of amide I and amide II (**Table SI 2**) were clearly assigned at 1661 cm^{-1} and 1554 cm^{-1}. Compared to the native BSA594 molecule, the characteristic peaks of the amide–NH groups are less intense for the magnetic nanobioconjugates, which may be an indication of an interaction between the BSA and the MNP surface by the –NH groups.[548] However, for a detailed determination of bonding type a Raman analysis needs to be performed.

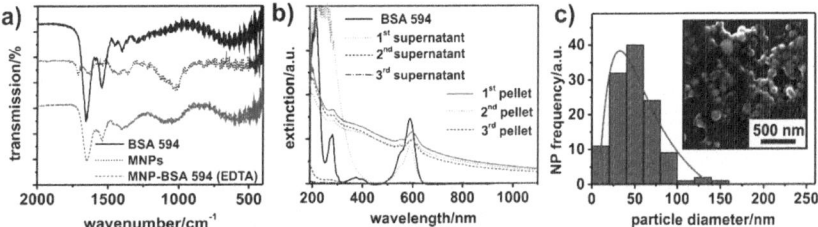

Figure 4.57. Characteristics of PLAL-generated magnetic nanobioconjugates. a) Fourier-Transform infrared spectra of MNPs in ddH$_2$O (purple dotted line), BSA with Alexa 594 fluorophore (black solid line) and the third pellet of MNP-BSA-Alexa594 conjugates after EDTA treatment (red dashed line). **b)** UV-vis spectra of BSA with Alexa594 fluorophore (black solid line) and the pellets (first pellet = green solid line, second pellet = blue dotted line, third pellet = red dashed line) and supernatants (first supernatant = orange dotted line, second supernatant = pink dashed line, third supernatant = purple dash-dotted line) of MNP-BSA-Alexa594 conjugates after three purification steps. **c)** Particle size distribution of MNP-BSA-Alexa594 conjugates after EDTA treatment and purification with LogNormal fitting and corresponding scanning electron micrograph in the inset. Adapted with permission from M. Meißner, copyright 2013 by Marita Meißner, Bachelor thesis.[415]

Successful conjugation and strength of binding was further explored using UV-vis spectrophotometry (**Figure 4.57b**). The extinction spectrum of pure BSA-Alexa594 solution exhibited both the conventional BSA protein peak at 280 nm and the fluorophore peaks at 380 and 594 nm. After *in situ* conjugation and EDTA treatment, the bioconjugate was washed threefold and spectra of both, pellets and supernatants were recorded (**Figure 4.57b**). After the 1st centrifugation, high amounts of removed EDTA were detected in the UV region (190–380 nm) of the supernatant. In addition, a distinct amount of BSA-Alexa594 was found, which is indicated by the fluorophore peak. However, in the pellet all peaks of BSA-Alexa594 (280, 380, 594 nm) were still detected and the high offset resulted from the MNP contribution. After the 2nd and 3rd washing steps, only trace amounts of EDTA/BSA-Alexa594 were found in the supernatant, while the pellet spectra including the same number of peaks, peak positions and intensities were almost unchanged. This result indicates a successful conjugation between MNPs and biomolecules and proves their strong, probably covalent connection. A final verification was performed with SEM analysis (**Figure 4.57c**). Initially, the size distribution (46±9 nm) was found to be insignificantly affected by bioconjugation compared to EDTA-treated MNPs (**Figure 4.57b**) and furthermore, the conjugation was clearly observed on the electron micrographs, which resulted in an organic layer on top of the singular particles (**Figure 4.57c**, inset).

The conjugation efficiency was calculated to be 59 % for BSA-Alexa594, which resulted in a surface coverage of 39.8 nmol per mg Fe (**Table 4.17**). This high value is 17-fold higher than that which was obtained with *ex situ* conjugation by Amendola et al.[546]

Table 4.17. Calculated conjugation efficiencies and surface coverage values for MNP-BSA-Alexa594 bioconjugates.

	MNP-BSA-Alexa-594
Conjugation Efficiency/%	59
Surface Coverage/nmol mg Fe^{-1}	39.8

To analyze the biological functionality of magnetic nanobioconjugates, their attachment to human fibroblasts was determined by fluorescence microscopy after co-incubation for 2 hours and thorough rinsing. In addition to the DAPI staining of the cell nuclei (**Figure 4.58a**) and to the autofluorescence of cytoplasm due to the absence of a mounting medium (**Figure 4.58b**), bright, spotted signals in the red channel (**Figure 4.58c**) were also detected and associated with the Alexa594 fluorophore of magnetic nanobioconjugates. The signals were co-localized (**Figure 4.58d**) with the outer cell cytoplasm and likely attached to the cell membrane. However, the signal appears from the fluorophor label and is no guarantee of an intact association between MNPs and biomolecules after incubation in cell culture medium.

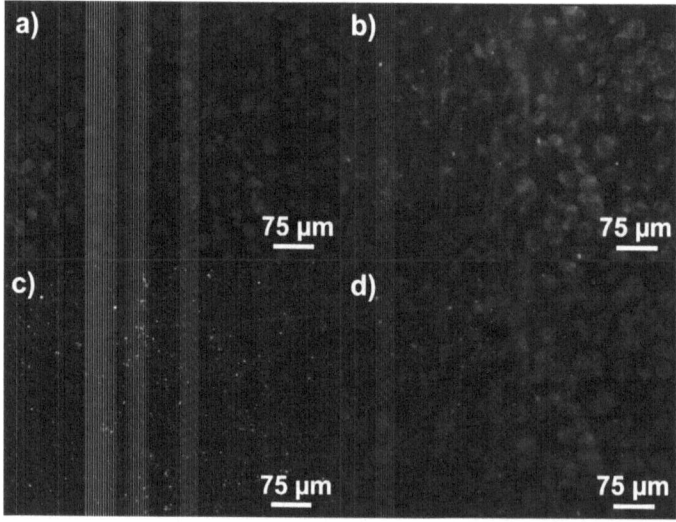

Figure 4.58. Immunolabeling of human fibroblasts with MNP-BSA-Alexa594 conjugates. a) Cell nuclei = blue   color.   **b)** Cellular   autofluorescence = green   color.   **c)** Fluorophore   signal   of MNP-BSA-Alexa594 bioconjugates = red color. **d)** Overlay of channels a)–c). Adapted with permission from M. Merkle, copyright 2013 by Matthias Merkle, Bachelor thesis.[549]

Therefore, the incubated fibroblasts were further applied to a magnetic attraction assay as schematically illustrated in **Figure 4.59**. A cell culture plate was modified with induction discs at the bottomside and then magnetic nanobioconjugate-incubated cells were inserted and homogeneous shaked for 30 minutes. After two days, photographs documented the cell growth at the position of NdFeB discs and in the peripherie (**Figure 4.59**).

The cell growth was clearly amplified and denser at the induction disc position than anywhere else in the periphery (**Figure 4.59**), which indicates the magnetic attraction force on cell-attached MNPs. Thus, considering both experiments in context and also consulting the results of the FT-IR and UV-vis analyses, it is possible to state that the conjugation between MNPs and BSA-Alexa594 biomolecules is a significant covalent manner, because it is not separated by centrifugal force of purification, or by salt-shielding forces of cell culture medium incubation. In addition, the conjugates are able to image cells, and manipulate them with magnetic attraction.

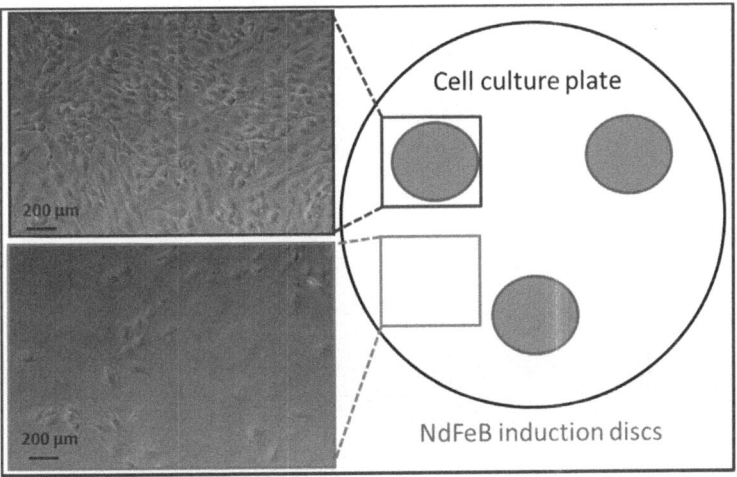

Figure 4.59. Schematic overview of the magnetic cell manipulation experiment. The cell culture plate with positions of NdFeB induction discs is indicated and photographs of cell growth at the induction disc position and in the periphery are presented. Adapted with permission from M. Merkle, copyright 2013 by Matthias Merkle, Bachelor thesis.[549]

In ongoing experiments, the bioconjugation to other molecules, including fluorophore-labeled antibodies and drugs was verified for MNPs and can be found in the bachelor theses from M. Meißner[415] and M. Merkle.[549]

However, due to the mixture of fabricated iron-oxides, solution aging and the modification of oxidation states, the exact binding mechanism could not be clearly determined thus far and the reproducible fabrication of nanobioconjugates is challenging.

4.3.5.   Summary and discussion

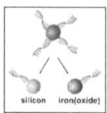 In summary, the transferability of the *in situ* bioconjugation method to other biomedically relevant materials such as silicon and iron was successfully demonstrated. The fabricated nanobioconjugates were stable in water and featured biological functionality for imaging and magnetic manipulation applications. Although physisorption appear to be predominant for conjugate formation, the detailed binding mechanisms have not been fully discovered.

To compare the surface coverages of *in situ* conjugated AuNP bioconjugates with *in situ* conjugated SiNP and MNP bioconjugates, the determined number of biomolecules per particle was recalculated for a 10 nm standard particle and matched with the literature data (**Table 4.18**).

Table 4.18. Comparison of surface coverage values for *in situ* bioconjugated gold, silicon and iron-based nanoparticles. The number (#) of biomolecules (Bio) per particle was recalculated for a standard 10 nm particle.[a,550]

Nanobio-conjugates	MW of Ligands /g mol^{-1}	#Bio per NP	Modal Particle Diameter/nm	Particle Area/nm^2	#Bio per 10 nm NP
AuNP-ssDNA(SH)	5,770	163	9	254	202
AuNP-BSA	66,000	6602	25	1960	1058
SiNP-ssDNA(SH)	5,770	1.5	3.5	38.5	12
SiNP-Protein A	~ 42,000	5	8	201	8
MNP-BSA-Alexa594	~ 66,000	7570	46	6650	357
AuNP-OES(SH)[a]	470	2410	17	908	833
AuNP-OES(SH)-P1[a]	~ 1.050	~ 550 P1	17	908	190 P1

Although several publications indicate the surface coverage of AuNPs with (bio)molecules, only a few are comparable with each other because coverage units differ widely and biomolecule specifications such as their molecular weight are not declared.

Bartczak et al. conjugated CRM-synthesized AuNPs with monocarboxy-(1-mercaptoundec-11-yl)hexaethylene glycol (OES), featuring a thiol function. A subsequent EDC/NHS treatment allowed for the functionalization of AuNP-OES conjugates with a peptide (P1) and the numbers of OES and P1 ligands per 10 nm particle were calculated to be 833 and 190, respectively.

Compared to these data, the *in situ* bioconjugation of ps-PLAL-fabricated AuNPs (**see Chapter 4.1.1.3**) resulted in surface coverages of 202 and 1058 biomolecules per 10 nm particle for AuNP-ssDNA and AuNP-BSA bioconjugates, respectively. At first sight, the number of attached ssDNA molecules seems relatively low compared to the 4-fold number of attached OES molecules. However, the MW of the biomolecules must be considered which is by a factor of > 10 lower for OES than for ssDNA. Thus, the three-dimensional size is also reduced, so that several OES molecules have the same volume as an ssDNA molecule. From this, it follows that at least a similar coverage can be assumed for both conjugates.

On the other hand, a number of 1058 BSA molecules per 10 nm particle were calculated for AuNP-BSA bioconjugates, which is by a factor of > 5 higher than for AuNP-OES-P1 bioconjugates. In this case, the MW of the BSA molecule is 60-fold higher than for P1 molecules, which verifies a superior amount of BSA coupled to the gold. However, it was already explained in **Chapter 4.1.1.3**, that BSA tend to form multilayers on the particle, yielding extremely high coverage values while monolayer formation would yield much lower amounts.

In further comparing the similar coverage values of AuNP-ssDNA and AuNP-OES-P1 bioconjugates while concerning the 5-fold higher MW of ssDNA, the nucleotides seem to bind more efficiently to AuNPs. However, in this case, the form and volume of the molecules must be considered. While ssDNA features an elongated chain structure, the P1 is a globular peptide with increased steric dimensions, which reduce the conjugable amount significantly.

Focusing next on SiNP bioconjugates, a much lower number of approximately 10 biomolecules per 10 nm particle was determined for both conjugates, although the same ssDNA as for AuNP-ssDNA bioconjugates fabrication and a similar-sized protein as for AuNP-BSA bioconjugate functionalization were adopted. These reduced coverages could result from an inferior binding mechanism such as physisorption, compared to the strong dative bond formation between thiol and gold. The physisorption of proteins onto silicon surfaces has actually been described in the literature.[522]

In this regard, an important benefit of PLAL is the ability to fabricate partially oxidized nanoparticles with a distinct surface charge and zeta potential in water promptly. This charge allows for electrostatic interactions with oppositely charged molecules and eases the adhesion/adsorption process of biomolecules onto the particle surface. Up to the present, several other SiNP bioconjugates had successfully been fabricated with *in situ* conjugation (unpublished data), supporting the physisorption theory.

Despite the low coverage value, the biological functionality of SiNP-Protein A bioconjugates was further proven, indicating that for a targeting application, even a low number of 10 biomolecules per particle could be sufficient. Theoretically, even a single biomolecule per particle could be sufficient if incorrect binding orientation and biomolecule degradation are avoided.

Finally, discussing the MNP-BSA-Alexa594 bioconjugates, a biomolecule number of 357 per 10 nm particle was calculated, which can be directly compared to 1058 conjugated BSA molecules per particle for AuNP-BSA bioconjugates. As described previously, the surface atoms of the nanoparticles have a higher surface energy than the core atoms and try to reduce this by binding to other molecules. However, for metals that are susceptible to corrosion such as iron, the free energy is lowered which strongly affects protein attachment and results in lower coverage values as in the case of MNP-BSA-Alexa594. Nevertheless, if the 357 BSA molecules are compared to the 202 ssDNA molecules attached to AuNP-ssDNA bioconjugates and a > 10 fold higher MW of BSA molecules is considered, then a multilayer formation still seems very likely. This assumption is supported by the SEM images of MNP-BSA-Alexa594 bioconjugates, illustrating a high amount of organic material covering the particles (**Figure 4.57c**).

The conjugation appears to be physisorption-driven, although other proteins with an adsorptive character such as Cytochrome C or Protein A were not always conjugable to the MNPs. This finding may indicate that the development of iron phases is in sensitive correlation to the applied aqueous medium and that an iron-oxide layer has most likely not been formed. This hypothesis should be confirmed with continuative experiments. However, the biological functionality of magnetic, iron-based BSA-Alexa594 bioconjugates was further proven, allowing for the targeting and magnetic manipulation of cells.

Because the focus of this thesis was on gold nanoparticles, the extent of the transferability study was kept moderate and thus it was not possible to resolve all questions regarding the bond type of conjugation and the exact driving forces of nanobioconjugate formation.

However, the aim of the study to transfer the *in situ* bioconjugation technique to silicon and iron-based materials was successfully reached. Reasonable surface coverage values were gained, depending on the conjugation mechanism and applied biomolecule and the nanobioconjugates featured biological functionality.

Ongoing experiments should be conducted to compare the efficiency of the method with the *ex situ* conjugation technique and to solve the aforementioned open questions. In addition, other biologically interesting materials such as silver as an anti-bacterial metal and cerium-oxide as a ROS-scavenger could be included in the analysis.

5. Summary and Conclusion

5.1. Summary

Nanoparticles are being developed for a multitude of applications in the fields of biomedicine and reproductive biology to date. An appealing fabrication method is the pulsed laser ablation in liquids (PLAL), which enables the production of gold nanoparticles and also their *in situ* bioconjugation with biomolecules on the timescale of minutes.

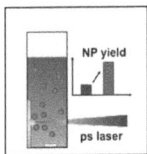 A crucial drawback of the PLAL process is generally that there is a mismatch between an efficient production yield and the maintenance of optimal conditions for the fabrication of functional nanobioconjugates. For a long time, only a maximum nanobioconjugate yield of approximately 11 µg min⁻¹ had been achieved using femtosecond-pulsed LAL that resulted in nearly 100 % integrity preservation of biomolecules.[39] That was the basis for the development of this thesis. To enhance the nanoparticle yield, it was studied whether longer pulse duration could increase the ablated gold mass per time. In fact, using picosecond pulses for PLAL instead of femtosecond pulses, the nanobioconjugate yield was significantly increased by a factor of 15 to 168 µg min⁻¹. Moreover, the produced nanobioconjugates featured nearly 100 % integrity preservation when fabricated with strictly defined process parameters. Interestingly, the nanoparticle concentration could be further increased by the post-processing techniques of ultrafiltration and solvent evaporation. A maximum concentration factor of 2-3 was reached with ultrafiltration. The up-concentrated nanobioconjugates were functional; however the efficiency of ultrafiltration was highly dependent on the material of the filter membrane. Moreover, high particle losses of approximately 40 % had to be accepted. Conversely, the concentration increase of nanobiohybrids by solvent evaporation was highly efficient by a factor of 13. However, because of long processing times at room temperature the risk of biomolecule inactivation is quite high and should be considered carefully.

In summary, the adoption of ps-PLAL for AuNP and AuNP bioconjugate fabrication could allow for competitiveness of the PLAL technique on the NP fabrication market, especially if it is combined with an additional post-processing step of ultrafiltration or solvent evaporation, yielding mg mL⁻¹ concentration scale.

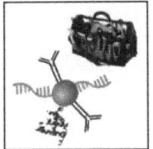 When starting the work on this thesis, there had been no comprehensive guideline for the laser-based fabrication of gold nanoparticle bioconjugates, especially regarding the specific demand on structure-function relationship. During the thesis workout, it turned out to be a particular challenge to include all relevant process parameters because of their diversity. Moreover, the parameters did not only influence the conjugation process but they

could also amplify or erase the benefit and function of each other. However, four consideration areas were subdivided for the discussion of the process parameters.

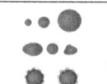 (I) The effects of nanoparticles' intrinsic parameters were studied on the examples of particle size and surface charge.

The methods of *in situ* photofragmentation and *ex situ* centrifugation were successfully applied to modify the particle size distribution and to separate distinct particle size classes. Moreover, the fabrication of AuNPs with ps-PLAL was found to generate partially oxidized surfaces (~ 5 % of atoms) with an Au^+ configuration. Compared to other studies, the extent of surface oxidation seems to be strongly dependent on the laser parameters such as pulse length, pulse energy, repetition rate, fluence and wavelength.

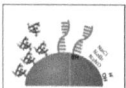 (II) Choice of binding stability and functional group, of ligand amount and ligand charge and of the surrounding medium.

The solvent for ablation should provide optimal conditions for the electrostatic stability of nanoparticles and biomolecules and it should allow for the dilution or transfer of nanobioconjugates into biological relevant media.

For optimal binding stability, the covalent attachment of ligands with a thiol or disulfide function should be aimed. There was no difference for the conjugation by thiol or disulfide function determined. However, the molecule structure should be considered carefully, because spontaneous fragmentation/dissolution of AuNPs by electron transfer from electron-donor-containing moieties could occur. To achieve optimal nanobioconjugate formation and functionality without precipitation or multilayer formation, the adopted ligand concentration should be within a concentration window that is defined by two thresholds termed minimum ligand concentration and maximum ligand concentration. The effect of charge compensation between net-charge negative AuNPs and net-charge positive ligands should be avoided, because it induces the reduction of the interparticle distance and allows for particle agglomeration.

 (III) Ligand characteristics as their length, dimension, binding orientation or amphiphilic nature and the adoption of diverse ligands for bivalent functionalization and surface saturation.

The chain of linear ligands should be kept short to avoid enhanced coiling and wrapping effects of the flexible ligands around the AuNP surface, which significantly limits the surface coverage. Moreover, it should be considered that the molecular size is in direct relation to a large molecular footprint and thus to the number of attachable ligands. For ligands that have active centers, a correct orientation on the nanoparticles should be enabled by the use of specific linker molecules. In addition, it should be considered for nucleotides that the insertion position of a sulphuric function at the strand end will either yield a high amount of full-length and capped failure nucleotides with the modification or only a low amount of the full-length product with the modification. The fabrication of bivalent AuNP bioconjugates can be enabled with three different approaches, including the *in situ* conjugation with ligand A and the *ex situ* conjugation with ligand B, the *in situ* conjugation with ligand B and the *ex situ* conjugation with ligand A and the *in situ* co-conjugation with both ligands at once. All approaches will result in bivalent conjugates. However, the co-conjugation is most effective due to a single-step process. A non-functional dummy ligand might be applied to pre-saturate the particle surface in order to control the surface coverage of a functional secondary ligand. However, it may also be applied to post-saturate the particle in order to increase the nanobioconjugate stability even in highly saline media. Moreover, the amphiphilic nature of the biomolecule may influence the biological functionality of the nanobioconjugates. Depending on the application, polycationic, bivalent bioconjugates should feature different properties than cationic-neutral, bivalent bioconjugates and amphiphilic bioconjugates with a twisted anionic-neutral-cationic composition and a high hydrophobic content.

(IV) The biological functionality of nanobioconjugates is their most crucial quality. Thus, the PLAL-generated conjugates that were fabricated in the framework of this thesis were analyzed regarding functionality in various laboratory assays such as immunoblotting and they were applied for *in vitro* tests such as cellular uptake studies, cytotoxicity screening or the specific immunolabeling. Within all those studies, the PLAL-generated nanobioconjugates were highly functional and featured the same or even a better quality than commercial products.

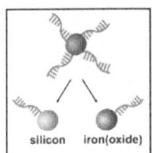

The transferability of an established technique e.g. from one to another materials is an important factor for the broadband-compatibility of a method. When the structure of this thesis was outlined, the laser-based *in situ* bioconjugation was solely used for the functionalization of gold nanoparticles with (thiolized) ligands in a single-step process. However, the adoption of the technique to other biologically relevant materials such as silicon and iron was of high interest. Within this thesis, the fabrication of silicon nanobioconjugates and magnetic, iron-based nanobioconjugates with *in situ* bioconjugation technique during PLAL was successfully demonstrated. The obtained nanobioconjugates were directly fabricated in ultrapure water and featured a high colloidal stability. This is outstanding, compared to the conventional biofunctionalization approaches of nanoparticles from those materials, which are generally performed in organic solvents with high amounts of stabilizers and which comprise complex purification procedures. However, with *in situ* bioconjugation method, reasonable surface coverage values were gained, depending on the conjugation mechanism and applied biomolecule. Moreover, the nanobioconjugates featured biological functionality for bio-imaging and magnetic manipulation applications and are highly promising for biomedical applications.

5.2.   Conclusion and Outlook

Golden bioperspective

Thus far, the nanotechnology has transcended the traditional boundaries between common research areas such as physics, chemistry and biology/medicine and has been characterized by its capacity to revolutionize nearly everything. This has made the work on this thesis highly challenging but also very exciting.

This thesis deals with the complex issue of functional nanobioconjugate fabrication with *in situ* bioconjugation during PLAL and discusses in three chapters the options to increase the yield of nanoparticles and nanobioconjugates, the intrinsic and extrinsic factors that influence the complex bioconjugation process and the transferability of the *in situ* conjugation method onto materials other than gold.

As a main outcome of this study, a guideline has been established for the laser-based fabrication of distinct AuNP bioconjugates with a focus on customer' needs for specific biological applications and on the limitations of the fabrication process.

An important benefit of PLAL is the ability to fabricate stable metal nanoparticles of nearly all materials in pure water promptly, yielding *clean* surfaces that are free of organic species. The surfactant-free surface is highly important for certain colloidal applications such as for surface-enhanced Raman scattering or catalysis. Thus, the competitiveness of PLAL method with respect to existing particle synthesis techniques and the global market demand is a main issue and yield enhancement will further be an important topic. Novel approaches such as PLAL in liquid flow[354] or wire ablation[304] may be alternative routes for the future to gain production up and beyond the gram scale.

The modulation of particles' intrinsic parameters and the main impacts of the nanoenvironment and the ligand characteristics were studied. However, only a small insight was provided when compared to the broad spectrum of parameters that have an impact on nanobioconjugate formation and on structure-function-relationship. Moreover, due to the sensitivity of PLAL method on the used process parameters, already slight modifications in wavelength or fluence could change the whole system. Thus a study with focus on the physical aspects could significantly continue this thesis.

Finally, because the *in situ* bioconjugation technique was successfully transferred to silicon and iron-based nanoparticles, it is highly probable that other materials are also applicable for the method. In these terms, biomedically relevant materials such as anti-bacterial silver or ROS scavenging nanoceria could be tested in order to broaden the fabrication line-up and to meet more customer's needs.

Overall, the fundamental results of this thesis may provide a basis for other manufacturers using the PLAL technique to produce customized nanobioconjugates for various applications. Specifically in the biomedical sector, the demand for nontoxic, bioactive and multivalent gold nanoparticles for diagnostics and disease treatment will presumably continue to grow in the next decades, which could announce a literally *golden bioperspective*.

6. References

[1] WHO, *World Health Statistics 2014 - A Wealth of Information on Global Public Health.* **2014**, Geneva.

[2] Jain K K, *Trends Biotechnol,* **2006** *24,* 143-145.

[3] Wang M, Thanou M, *Pharmacol Res,* **2010** *62,* 90-99.

[4] Saha K, Agasti S S, Kim C, Li X, Rotello V M, *Chem Rev,* **2012** *112,* 2739-2779.

[5] Gannon C J, Patra C R, Bhattacharya R, Mukherjee P, Curley S A, *J Nanobiotechnology,* **2008** *6,* 2.

[6] Allen T M, Cullis P R, *Science,* **2004** *303,* 1818-1822.

[7] Panyam J, Labhasetwar V, *Adv Drug Delivery Rev,* **2003** *55,* 329-347.

[8] Lavitrano M, Busnelli M, Cerrito M G, Giovannoni R, Manzini S,Vargiolu A, *Reprod Fertil Dev,* **2006** *18,* 19-23.

[9] Sarrate Z, Blanco J, Anton E, Egozcue S, Egozcue J, Vidal F, *Asian J Androl,* **2005** *7,* 227-236.

[10] Johnson L A, Welch G R, Keyvanfar K, Dorfmann A, Fugger E F,Schulman J D, *Hum Reprod,* **1993** *8,* 1733-1739.

[11] Ogilvie C M, Braude P R, Scriven P N, *J Histochem Cytochem,* **2005** *53,* 255-260.

[12] Rath D, Barcikowski S, de Graaf S, Garrels W, Grossfeld R, Klein S, Knabe W, Knorr C, Kues W, Meyer H, Michl J, Moench-Tegeder G, Rehbock C, Taylor U, Washausen S, *Reproduction,* **2013** *145,* R15-30.

[13] Wagner V, Dullaart A, Bock A-K, Zweck A, *Nat Biotech,* **2006** *24,* 1211-1217.

[14] Santos M I, Tuzlakoglu K, Fuchs S, Gomes M E, Peters K, Unger R E, Piskin E, Reis R L, Kirkpatrick C J, *Biomaterials,* **2008** *29,* 4306-4313.

[15] Buyukhatipoglu K, Chang R, Sun W, Clyne A M, *Tissue Eng Part C Methods,* **2010** *16,* 631-642.

[16] Bigdeli A K, Lyer S, Detsch R, Boccaccini A R, Beier J P, Kneser U, Horch R E, Arkudas A, *Nanotech Rev,* **2013** *2.*

[17] Neumeister A, Bartke D, Barsch N, Weingartner T, Guetaz L, Montani A, Compagnini G,Barcikowski S, *Langmuir,* **2012** *28,* 12060-12066.

[18] Lavenus S, Ricquier J C, Louarn G, Layrolle P, *Nanomedicine (Lond),* **2010** *5,* 937-947.

[19] Bressan E, Sbricoli L, Guazzo R, Tocco I, Roman M, Vindigni V, Stellini E, Gardin C, Ferroni L, Sivolella S, Zavan B, *Int J Mol Sci,* **2013** *14,* 1918-1931.

[20] Scampicchio M, Arecchi A, Mannino S, *Nanotechnology,* **2009** *20,* 135501.

[21] Zanello L P, Zhao B, Hu H, Haddon R C, *Nano Lett,* **2006** *6,* 562-567.

[22] Patolsky F, Weizmann Y, Willner I, *Nat Mater,* **2004** *3,* 692-695.

[23] Penner R M, *Annu Rev Anal Chem (Palo Alto Calif),* **2012** *5,* 461-485.

[24] Salata O, *J. Nanobiotech,* **2004** *2,* 3.

[25] De M, Ghosh P S, Rotello V M, *Adv Mater,* **2008** *20,* 4225-4241.

[26] Boisselier E, Astruc D, *Chem Soc Rev,* **2009** *38,* 1759-1782.

[27] Giljohann D A, Seferos D S, Daniel W L, Massich M D, Patel P C, Mirkin C A, *Angew Chem Int Edit,* **2010** *49,* 3280-3294.

[28] Dreaden E C, Alkilany A M, Huang X, Murphy C J, El-Sayed M A, *Chem Soc Rev*, **2012** *41*, 2740-2779.

[29] Schmid G, *Nanoparticles: From Theory to Application*, 2nd ed. Wiley-VCH, **2004**, Weinheim.

[30] BBC Research, *Nanotechnology: A Realistic Market Assessment*. **2012**, pp. 226.

[31] Kumar C S S R, *Nanomaterials for Medical Diagnosis and Therapy*, Wiley-VCH, **2007**, Weinheim.

[32] Kelly K L, Coronado E, Zhao L L, Schatz G C, *J Phys Chem B*, **2002** *107*, 668-677.

[33] Hu M, Chen J, Li Z Y, Au L, Hartland G V, Li X, Marquez M, Xia Y, *Chem Soc Rev*, **2006** *35*, 1084-1094.

[34] Alkilany A M, Murphy C J, *J Nanopart Res*, **2010** *12*, 2313-2333.

[35] Zhou J, Ralston J, Sedev R, Beattie D A, *J Colloid Interface Sci*, **2009** *331*, 251-262.

[36] Qiao F-Y, Liu J, Li F-R, Kong X-L, Zhang H-L, Zhou H-X, *Appl Surf Sci*, **2008** *254*, 2941-2946.

[37] Brust M, Walker M, Bethell D, Schiffrin D J, Whyman R, *J Chem Soc Chem Comm*, **1994**, 801-802.

[38] Turkevich J, Stevenson P C, Hillier J, *Discuss Faraday Soc*, **1951** *11*, 55-75.

[39] Petersen S, Barcikowski S, *Adv Funct Mater*, **2009** *19*, 1167-1172.

[40] Amendola V, Meneghetti M, *Phys Chem Chem Phys*, **2009** *11*, 3805-3821.

[41] Zeng H, Du X-W, Singh S C, Kulinich S A, Yang S, He J, Cai W, *Adv Funct Mater*, **2012** *22*, 1333-1353.

[42] Barcikowski S, Devesa F, Moldenhauer K, *J Nanopart Res*, **2009** *11*, 1883-1893.

[43] Niece S L, *Gold*, Harvard University Press, **2009**, Cambridge/Massachusetts.

[44] Paracelsus S, F., *Volumen Paramirum und Opus Paramirum*, Eugen Diederichs Verlag, **1904**, Jena.

[45] World Gold Council, http://www.gold.org, September 25th, **2014**.

[46] Bhushan B, *Principles and Applications of Tribology*, 2nd ed. John Wiley & Sons **2013**, New York.

[47] Schlesinger M, Paunovic M, *Modern Electroplating*, 5th ed. John Wiley & Sons, **2011**, New York.

[48] Greenwood N N, Earnshaw A, *Chemistry of the Elements*, 2nd ed. Pergamon Press, **1984**, Oxford.

[49] Haynes W M, *CRC Handbook of Chemistry and Physics, 93rd Edition*, Taylor & Francis, **2012**, Florida.

[50] Kaye G W C, Laby T H, *Tables of Physical and Chemical Constants: And Some Mathematical Functions*, 16th ed. John Wiley & Sons, **1966**, New York.

[51] Müller M, *Strukturelle und Chemische Charakterisierung von Selbst-Assemblierten Monolagen Organischer Moleküle auf Oberflächen*, Hochschulbibliothek der Rheinisch-Westfälischen Technischen Hochschule Aachen, **2013**.

[52] Beardmore K M, Kress J D, Grønbech-Jensen N, Bishop A R, *Chem Phys Lett*, **1998** *286*, 40-45.

[53] Miessler G L, Fischer P J, Tarr D A, *Inorganic Chemistry*, 5th ed. Pearson Education, **2013**, New York.

[54] Grönbeck H, Curioni A, Andreoni W, *J Am Chem Soc*, **2000** *122*, 3839-3842.

[55] Crozier T E, Yamamoto S, *J Chem Eng Data*, **1974** *19*, 242-244.

[56] Nagahara T, Suemasu T, Aida M, Ishibashi T-a, *Langmuir*, **2009** *26*, 389-396.

[57] Valkenier H, Huisman E H, van Hal P A, de Leeuw D M, Chiechi R C, Hummelen J C, *J Am Chem Soc*, **2011** *133*, 4930-4939.

[58] Garg N, Friedman J M, Lee T R, *Langmuir*, **2000** *16*, 4266-4271.

[59] Peterlinz K A, Georgiadis R, *Langmuir*, **1996** *12*, 4731-4740.

[60] Faraday M, *Philos Trans R Soc Lond B Biol Sci*, **1857** *147*, 145-181.

[61] Jain P K, Huang X, El-Sayed I H, El-Sayed M A, *Plasmonics*, **2007** *2*, 107-118.

[62] Sannomiya T, Hafner C, Voros J, *J Biomed Opt*, **2009** *14*, 064027.

[63] Barchanski A, Sajti L C, Sehring C, Petersen S, Barcikowski S, *Design of Bi-Functional Bioconjugated Gold Nanoparticles by Pulsed Laser Ablation with Minimized Degradation*, LPM 2010: 11th International Symposium on Laser Precision MicrofabricationJune 7-10, 2010, Stuttgart, Germany. **2010**.

[64] Mody V, Siwale R, Singh A, Mody H, *J Pharm Bioallied Sci*, **2010** 2, 282-289.

[65] Kreibig U, Vollmer M, *Optical Properties of Metal Clusters*, Springer-Verlag, **1995**, Berlin Heidelberg.

[66] Khlebtsov N G, *Quantum Electron*, **2008** *38*, 504.

[67] Ghosh S K, Pal T, *Chem Rev*, **2007** *107*, 4797-4862.

[68] Mie G, *Ann Phys*, **1908** *330*, 377-445.

[69] Besner S, Kabashin A V, Meunier M, *Appl Phys A*, **2007** *88*, 269-272.

[70] Jain P K, Lee K S, El-Sayed I H, El-Sayed M A, *J Phys Chem B*, **2006** *110*, 7238-7248.

[71] Link S, El-Sayed M A, *J Phys Chem B*, **1999** *103*, 4212-4217.

[72] Huang X, El-Sayed I H, Qian W, El-Sayed M A, *J Am Chem Soc*, **2006** *128*, 2115-2120.

[73] Vigderman L, Zubarev E R, *Chem Mater*, **2013** *25*, 1450-1457.

[74] Nedyalkov N N, Imamova S, Atanasov P A, Tanaka Y, Obara M, *J Nanopart Res*, **2011** *13*, 2181-2193.

[75] Mulvaney P, *Langmuir*, **1996** *12*, 788-800.

[76] Raschke G, Kowarik S, Franzl T, Sönnichsen C, Klar T A, Feldmann J, Nichtl A, Kürzinger K, *Nano Lett*, **2003** *3*, 935-938.

[77] Muto H, Yamada K, Miyajima K, Mafuné F, *J Phys Chem C*, **2007** *111*, 17221-17226.

[78] Link S, El-Sayed M A, *Annu Rev Phys Chem*, **2003** *54*, 331-366.

[79] He H, Xie C, Ren J, *Anal Chem*, **2008** *80*, 5951-5957.

[80] Thurn K T, Brown E, Wu A, Vogt S, Lai B, Maser J, Paunesku T, Woloschak G E, *Nanoscale Res Lett*, **2007** 2, 430-441.

[81] Slouf M, Kuzel R, Matej Z, *Z Kristallogr*, **2006**, 319-324.

[82] Chithrani B D, Ghazani A A, Chan W C W, *Nano Lett*, **2006** *6*, 662-668.

[83] Nativo P, Prior I A, Brust M, *ACS Nano*, **2008** *2*, 1639-1644.

[84] Murphy C J, Gole A M, Stone J W, Sisco P N, Alkilany A M, Goldsmith E C, Baxter S C, *Accounts Chem Res*, **2008** *41*, 1721-1730.

[85] Parab H J, Chen H M, Lai T-C, Huang J H, Chen P H, Liu R-S, Hsiao M, Chen C-H, Tsai D-P, Hwu Y-K, *J Phys Chem C*, **2009** *113*, 7574-7578.

[86] Lindfors K, Kalkbrenner T, Stoller P, Sandoghdar V, *Phys Rev Lett*, **2004** *93*.

[87] van Dijk M A, Tchebotareva A L, Orrit M, Lippitz M, Berciaud S, Lasne D, Cognet L, Louis B, *Phys Chem Chem Phys*, **2006** *8*, 3486-3495.

[88] Sandoghdar V, Klotzsch E, Jacobsen V, Renn A, Håkanson U, Agio M, Gerhardt I, Seelig J, Wrigge G, *CHIMIA Int J Chem*, **2006** *60*, 761-764.

[89] Zagaynova E V, Shirmanova M V, Kirillin M Y, Khlebtsov B N, Orlova A G, Balalaeva I V, Sirotkina M A, Bugrova M L, Agrba P D, Kamensky V A, *Phys Med Biol*, **2008** *53*, 4995-5009.

[90] Ponce de Leon Y P, Pichardo-Molina J L, Ochoa N A, Luna-Moreno D, *J Nanomater*, **2012**, 9.

[91] Schmelzeisen M, Austermann J, Kreiter M, *Opt Express*, **2008** *16*, 17826-17841.

[92] Qian X, Peng X-H, Ansari D O, Yin-Goen Q, Chen G Z, Shin D M, Yang L, Young A N, Wang M D, Nie S, *Nat Biotech*, **2008** *26*, 83-90.

[93] Klein S, Petersen S, Taylor U, Rath D, Barcikowski S, *J Biomed Opt*, **2010** *15*, 036015.

[94] Sezgin E, Karatas Ö F, Cam D, Sur I, Sayin I, Avci E, Keseroglu K, Sülek S, Culha M, *J Eng Nat Sci (Sigma)*, **2008** *26*, 227-246.

[95] Boyer D, Tamarat P, Maali A, Lounis B, Orrit M, *Science*, **2002** *297*, 1160-1163.

[96] Terentyuk G S, Maslyakova G N, Suleymanova L V, Khlebtsov B N, Kogan B Y, Akchurin G G, Shantrocha A V, Maksimova I L, Khlebtsov N G, Tuchin V V, *J Biophotonics*, **2009** *2*, 292-302.

[97] Werner D, Hashimoto S, Uwada T, *Langmuir*, **2010** *26*, 9956-9963.

[98] Lasne D, Blab G A, Berciaud S, Heine M, Groc L, Choquet D, Cognet L, Lounis B, *Biophys J*, **2006** *91*, 4598-4604.

[99] Wang B, Joshi P P, Sapozhnikova V, Amirian J, Litovsky S H, *Proc SPIE*, **2010** *7564*, 75640A.

[100] Mallidi S, Larson T, Tam J, Joshi P P, Karpiouk A, Sokolov K, Emelianov S, *Nano Lett*, **2009** *9*, 2825-2831.

[101] Khlebtsov N, Dykman L, *Chem Soc Rev*, **2011** *40*, 1647-1671.

[102] OECD, *Guidance on Sample Preparation and Dosimetry for the Safety Testing of Manufactured Nanomaterials*, Series on the Safety of Manufactured Nanomaterials. OECD Publication Service, **2012**, Paris, France.

[103] Faux S P, Tran C L, Miller A D, Jones, Montellier C, Donaldson K, *In-Vitro Determinants of Particulate Toxicity: The Dose Metric for Poorly Soluble Dusts*, Health and Safety Executive Report 154, **2003**.

[104] Oberdorster G, Oberdorster E, Oberdorster J, *Environ Health Perspect*, **2005** *113*, 823-839.

[105] Nel A E, Madler L, Velegol D, Xia T, Hoek E M, Somasundaran P, Klaessig F, Castranova V, Thompson M, *Nat Mater*, **2009** *8*, 543-557.

[106] Alkilany A M, Lohse S E, Murphy C J, *Accounts Chem Res*, **2012** *46*, 650-661.

[107] Hühn D, Kantner K, Geidel C, Brandholt S, De Cock I, Soenen S J H, Rivera-Gil P, Montenegro J-M, Braeckmans K, Müllen K, Nienhaus G U, Klapper M, Parak W J, *ACS Nano*, **2013** *7*, 3253-3263.

[108] Leifert A, Pan Y, Kinkeldey A, Schiefer F, Setzler J, Scheel O, Lichtenbeld H, Schmid G, Wenzel W, Jahnen-Dechent W, Simon U, *Proc Natl Acad Sci USA*, **2013** *110*, 8004-8009.

[109] Liu Y, Meyer-Zaika W, Franzka S, Schmid G, Tsoli M, Kuhn H, *Angew Chem Int Ed*, **2003** *42*, 2853-2857.

[110] Singh N, Manshian B, Jenkins G J S, Griffiths S M, Williams P M, Maffeis T G G, Wright C J, Doak S H, *Biomaterials*, **2009** *30*, 3891-3914.

[111] Zhang H, Ji Z, Xia T, Meng H, Low-Kam C, Liu R, Pokhrel S, Lin S, Wang X, Liao Y-P, Wang M, Li L, Rallo R, Damoiseaux R, Telesca D, Mädler L, Cohen Y, Zink J I, Nel AE, *ACS Nano*, **2012** *6*, 4349-4368.

[112] Albanese A, Tang P S, Chan W C, *Annu Rev Biomed Eng*, **2012** *14*, 1-16.

[113] Taylor U, Barchanski A, Kues W, Barcikowski S, Rath D, *Reprod Domest Anim*, **2012** *47*, 359-368.

[114] Taylor U, Barchanski A, Garrels W, Klein S, Kues W, Barcikowski S, Rath D, *Toxicity of Gold Nanoparticles on Somatic and Reproductive Cells*, in: Nano-Biotechnology for Biomedical and Diagnostic Research. Springer Netherlands, **2012**, pp. 125-133.

[115] Taylor U, Rehbock C, Streich C, Rath D, Barcikowski S, *Nanomedicine (Lond)*, **2014** *9*, 1971-1989.

[116] Shenoy D, Fu W, Li J, Crasto C, Jones G, DiMarzio C, Sridhar S, Amiji M, *Int J Nanomedicine*, **2006** *1*, 51-57.

[117] Salmaso S, Caliceti P, Amendola V, Meneghetti M, Magnusson J P, Pasparakis G, Alexander C, *J Mater Chem*, **2009** *19*, 1608-1615.

[118] Fu W, Shenoy D, Li J, Crasto C, Jones G, Dimarzio C, Sridhar S, Amiji M, *MRS Proceedings*, **2004** *845*, AA5.4.

[119] Thomas M, Klibanov A M, *P Natl A Sci USA*, **2003** *100*, 9138-9143.

[120] Connor E E, Mwamuka J, Gole A, Murphy C J, Wyatt M D, *Small*, **2005** *1*, 325-327.

[121] Massich M D, Giljohann D A, Schmucker A L, Patel P C, Mirkin C A, *ACS Nano*, **2010** *4*, 5641-5646.

[122] Taylor U, Klein S, Petersen S, Kues W, Barcikowski S, Rath D, *Cytom Part A*, **2010** *77A*, 439-446.

[123] Tkachenko A G, Xie H, Liu Y, Coleman D, Ryan J, Glomm W R, Shipton M K, Franzen S, Feldheim D L, *Bioconj Chem*, **2004** *15*, 482-490.

[124] Pan Y, Neuss S, Leifert A, Fischler M, Wen F, Simon U, Schmid G, Brandau W, Jahnen-Dechent W, *Small*, **2007** *3*, 1941-1949.

[125] Pan Y, Leifert A, Ruau D, Neuss S, Bornemann J, Schmid G, Brandau W, Simon U, Jahnen-Dechent W, *Small*, **2009** *5*, 2067-2076.

[126] Patra H K, Banerjee S, Chaudhuri U, Lahiri P, Dasgupta A K, *Nanomed Nanotech Biol Med*, **2007** *3*, 111-119.

[127] Ding Y, Bian X, Yao W, Li R, Ding D, Hu Y, Jiang X, Hu Y, *ACS Appl Mater Interfaces*, **2010** *2*, 1456-1465.

[128] Hanley C, Layne J, Punnoose A, Reddy K M, Coombs I, Coombs A, Feris K, Wingett D, *Nanotech*, **2008** *19*, 295103.

[129] Bregoli L, Chiarini F, Gambarelli A, Sighinolfi G, Gatti A M, Santi P, Martelli A M, Cocco L, *Toxicology*, **2009** *262*, 121-129.

[130] Goodman C M, McCusker C D, Yilmaz T, Rotello V M, *Bioconj Chem*, **2004** *15*, 897-900.

[131] Bartneck M, Keul H A, Singh S, Czaja K, Bornemann J, Bockstaller M, Moeller M, Zwadlo-Klarwasser G, Groll J, *ACS Nano*, **2010** *4*, 3073-3086.

[132] Paracelsus, Huser J, *Bücher Und Schrifften*, Konrad von Waldkirch, **1589**, Basel.

[133] Balasubramanian S K, Jittiwat J, Manikandan J, Ong C-N, Yu L E, Ong W-Y, *Biomaterials*, **2010** *31*, 2034-2042.

[134] Ema M, Kobayashi N, Naya M, Hanai S, Nakanishi J, *Reprod Toxicol*, **2010** *30*, 343-352.

[135] Wiwanitkit V, Sereemaspun A, Rojanathanes R, *Fertil Steril*, **2009** *91*, e7-e8.

[136] Tiedemann D, Taylor U, Rehbock C, Jakobi J, Klein S, Kues W A, Barcikowski S, Rath D, *Analyst*, **2014** *139*, 931-942.

[137] Taylor U, Barchanski A, Petersen S, Kues W A, Baulain U, Gamrad L, Sajti L, Barcikowski S, Rath D, *Nanotoxicology*, **2013**.

[138] Moretti E, Terzuoli G, Renieri T, Iacoponi F, Castellini C, Giordano C, Collodel G, *Andrologia*, **2013** *45*, 392-396.

[139] Zakhidov S T, Marshak T L, Malolina E A, Kulibin A Y, Zelenina I A, Pavluchenkova S M, Rudoi V M, Dement'eva O V, Skuridin S G, Evdokimov Y M, *Biochem (Mosc) Suppl Ser A: Membr Cell Biol*, **2010** *4*, 293-296.

[140] Takahashi S, Matsuoka O, *J Rad Res*, **1981** *22*, 242-249.

[141] Semmler-Behnke M, Kreyling W G, Lipka J, Fertsch S, Wenk A, Takenaka S, Schmid G, Brandau W, *Small*, **2008** *4*, 2108-2111.

[142] Challier J C, Panigel M, Meyer E, *Int J Nucl Med Biol*, **1973** *1*, 103-106.

[143] Sadauskas E, Wallin H, Stoltenberg M, Vogel U, Doering P, Larsen A, Danscher G, *Part Fibre Toxicol*, **2007** *4*, 1-7.

[144] Myllynen P K, Loughran M J, Howard C V, Sormunen R, Walsh A A, Vähäkangas K H, *Reprod Toxicol*, **2008** *26*, 130-137.

[145] Bar-Ilan O, Albrecht R M, Fako V E, Furgeson D Y, *Small*, **2009** *5*, 1897-1910.

[146] Browning L M, Lee K J, Huang T, Nallathamby P D, Lowman J E, Nancy Xu X-H, *Nanoscale*, **2009** *1*, 138-152.

[147] Zielinska A K, Sawosz E, Grodzik M, Chwalibog A, Kamaszewski M, *Ann Warsaw Univ Life Sci SGGW, Anim Sci*, **2009** *46*, 249-253.

[148] Sawosz E, Chwalibog A, Szeliga J, Sawosz F, Grodzik M, Rupiewicz M, Niemiec T, Kacprzyk K, *Int J Nanomedicine*, **2010** *5*, 631-637.

[149] Taylor U, Garrels W, Barchanski A, Peterson S, Sajti L, Lucas-Hahn A, Gamrad L, Baulain U, Klein S, Kues W A, Barcikowski S, Rath D, *Beilstein J Nanotech*, **2014** *5*, 677-688.

[150] Pan Y, Leifert A, Graf M, Schiefer F, Thoröe-Boveleth S, Broda J, Halloran M C, Hollert H, Laaf D, Simon U, Jahnen-Dechent W, *Small*, **2013** *9*, 863-869.

[151] Chen Y S, Hung Y C, Liau I, Huang G S, *Nanoscale Res Lett*, **2009** *4*, 858-864.

[152] Cho W-S, Cho M, Jeong J, Choi M, Cho H-Y, Han B S, Kim S H, Kim H O, Lim Y T, Chung B H, Jeong J, *Toxicol Appl Pharmacol*, **2009** *236*, 16-24.

[153] Tedesco S, Doyle H, Redmond G, Sheehan D, *Mar Environ Res*, **2008** *66*, 131-133.

[154] Fent G M, Casteel S W, Kim D Y, Kannan R, Katti K, Chanda N, Katti K, *Nanomed Nanotech Biol Med*, **2009** *5*, 128-135.

[155] Hainfeld J F, Slatkin D N, Smilowitz H M, *Phys Med Biol*, **2004** *49*, N309.

[156] Kattumuri V, Katti K, Bhaskaran S, Boote E J, Casteel S W, Fent G M, Robertson D J, Chandrasekhar M, Kannan R, Katti K V, *Small*, **2007** *3*, 333-341.

[157] Kim J H, Kim J H, Kim K-W, Kim M H, Yu Y S, *Nanotechnology*, **2009** *20*, 505101.

[158] Lasagna-Reeves C, Gonzalez-Romero D, Barria M A, Olmedo I, Clos A, Sadagopa Ramanujam V M, Urayama A, Vergara L, Kogan M J, Soto C, *Biochem Biophys Rese Co*, **2010** *393*, 649-655.

[159] Abraham G E, Himmel P B, *J Nutr Environ Med*, **1997** *7*, 295-305.

[160] Zhang X D, Wu H Y, Wu D, Wang Y Y, Chang J H, Zhai Z B, Meng A M, Liu P X, Zhang L A, Fan F Y, *Int J Nanomedicine*, **2010** *5*, 771-781.

[161] Silverstein S C, Steinman R M, Cohn Z A, *Annu Rev Biochem*, **1977** *46*, 669-722.

[162] Wilson D B, *Annu Rev Biochem*, **1978** *47*, 933-965.

[163] Schatz G, Dobberstein B, *Science*, **1996** *271*, 1519-1526.

[164] Jahn R, Südhof T C, *Annu Rev Biochem*, **1999** *68*, 863-911.

[165] Cheng L C, Jiang X, Wang J, Chen C, Liu R S, *Nanoscale*, **2013** *5*, 3547-3569.

[166] Chithrani D B, *Insciences J*, **2011**, 115-135.

[167] Zhao F, Zhao Y, Liu Y, Chang X, Chen C, Zhao Y, *Small*, **2011** *7*, 1322-1337.

[168] Chithrani B D, Chan W C W, *Nano Lett*, **2007** *7*, 1542-1550.

[169] Jiang W, Kim B Y, Rutka J T, Chan W C, *Nat Nanotechnol*, **2008** *3*, 145-150.

[170] Zhang K, Fang H, Chen Z, Taylor J-S A, Wooley K L, *Bioconj Chem*, **2008** *19*, 1880-1887.

[171] Arvizo R R, Miranda O R, Thompson M A, Pabelick C M, Bhattacharya R, Robertson J D, Rotello V M, Prakash Y S, Mukherjee P, *Nano Lett*, **2010** *10*, 2543-2548.

[172] Gao H, Shi W, Freund L B, *Proc Natl Acad Sci USA*, **2005** *102*, 9469-9474.

[173] Ma X, Wu Y, Jin S, Tian Y, Zhang X, Zhao Y, Yu L, Liang X-J, *ACS Nano*, **2011** *5*, 8629-8639.

[174] Wang S-H, Lee C-W, Chiou A, Wei P-K, *J. Nanobiotech*, **2010** *8*, 1-13.

[175] de la Fuente J M, Berry C C, *Bioconj Chem*, **2005** *16*, 1176-1180.

[176] Gu Y J, Cheng J, Lin C C, Lam Y W, Cheng S H, Wong W T, *Toxicol Appl Pharmacol*, **2009** *237*, 196-204.

[177] Kong W H, Bae K H, Jo S D, Kim J S, Park T G, *Pharm Res*, **2012** *29*, 362-374.

[178] Oh E, Delehanty J B, Sapsford K E, Susumu K, Goswami R, Blanco-Canosa J B, Dawson P E, Granek J, Shoff M, Zhang Q, Goering P L, Huston A, Medintz I L, *ACS Nano*, **2011** *5*, 6434-6448.

[179] Ryan J A, Overton K W, Speight M E, Oldenburg C N, Loo L, Robarge W, Franzen S, Feldheim D L, *Anal Chem*, **2007** *79*, 9150-9159.

[180] Cho E C, Xie J, Wurm P A, Xia Y, *Nano Letters*, **2009** *9*, 1080-1084.

[181] Freese C, Gibson M I, Klok H A, Unger R E, Kirkpatrick C J, *Biomacromolecules*, **2012** *13*, 1533-1543.

[182] Liang M, Lin I C, Whittaker M R, Minchin R F, Monteiro M J, Toth I, *ACS Nano*, **2009** *4*, 403-413.

[183] Zhu Z-J, Wang H, Yan B, Zheng H, Jiang Y, Miranda O R, Rotello V M, Xing B, Vachet RW, *Environ Sci Technol*, **2012** *46*, 12391-12398.

[184] Weiss L, Zeigel R, Jung O S, Bross I D J, *Exp Cell Res*, **1972** *70*, 57-64.

[185] Mutsaers S E, Papadimitriou J M, *J Leukoc Biol*, **1988** *44*, 17-26.

[186] Lynch I, Dawson K A, *Nano Today*, **2008** *3*, 40-47.

[187] Panyam J, Zhou W Z, Prabha S, Sahoo S K, Labhasetwar V, *Faseb J*, **2002** *16*, 1217-1226.

[188] Stewart K M, Horton K L, Kelley S O, *Org Biomol Chem*, **2008** *6*, 2242-2255.

[189] Verma A, Uzun O, Hu Y, Hu Y, Han H-S, Watson N, Chen S, Irvine DJ, Stellacci F, *Nat Mater*, **2008** *7*, 588-595.

[190] Dekiwadia C D, Lawrie A C, Fecondo J V, *J Pept Sci*, **2012** *18*, 527-534.

[191] Everts M, Saini V, Leddon J L, Kok R J, Stoff-Khalili M, Preuss M A, Millican C L, Perkins G, Brown J M, Bagaria H, Nikles D E, Johnson D T, Zharov V P, Curiel D T, *Nano Letters*, **2006** *6*, 587-591.

[192] Shi X, Wang S, Meshinchi S, Van Antwerp M E, Bi X, Lee I, Baker J R, *Small*, **2007** *3*, 1245-1252.

[193] Green M, Loewenstein P M, *Cell*, **1988** *55*, 1179-1188.

[194] Derossi D, Calvet S, Trembleau A, Brunissen A, Chassaing G, Prochiantz A, *J Biol Chem*, **1996** *271*, 18188-18193.

[195] Richard J P, Melikov K, Vives E, Ramos C, Verbeure B, Gait M J, Chernomordik L V, Lebleu B, *J Biol Chem*, **2003** *278*, 585-590.

[196] Mandal D, Maran A, Yaszemski M J, Bolander M E, Sarkar G, *J Mater Sci Mater Med*, **2009** *20*, 347-350.

[197] Petersen S, Barchanski A, Taylor U, Klein S, Rath D, Barcikowski S, *J Phys Chem C*, **2010** *115*, 5152-5159.

[198] Barchanski A, Taylor U, Sajti C L, Gamrad L, Kues W A, Rath D, Barcikowski S, *J Biomed Nanotechnol*, **2015** *11*, 1-11.

[199] Pujals S, Bastús N G, Pereiro E, López-Iglesias C, Puntes VF, Kogan M J, Giralt E, *ChemBioChem*, **2009** *10*, 1025-1031.

[200] Berry C C, De la Fuente J M, Mullin M, Wai Ling Chu S, Curtis A S G, *IEEE Trans Nanobioscience*, **2007** *6*, 262-269.

[201] Krpetic Z, Nativo P, Porta F, Brust M, *Bioconjug Chem*, **2009** *20*, 619-624.

[202] De Jong W H, Hagens W I, Krystek P, Burger M C, Sips A J, Geertsma RE, *Biomaterials*, **2008** *29*, 1912-1919.

[203] Lipka J, Semmler-Behnke M, Sperling R A, Wenk A, Takenaka S, Schleh C, Kissel T, Parak W J, Kreyling W G, *Biomaterials*, **2010** *31*, 6574-6581.

[204] Sonavane G, Tomoda K, Makino K, *Colloid Surface B*, **2008** *66*, 274-280.

[205] Hainfeld J F, Slatkin D N, Focella T M, Smilowitz H M, *Br J Radiol*, **2006** *79*, 248-253.

[206] Zhang X D, Wu D, Shen X, Liu P X, Fan F Y, Fan S J, *Biomaterials*, **2012** *33*, 4628-4638.

[207] Zhou C, Long M, Qin Y, Sun X, Zheng J, *Angew Chem Int Ed*, **2011** *50*, 3168-3172.

[208] Sokolov K, Follen M, Aaron J, Pavlova I, Malpica A, Lotan R, Richards-Kortum R, *Cancer Res*, **2003** *63*, 1999-2004.

[209] Wang Z, Ma L, *Coordin Chem Rev*, **2009** *253*, 1607-1618.

[210] Han G, Ghosh P, De M, Rotello V M, *Nanobiotechnol*, **2007** *3*, 40-45.

[211] Anker J N, Hall W P, Lyandres O, Shah N C, Zhao J, Van Duyne R P, *Nat Mater*, **2008** *7*, 442-453.

[212] Xiao L, Wei L, He Y, Yeung E S, *Anal Chem*, **2010** *82*, 6308-6314.

[213] de la Escosura-Muñiz A, Parolo C, Merkoçi A, *Materials Today*, **2010** *13*, 24-34.

[214] Lou S, Ye J-y, Li K-q, Wu A, *Analyst*, **2012** *137*, 1174-1181.

[215] Sokolov K, Aaron J, Hsu B, Nida D, Gillenwater A, Follen M, MacAulay C, Adler-Storthz K, Korgel B, Descour M, Pasqualini R, Arap W, Lam W, Richards-Kortum R, *Technol Cancer Res Treat*, **2003** *2*, 491-504.

[216] Rosi N L, Mirkin C A, *Chem Rev*, **2005** *105*, 1547-1562.

[217] Graham D, Faulds K, Thompson D, McKenzie F, Stokes R, Dalton C, Stevenson R, Alexander J, Garside P, McFarlane E, *Biochem Soc Trans*, **2009** *37*, 697-701.

[218] McKenzie F, Faulds K, Graham D, *Chem Commun (Camb)*, **2008**, 2367-2369.

[219] Hill H D, Mirkin C A, *Nat Protoc*, **2006** *1*, 324-336.

[220] Jia C P, Zhong X Q, Hua B, Liu M Y, Jing F X, Lou X H, Yao S H, Xiang J Q, Jin QH, Zhao J L, *Biosens Bioelectron*, **2009** *24*, 2836-2841.

[221] Liu Y, Liu Y, Mernaugh R L, Zeng X, *Biosens Bioelectron*, **2009** *24*, 2853-2857.

[222] Aslan K, Lakowicz J R, Geddes C D, *Anal Biochem*, **2004** *330*, 145-155.

[223] Wilson R, *Chem Soc Rev*, **2008** *37*, 2028-2045.

[224] Medley C D, Smith J E, Tang Z, Wu Y, Bamrungsap S, Tan W, *Anal Chem*, **2008** *80*, 1067-1072.

[225] Yang J, Eom K, Lim E-K, Park J, Kang Y, Yoon D S, Na S, Koh E K, Suh J-S, Huh Y-M, Kwon T Y, Haam S, *Langmuir*, **2008** *24*, 12112-12115.

[226] Rojanathanes R, Sereemaspun A, Pimpha N, Buasorn V, Ekawong P, Wiwanitkit V, *Taiwan J Obstet Gynecol*, **2008** *47*, 296-299.

[227] Tang L, Casas J, *Biosens Bioelectron*, **2014** *61*, 70-75.

[228] Zhang J, Wang L, Pan D, Song S, Boey F Y C, Zhang H, Fan C, *Small*, **2008** *4*, 1196-1200.

[229] El-Sayed I H, Huang X, El-Sayed M A, *Nano Lett*, **2005** *5*, 829-834.

[230] Kah J C, Kho K W, Lee C G, James C, Sheppard R, Shen Z X, Soo K C, Olivo M C, *Int J Nanomedicine*, **2007** *2*, 785-798.

[231] Jain P K, El-Sayed I H, El-Sayed M A, *Nano Today*, **2007** *2*, 18-29.

[232] Qian X M, Nie S M, *Chem Soc Rev*, **2008** *37*, 912-920.

[233] Huang Y-F, Lin Y-W, Lin Z-H, Chang H-T, *J Nanopart Res*, **2008** *11*, 775-783.

[234] Tanev S, Sun W, Pond J, Tuchin V V, Zharov V P, *J Biophotonics*, **2009** *2*, 505-520.

[235] Crow M J, Marinakos S M, Cook J M, Chilkoti A, Wax A, *Cytom Part A*, **2011** *79A*, 57-65.

[236] Van de Broek B, Devoogdt N, D'Hollander A, Gijs H-L, Jans K, Lagae L, Muyldermans S, Maes G, Borghs G, *ACS Nano*, **2011** *5*, 4319-4328.

[237] Rodriguez-Lorenzo L, Fytianos K, Blank F, vonGarnier C, Rothen-Rutishauser B, Petri-Fink A, *Small*, **2014** *10*, 1341-1350.

[238] Baronzio G F, Hager E D, *Hyperthermia in Cancer Treatment: A Primer*, Landes Bioscience & Springer **2008**, USA.

[239] Friedman M, Mikityansky I, Kam A, Libutti SK, Walther M M, Neeman Z, Locklin J K, Wood B J, *Cardiovasc Inter Rad*, **2004** *27*, 427-434.

[240] Jordan A, Scholz R, Wust P, Fahling H, Roland F, *J Magn Magn Mater*, **1999** *201*, 413-41.

[241] Laurent S, Dutz S, Hafeli U O, Mahmoudi M, *Adv Colloid Interface Sci*, **2011** *166*, 8-23.

[242] Thomas L A, Dekker L, Kallumadil M, Southern P, Wilson M, Nair S P, Pankhurst Q A, Parkin IP, *J Mater Chem*, **2009** *19*, 6529.

[243] Overgaard J, *Cancer*, **1977** *39*, 2637-2646.

[244] Dennis C L, Jackson A J, Borchers J A, Hoopes P J, Strawbridge R, Foreman A R, van Lierop J, Gruttner C, Ivkov R, *Nanotechnology*, **2009** *20*, 395103.

[245] Pissuwan D, Valenzuela S M, Cortie M B, *Trends Biotechnol*, **2006** *24*, 62-67.

[246] Visaria R K, Griffin R J, Williams B W, Ebbini E S, Paciotti G F, Song C W, Bischof JC, *Mol Cancer Ther*, **2006** *5*, 1014-1020.

[247] Lapotko D O, Lukianova E, Oraevsky A A, *Lasers Surg Med*, **2006** *38*, 631-642.

[248] Loo C, Lin A, Hirsch L, Lee MH, Barton J, Halas N, West J, Drezek R, *Technol Cancer Res Treat*, **2004** *3*, 33-40.

[249] Huang X, Jain P K, El-Sayed I H, El-Sayed M A, *Lasers Med Sci*, **2008** *23*, 217-228.

[250] Pitsillides C M, Joe E K, Wei X, Anderson R R, Lin C P, *Biophys J*, **2003** *84*, 4023-4032.

[251] Kalies S, Birr T, Heinemann D, Schomaker M, Ripken T, Heisterkamp A, Meyer H, *J Biophotonics*, **2014** *7*, 474-482.

[252] Remaut K, Sanders N N, De Geest B G, Braeckmans K, Demeester J, De Smedt S C, *Mater Sci Eng Rep*, **2007** *58*, 117-161.

[253] Howarth J, Lee Y, Uney J, *Cell Biol Toxicol*, **2010** *26*, 1-20.

[254] Puvanakrishnan P, Park J, Chatterjee D, Krishnan S, Tunnell J W, *Int J Nanomedicine*, **2012** *7*, 1251-1258.

[255] Liu J, Yu M, Zhou C, Yang S, Ning X, Zheng J, *J Am Chem Soc*, **2013** *135*, 4978-4981.

[256] Hong R, Han G, Fernández J M, Kim B-j, Forbes N S, Rotello V M, *J Am Chem Soc*, **2006** *128*, 1078-1079.

[257] Braun G B, Pallaoro A, Wu G, Missirlis D, Zasadzinski J A, Tirrell M, Reich N O, *ACS Nano*, **2009** *3*, 2007-2015.

[258] Gibson J D, Khanal B P, Zubarev E R, *J Am Chem Soc*, **2007** *129*, 11653-11661.

[259] Farokhzad O C, Cheng J, Teply B A, Sherifi I, Jon S, Kantoff P W, Richie J P, Langer R, *Proc Natl Acad Sci USA*, **2006** *103*, 6315-6320.

[260] Wijaya A, Schaffer S B, Pallares I G, Hamad-Schifferli K, *ACS Nano*, **2008** *3*, 80-86.

[261] Han G, You C-C, Kim B-j, Turingan R S, Forbes N S, Martin C T, Rotello V M, *Angew Chem*, **2006** *118*, 3237-3241.

[262] Poon L, Zandberg W, Hsiao D, Erno Z, Sen D, Gates B D, Branda N R, *ACS Nano*, **2010** *4*, 6395-6403.

[263] Panchapakesan B, Book-Newell B, Sethu P, Rao M, Irudayaraj J, *Nanomedicine (Lond)*, **2011** *6*, 1787-1811.

[264] Lux Research, *How to Evaluate Emerging Targeted Delivery Technologies*, November 14th **2008**.

[265] Zhao W, Lee T M H, Leung S S Y, Hsing I M, *Langmuir*, **2007** *23*, 7143-7147.

[266] Basu S, Harfouche R, Soni S, Chimote G, Mashelkar R A, Sengupta S, *Proc Natl Acad Sci USA*, **2009** *106*, 7957-7961.

[267] Fischler M, Simon U, *J Mater Chem*, **2009** *19*, 1518.

[268] Wang J, Munir A, Zhou H S, *Talanta*, **2009** *79*, 72-76.

[269] Walter J G, Petersen S, Stahl F, Scheper T, Barcikowski S, *J Nanobiotech*, **2010** *8*, :21.

[270] Kumar S, Aaron J, Sokolov K, *Nat Protoc*, **2008** *3*, 314-320.

[271] Rezaei B, Khayamian T, Majidi N, Rahmani H, *Biosens Bioelectron*, **2009** *25*, 395-399.

[272] Krasovskii V I, Nagovitsyn I A, Chudinova G K, Savranskii V V, Karavanskii V A, *B Lebedev Phys Inst*, **2007** *34*, 321-324.

[273] ASTM International, E-245-06 Terminology for nanotechnology, September 25th **2014**.

[274] Gardea-Torresdey J L, Parsons J G, Gomez E, Peralta-Videa J, Troiani H E, Santiago P, Yacaman M *Nano Lett*, **2002** *2*, 397-401.

[275] Agnihotri M, Joshi S, Kumar A R, Zinjarde S, Kulkarni S, *Mater Lett*, **2009** *63*, 1231-1234.

[276] Pimpang P, Sutham W, Mangkorntong N, Mangkorntong P, Choopun S, *Chiang Mai J Sci*, **2008** *35*, 250-257.

[277] Cho K, Chang H, Kil D S, Kim B-G, Jang H D, *J Ind Eng Chem*, **2009** *15*, 243-246.

[278] Khomutov G B, *Colloid Surface A*, **2002** *202*, 243-267.

[279] Nakamoto M, Yamamoto M, Fukusumi M, *Chem Commun (Camb)*, **2002**, 1622-1623.

[280] Shimizu T, Teranishi T, Hasegawa S, Miyake M, *J Phys Chem B*, **2003** *107*, 2719-2724.

[281] Chen F, Xu G-Q, Hor T S A, *Mater Lett*, **2003** *57*, 3282-3286.

[282] Sohn B-H, Choi J-M, Yoo S I, Yun S-H, Zin W-C, Jung J C, Kanehara M, Hirata T, Teranishi T, *J Am Chem Soc*, **2003** *125*, 6368-6369.

[283] Sau T, Pal A, Jana N R, Wang Z L, Pal T, *J Nanopart Res*, **2001** *3*, 257-261.

[284] Watzky M A, Finke R G, *J Am Chem Soc*, **1997** *119*, 10382-10400.

[285] Daniel M-C, Astruc D, *Chem Rev*, **2003** *104*, 293-346.

[286] Kumar S, Gandhi K S, Kumar R, *Ind Eng Chem Res*, **2006** *46*, 3128-3136.

[287] Noyong M, *Synthese und Organisation von Gold-Nanopartikeln mittels DNA*. Technische Hochschule Aachen, Thesis, **2005**, Aachen.

[288] Templeton A C, Wuelfing W P, Murray R W, *Accounts Chem Res*, **1999** *33*, 27-36.

[289] Kassam A, Bremner G, Clark B, Ulibarri G, Lennox R B, *J Am Chem Soc*, **2006** *128*, 3476-3477.

[290] Woehrle G H, Brown L O, Hutchison J E, *J Am Chem Soc*, **2005** *127*, 2172-2183.

[291] Brewer S H, Glomm W R, Johnson M C, Knag M K, Franzen S, *Langmuir*, **2005** *21*, 9303-9307.

[292] Barcikowski S, Hahn A, Kabashin A V, Chichkov B N, *Appl Phys A*, **2007** *87*, 47-55.

[293] Patil P P, Phase D M, Kulkarni S A, Ghaisas S V, Kulkarni S K, Kanetkar S M, Ogale S B, Bhide V G, *Phys Rev Lett*, **1987** *58*, 238-241.

[294] Henglein A, *J Phys Chem*, **1993** *97*, 5457-5471.

[295] Merk V, Rehbock C, Becker F, Hagemann U, Nienhaus H, Barcikowski S, *Langmuir*, **2014** *30*, 4213-4222.

[296] Barcikowski S, Walter J, Hahn A, Koch J, Haloui H, Herrmann T, Gatti A, *J Laser Micro/Nanoeng JLMN*, **2009** *4*, 159-164.

[297] Mafuné F, Kohno J-y, Takeda Y, Kondow T, Sawabe H, *J Phys Chem B*, **2000** *104*, 9111-9117.

[298] Kabashin A V, Meunier M, *J Appl Phys*, **2003** *94*, 7941-7943.

[299] Lee I, Han S W, Kim K, *Chem Commun*, **2001**, 1782-1783.

[300] Jakobi J, Menendez-Manjon A, Chakravadhanula VS, Kienle L, Wagener P, Barcikowski S, *Nanotechnology*, **2011** *22*, 145601.

[301] Baersch N, Jakobi J, Weiler S, Barcikowski S, *Nanotechnology*, **2009** *20*, 445603.

[302] Sajti C L s, Sattari R, Chichkov B N, Barcikowski S, *J Phys Chem C*, **2010** *114*, 2421-2427.

[303] Semaltianos N G, *Crc Cr Rev Sol State*, **2010** *35*, 105-124.

[304] Messina GC, Wagener P, Streubel R, De Giacomo A, Santagata A, Compagnini G, Barcikowski S, *Phys Chem Chem Phys*, **2013** *15*, 3093-3098.

[305] Petersen S, Barcikowski S, *J Phys Chem C*, **2009** *113*, 19830-19835.

[306] Yan Z, Compagnini G, Chrisey D B, *J Phys Chem C*, **2010** *115*, 5058-5062.

[307] Itina T E, *J Phys Chem C*, **2011** *115*, 5044-5048.

[308] Dolgaev S I, Simakin A V, Voronov V V, Shafeev G A, Bozon-Verduraz F, *Appl Surf Sci*, **2002** *186*, 546-551.

[309] Sylvestre J P, Kabashin A V, Sacher E, Meunier M, *Appl Phys A*, **2005** *80*, 753-758.

[310] Noël S, Hermann J, Itina T, *Appl Surf Sci*, **2007** *253*, 6310-6315.

[311] Amendola V, Meneghetti M, *Phys Chem Chem Phys*, **2013** *15*, 3027-3046.

[312] Nichols W T, Sasaki T, Koshizaki N, *J Appl Phys*, **2006** *100*, 114912.

[313] Singh S C, Zeng H B, Guo C, Cai W, *Nanomaterials: Processing and Characterization with Lasers*, Wiley, **2012**, Weinheim.

[314] Peyre P, Fabbro R, *Opt Quant Electron*, **1995** *27*, 1213-1229.

[315] Gökce B, van't Zand D D, Menéndez-Manjón A, Barcikowski S, *Chem Phys Lett*, **2015** *626*, 96-101.

[316] Mafuné F, Kohno J-y, Takeda Y, Kondow T, *J Phys Chem B*, **2001** *105*, 9050-9056.

[317] Ibrahimkutty S, Wagener P, Menzel A, Plech A, Barcikowski S, *Appl Phys Lett*, **2012** *101*, 103104.

[318] Rehbock C, Merk V, Gamrad L, Streubel R, Barcikowski S, *Phys Chem Chem Phys*, **2013** *15*, 3057-3067.

[319] Mafuné F, Kohno J-y, Takeda Y, Kondow T, Sawabe H, *J Phys Chem B*, **2001** *105*, 5114-5120.

[320] Compagnini G, Scalisi A A, Puglisi O, *J Appl Phys*, **2003** *94*, 7874.

[321] Amendola V, Polizzi S, Meneghetti M, *J Phys Chem B*, **2006** *110*, 7232-7237.

[322] Zhang X, Servos M R, Liu J, *J Am Chem Soc*, **2012** *134*, 9910-9913.

[323] Liu Y, Zhang Y, Wang J, *CrystEngComm*, **2014** *16*, 5948-5967.

[324] Sylvestre J-P, Poulin S, Kabashin A V, Sacher E, Meunier M, Luong J H T, *J Phys Chem B*, **2004** *108*, 16864-16869.

[325] Sylvestre J-P, Kabashin A V, Sacher E, Meunier M, Luong J H T, *J Am Chem Soc*, **2004** *126*, 7176-7177.

[326] Li X, Lenhart J J, Walker H W, *Langmuir*, **2012** *28*, 1095-1104.

[327] Gebauer J S, Treuel L, *J Colloid Interface Sci*, **2011** *354*, 546-554.

[328] Cristoforetti G, Pitzalis E, Spiniello R, Ishak R, Giammanco F, Muniz-Miranda M, Caporali S, *Appl Surf Sci*, **2012** *258*, 3289-3297.

[329] Compagnini G, Scalisi A A, Puglisi O, Spinella C, *J Mater Res*, **2004** *19*, 2795-2798.

[330] Besner S b, Kabashin A V, Winnik F o M, Meunier M, *J Phys Chem C*, **2009** *113*, 9526-9531.

[331] Fong Y-Y, Gascooke J R, Visser B R, Metha G F, Buntine M A, *J Phys Chem C*, **2010** *114*, 15931-15940.

[332] Kabashin A V, Meunier M, Kingston C, Luong J H T, *J Phys Chem B*, **2003** *107*, 4527-4531.

[333] Marzun G, Nakamura J, Zhang X, Barcikowski S, Wagener P, *Appl Surf Sci.* 2015 348, 75-84.

[334] Hofmeister F, *Naunyn Schmiedebergs Arch Exp Pathol Pharmakol*, **1888** *24*, 247-260.

[335] Gamrad L, Rehbock C, Krawinkel J, Tumursukh B, Heisterkamp A, Barcikowski S, *J Phys Chem C*, **2014** 118, 10302-10313

[336] Kamat P V, Flumiani M, Hartland G V, *J Phys Chem B*, **1998** *102*, 3123-3128.

[337] Werner D, Furube A, Okamoto T, Hashimoto S, *J Phys Chem C*, **2011** *115*, 8503-8512.

[338] Procházka M, Štěpánek J, Vlčková B, Srnová I, Malý P, *J Mol Struct*, **1997** *410–411*, 213-216.

[339] Besner S, Kabashin A V, Meunier M, *Appl Phys Lett*, **2006** *89*, 233122.

[340] Amendola V, Meneghetti M, *J Mater Chem*, **2007** *17*, 4705-4710.

[341] Giusti A, Giorgetti E, Laza S, Marsili P, Giammanco F, *J Phys Chem C*, **2007** *111*, 14984-14991.

[342] Giorgetti E, Giammanco F, Marsili P, Giusti A, *J Phys Chem C*, **2011** *115*, 5011-5020.

[343] Giammanco F, Giorgetti E, Marsili P, Giusti A, *J Phys Chem C*, **2010** *114*, 3354-3363.

[344] Menéndez-Manjón A, Barcikowski S, *Appl Surf Sci*, **2011** *257*, 4285-4290.

[345] Maciulevičius M, Vinčiūnas A, Brikas M, Butsen A, Tarasenka N, Tarasenko N, Račiukaitis G, *Appl Phys A*, **2013** *111*, 289-295.

[346] Kim K K, Kwon H J, Shin S K, Song J K, Park S M, *Chem Phys Lett*, **2013** *588*, 167-173.

[347] Riabinina D, Chaker M, Margot J, *Nanotechnology*, **2012** *23*, 135603.

[348] Sobhan M A, Withford M J, Goldys E M, *Langmuir*, **2009** *26*, 3156-3159.

[349] Sobhan M A, Ams M, Withford M J, Goldys E M, *J Nanopart Res*, **2010** *12*, 2831-2842.

[350] Menéndez-Manjón A, Chichkov B N, Barcikowski S, *J Phys Chem C*, **2010** *114*, 2499-2504.

[351] Menéndez-Manjón A, Barcikowski S, Shafeev G A, Mazhukin V I, Chichkov B N, *Laser Part Beams*, **2010** *28*, 45.

[352] Ulman A, *Chem Rev*, **1996** *96*, 1533-1554.

[353] Mafuné F, Kohno J-y, Takeda Y, Kondow T, *J Phys Chem B*, **2002** *106*, 8555-8561.

[354] Sajti C L, Petersen S, Menéndez-Manjón A, Barcikowski S, *Appl Phys A*, **2010** *101*, 259-264.

[355] Takeda Y, Kondow T, Mafune F, *Nucleos Nucleot Nucl*, **2005** *24*, 1215-1225.

[356] Takeda Y, Kondow T, Mafune F, *J Phys Chem B*, **2006** *110*, 2393-2397.

[357] Takeda Y, Kondow T, Mafune F, *Phys Chem Chem Phys*, **2011** *13*, 586-592.

[358] Mutisya S, Franzel L, Barnstein B O, Faber T W, Ryan J J, Bertino M F, *Appl Surf Sci*, **2013** *264*, 27-30.

[359] Petersen S, Jakobi J, Barcikowski S, *Appl Surf Sci*, **2009** *255*, 5435-5438.

[360] Barcikowski S, Menéndez-Manjón A, Chichkov B, Brikas M, Račiukaitis G, *Appl Phys Lett*, **2007** *91*, 083113.

[361] Sajti C L s, Barchanski A, Wagener P, Klein S, Barcikowski S, *J Phys Chem C*, **2011** *115*, 5094-5101.

[362] Szymanski M S, Porter R A, *J Immunol Methods*, **2013** *387*, 262-269.

[363] Yan Z, Wang J, *Sci Rep*, **2012** *2*.

[364] Arruebo M, Valladares M, González-Fernández Á, *J Nanomaterials*, **2009** *2009*, 1-24.

[365] Loo C, Hirsch L, Lee M-H, Chang E, West J, Halas N, Drezek R, *Opt Lett*, **2005** *30*, 1012-1014.

[366] Pathak S, Davidson M C, Silva G A, *Nano Lett*, **2007** *7*, 1839-1845.

[367] Docter D, Distler U, Storck W, Kuharev J, Wünsch D, Hahlbrock A, Knauer S K, Tenzer S, Stauber R H, *Nat Protoc*, **2014** *9*, 2030-2044.

[368] Salvati A, Pitek A S, Monopoli M P, Prapainop K, Bombelli F B, Hristov D R, Kelly P M, Aberg C, Mahon E, Dawson K A, *Nat Nano*, **2013** *8*, 137-143.

[369] Tenzer S, Docter D, Kuharev J, Musyanovych A, Fetz V, Hecht R, Schlenk F, Fischer D, Kiouptsi K, Reinhardt C, Landfester K, Schild H, Maskos M, Knauer S K, Stauber R H, *Nat Nano*, **2013** *8*, 772-781.

[370] Monopoli M P, Aberg C, Salvati A, Dawson K A, *Nat Nanotechnol*, **2012** *7*, 779-786.

[371] He Q, Zhang J, Shi J, Zhu Z, Zhang L, Bu W, Guo L, Chen Y, *Biomaterials*, **2010** *31*, 1085-1092.

[372] Tam J M, Tam J O, Murthy A, Ingram D R, Ma L L, Travis K, Johnston K P, Sokolov K V, *ACS Nano*, **2010** *4*, 2178-2184.

[373] Faust C B, *Modern Chemical Techniques*, Royal Society of Chemistry, **1992**.

[374] Zhang P, Sham T K, *Phys Rev Lett*, **2003** *90*, 245502.

[375] Ghosh S K, Nath S, Kundu S, Esumi K, Pal T, *J Phys Chem B*, **2004** *108*, 13963-13971.

[376] Barchanski A, Funk D, Wittich O, Tegenkamp C, Chichkov B N, Sajti C L, *J Phys Chem C*, **2015** *119*, 9524-9533.

[377] Ball D W, *Field Guide to Spectroscopy*, Society of Photo Optical, **2006**, Washington.

[378] Barron A R, *Fourier Transform Infrared Spectroscopy of Metal Ligand Complexes*, Physical Methods in Chemistry and Nano Science. OpenStax CNX, **2014**.

[379] Barron A R, Payne C, *Raman and Surface-Enhanced Raman Spectroscopy*. OpenStax CNX, **2010**. Download for free at http://cnx.org/contents/73a1f8a3-32ba-4250-9277-ea639edafb80@1.

[380] Hollas J M, *Modern Spectroscopy*, John Wiley & Sons, **2004**, England.

[381] van der Heide P, *X-Ray Photoelectron Spectroscopy: An Introduction to Principles and Practices*, John Wiley & Sons, **2011**, USA.

[382] Barr T L, *Modern ESCA: The Principles and Practice of X-Ray Photoelectron Spectroscopy*, CRC Press, **1994**, Florida.

[383] Malvern Instruments Ltd., *Zetasizer Nano Series User Manual*, **2004**, Worcestershire.

[384] James A E, Driskell J D, *Analyst*, **2013** *138*, 1212-1218.

[385] Verwey E J W, *J Phys Colloid Chem*, **1947** *51*, 631-636.

[386] Stern O, *Z Elektrochem Angew P*, **1924** *30*, 508-516.

[387] Helmholtz H, *Ann Phys*, **1879** *243*, 337-382.

[388] Chapman D L, *Philos Mag Series 6*, **1913** *25*, 475-481.

[389] Gouy M, *J Phys Theor Appl*, **1910** *9*, 457-468.

[390] Pfeiffer C, Rehbock C, Hühn D, Carrillo-Carrion C, de Aberasturi DJ, Merk V, Barcikowski S, Parak W J, *J R Soc Interface*, **2014** *11*.

[391] Doane T L, Chuang C-H, Hill R J, Burda C, *Accounts Chem Res*, **2012** *45*, 317-326.

[392] von Smoluchowski M, *Ann Phys*, **1906** *326*, 756-780.

[393] Hückel E, *Phys Z*, **1924** *25*, 204-210.

[394] Henry D C, *P R Soc London*, **1931** 133, 106-129.

[395] Delgado A V, González-Caballero F, Hunter R J, Koopal L K, Lyklema J, *J Colloid Interface Sci*, **2007** *309*, 194-224.

[396] Slayter E M, Slayter H S, *Light and Electron Microscopy*, Cambridge University Press, **1992**, England.

[397] Bozzola JJ, Russell L D, *Electron Microscopy: Principles and Techniques for Biologists*, Jones and Bartlett, **1999**, USA.

[398] Amali A I, *High Angle Annular Dark Field Imaging in Scanning Transmission Electron Microscopy*, Arizona State University, **1995**.

[399] Fultz B, Howe J M, *Transmission Electron Microscopy and Diffractometry of Materials*, Springer Verlag **2012**, Berlin Heidelberg.

[400] Bell D, Garratt-Reed A, *Energy Dispersive X-Ray Analysis in the Electron Microscope*, BIOS Scientific, **2003**, Oxford.

[401] Herman B, *Fluorescence Microscopy*, BIOS Scientific, **1998**, Oxford.

[402] Mondal P P, Diaspro A, *Fundamentals of Fluorescence Microscopy: Exploring Life with Light*, Springer **2013**, The Netherlands.

[403] Moran-Mirabal J M, *Advanced-Microscopy Techniques for the Characterization of Cellulose Structure and Cellulose-Cellulase Interactions*, in: Cellulose - Fundamental Aspects. InTech, **2013**. Download for free at http://www.intechopen.com/books/cellulose-fundamental-aspects/advanced-microscopy-techniques-for-the-characterization-of-cellulose-structure-and-cellulose-cellula.

[404] Sheppard C, Shotton D, *Confocal Laser Scanning Microscopy*, BIOS Scientific, **1997**.

[405] Warren B E, *X-Ray Diffraction*, Dover Publications, **2012**, USA.

[406] Barron A R, Boyd T, *An Introduction to Single-Crystal X-Ray Crystallography*. OpenStax CNX, **2013**. Download for free at http://cnx.org/contents/ba27839d-5042-4a40-afcf-c0e6e39fb454@20.16.

[407] Yoshida Y, Langouche G, *Mössbauer Spectroscopy: Tutorial Book*, Springer, **2012**, Berlin Heidelberg.

[408] Guerra M, *57fe Mößbauerspektroskopie*. Helmholtz Zenrum Berlin, **2012**, Berlin. https://www.helmholtz-berlin.de/forschung/oe/ee/solare-brennstoffe/analytische-methoden/57 fe-moessbauerspektroskopie_de.html. Stand 16.10.2012.

[409] Dickson D P E, Berry F J, *Mössbauer Spectroscopy*, Cambridge University Press, **2005**, Cambridge.

[410] Watson J V, *Introduction to Flow Cytometry*, Cambridge University Press, **2004**, Cambridge.

[411] Tabll A, Ismail H, *The Use of Flow Cytometric DNA Ploidy Analysis of Liver Biopsies in Liver Cirrhosis and Hepatocellular Carcinoma*, in: Liver Biopsy. InTech, **2011**. Available from: http://www.intechopen.com/books/liver-biopsy/the-use-of-flow-cytometric-dna-ploidy-analysis-of-liver-biopsies-in-liver-cirrhosis-and-hepatocellul.

[412] Macey M G, *Flow Cytometry: Principles and Applications*, Humana Press, **2007**, New Jersey.

[413] Bonner W A, Hulett H R, Sweet R G, Herzenberg L A, *Rev Sci Instrum*, **1972** *43*, 404-409.

[414] Sehring C, *Bifunctionalisation of Laser-Generated Gold Nanoparticles for Biomedical Applications*. Fachhochschule Südwestfalen, Thesis, **2010**, Iserlohn.

[415] Meißner M, *Laserbasierte Generierung und Stabilisierung von Magnetischen Nanopartikel-Konjugaten für Biomedizinische Anwendungen*. Leibniz Universität Hannover, Thesis, **2013**, Hannover.

[416] Svedberg T, *Kolloid Z*, **1930** *51*, 10-24.

[417] Day E S, Bickford L R, Slater J H, Riggall N S, Drezek R A, West J L, *Int J Nanomedicine*, **2010** *5*, 445-454.

[418] Amendola V, Riello P, Polizzi S, Fiameni S, Innocenti C, Sangregorio C, Meneghetti M, *J Mater Chem*, **2011** *21*, 18665.

[419] Haiss W, Thanh N T K, Aveyard J, Fernig D G, *Anal Chem*, **2007** *79*, 4215-4221.

[420] Barchanski A, Hashimoto N, Petersen S, Sajti C, Barcikowski S, *Bioconj Chem*, **2012** *23*, 908-915.

[421] Hill H D, Millstone J E, Banholzer M J, Mirkin C A, *ACS Nano*, **2009** *3*, 418-424.

[422] Durán M, Willenbrock S, Barchanski A, Müller J-M V, Maiolini A, Soller J, Barcikowski S, Nolte I, Feige K, Murua Escobar H, *J Nanobiotech*, **2011** *9*, 1-11.

[423] Pirkmajer S, Chibalin A V, *Am J Physiol Cell Physiol*, **2011** 301, C272-C279.

[424] American Society for Testing and Materials (ASTM), *Standard Terminology Relating to Nanotechnology*, **2012**.

[425] Hackley V A, Ferraris C F, *The Use of Nomenclature in Dispersion Science and Technology. Nist Recommended Practice Guide*, in: National Technical Information Service (NIST), U.S. Department of Commerce, **2001**, Springfield.

[426] Riddick T M, Zeta-Meter Inc, *Control of Colloid Stability through Zeta Potential: With a Closing Chapter on Its Relationship to Cardiovascular Disease*, Published for Zeta-Meter, Inc., by Livingston Pub. Co., **1968**.

[427] Müller M, *Polyelectrolyte Complexes in the Dispersed and Solid State In: Application Aspects*, Springer **2013**, Berlin Heidelberg.

[428] Barchanski A, Funk D, Wittich O, Tegenkamp C, Chichkov B N, Sajti C L, *J Phys Chem C*, **2014**, 119, 9524-9533.

[429] Wright A K, Thompson M R, *Biophys J*, **1975** 15, 137-141.

[430] Jain P K, Qian W, El-Sayed M A, *J Am Chem Soc*, **2006** 128, 2426-2433.

[431] Song J Y, Jang H-K, Kim B S, *Process Biochem*, **2009** 44, 1133-1138.

[432] Babu P J, Sharma P, Kalita M, Bora U, *Front Mater Sci*, **2011** 5, 379-387.

[433] Seuvre A-M, Mathlouthi M, *Carbohyd Res*, **1987** 169, 83-103.

[434] Mady M M, Mohammed W A, El-Guendy N M, El-Sayed A A, *Int J Phys Sci*, **2011** 6, 7328-7334.

[435] Hackl E V, Kornilova S V, Blagoi Y P, *Int J Biol Macromol*, **2005** 35, 175-191.

[436] Grdadolnik J, Maréchal Y, *Biopolymers*, **2001** 62, 40-53.

[437] Tsai D-H, DelRio F W, Keene A M, Tyner K M, MacCuspie R I, Cho TJ, Zachariah M R, Hackley V A, *Langmuir*, **2011** 27, 2464-2477.

[438] Roach P, Farrar D, Perry C C, *J Am Chem Soc*, **2005** 127, 8168-8173.

[439] Kaiden K, Matsui T, Tanaka S, *Appl Spectroscopy*, **1987** 41, 180-184.

[440] Servagent-Noinville S, Revault M, Quiquampoix H, Baron M H, *J Colloid Interface Sci*, **2000** 221, 273-283.

[441] In S I W, *Water Security for the 21st Century - Innovative Approaches: Proceedings of the 10th Stockholm Water Symposium "Water Security for the 21st Century- Innovate Approaches"*, held in Stockholm, Sweden, 14-17th August 2000, IWA Pub., **2001**.

[442] Kim K S, Lee K H, Cho K, Park C E, *J Membrane Sci*, **2002** 199, 135-145.

[443] Jucker C, Clark M M, *J Membrane Sci*, **1994** 97, 37-52.

[444] Schafer A, *Natural Organics Removal Using Membranes: Principles, Performance, and Cost*, Taylor & Francis, **2001**, Florida.

[445] Ariga K, Hu X, Mandal S, Hill J P, *Nanoscale*, **2010** 2, 198-214.

[446] Sastry M, Swami A, Mandal S, Selvakannan P R, *J Mater Chem*, **2005** 15, 3161-3174.

[447] Zhang Y X, Zeng H C, *J Phys Chem C*, **2007** 111, 6970-6975.

[448] Pienpinijtham P, Han X X, Ekgasit S, Ozaki Y, *Phys Chem Chem Phys*, **2012** 14, 10132-10139.

[449] Liu S, Zhu T, Hu R, Liu Z, *Phys Chem Chem Phys*, **2002** 4, 6059-6062.

[450] Cho K-S, Talapin D V, Gaschler W, Murray C B, *J Am Chem Soc*, **2005** 127, 7140-7147.

[451] Patla I, Acharya S, Zeiri L, Israelachvili J, Efrima S, Golan Y, *Nano Lett*, **2007** 7, 1459-1462.

[452] Murray C B, Kagan C R, Bawendi M G, *Annu Rev Mater Sci*, **2000** 30, 545-610.

[453] Nikoobakht B, Wang Z L, El-Sayed M A, *J Phys Chem B*, **2000** 104, 8635-8640.

[454] Mortimer C E, Müller U,Beck J, *Chemie: Das Basiswissen Der Chemie*, Thieme, **2014**, Stuttgargt.

[455] Nanda K K, Maisels A, Kruis F E, Fissan H, Stappert S, *Phys Rev Lett*, **2003** 91, 106102.

[456] Nanda K K, *Pramana*, **2009** 72, 617-628.

[457] Peng Z, Walther T, Kleinermanns K, *J Phys Chem B*, **2005** 109, 15735-15740.

[458] Lau M, Haxhiaj I, Wagener P, Intartaglia R, Brandi F, Nakamura J, Barcikowski S, *Chem Phys Lett*, **2014** 610–611, 256-260.

[459] Eckstein H, Kreibig U, Z Phys, **1993** 26, 239-241.

[460] Petersen S, Soller J, Wagner S, Richter A, Bullerdiek J, Nolte I, Barcikowski S, Escobar H, J Nanobiotech, **2009** 7, 1-6.

[461] Moulder J F, Chastain J, Handbook of X-Ray Photoelectron Spectroscopy: A Reference Book of Standard Spectra for Identification and Interpretation of XPS Data, Physical Electronics Division, Perkin-Elmer Corporation, **1992**.

[462] Stöhr J, Siegmann H C, Magnetism: From Fundamentals to Nanoscale Dynamics, Springer, **2007**, Berlin.

[463] Fong Y Y, Gascooke J R, Visser B R, Harris H H, Cowie B C, Thomsen L, Metha G F, Buntine M A, Langmuir, **2013** 29, 12452-12462.

[464] Southam G, Fyfe W, Beveridge T, Miner Metall Proc, **2000** 17, 129-132.

[465] Alberts B, Johnson A, Lewis J, Walter P, Raff M, Roberts K, Molecular Biology of the Cell, 4th Edition, International Student Edition, Garland Science, **2002**, New York.

[466] Campbell N A, Reece J B, Markl J, Biologie, Spektrum, Akad. Verlag, **1997**, Heidelberg.

[467] Lund M, Vrbka L, Jungwirth P, J Am Chem Soc, **2008** 130, 11582-11583.

[468] Thompson D W, Collins I R, J Colloid Interface Sci, **1992** 152, 197-204.

[469] Gamrad L, Struktur-Funktions-Beziehung Lasergenerierter Nanopartikel-Peptid-Konjugate, Technical Chemistry I, University Dusiburg-Essen, Thesis, **2012**, Essen.

[470] Janeway C A,Travers P, Immunologie, Spektrum Akad. Verlag, **1995**, Heidelberg.

[471] Lopez-Acevedo O, Akola J, Whetten R L, Grönbeck H, Häkkinen H, J Phys Chem C, **2009** 113, 5035-5038.

[472] Wang T, Hu X, Dong S, Chem Commun, **2008**, 4625-4627.

[473] Wang F, He C, Han M-Y, Wu J H, Xu G Q, Chem Commun, **2012** 48, 6136-6138.

[474] Barchanski A, Lasergenerierte Nanopartikel-Biokonjugate zur Penetration von Tumorzellen, Institut für Zell- und Molekularpathologie, Medizinische Hochschule Hannover, Thesis, **2009**, Hannover.

[475] Steel A B, Levicky R L, Herne T M, Tarlov M J, Biophys J, **2000** 79, 975-981.

[476] Herne T M, Tarlov M J, J Am Chem Soc, **1997** 119, 8916-8920.

[477] Levicky R, Herne T M, Tarlov M J, Satija S K, J Am Chem Soc, **1998** 120, 9787-9792.

[478] Park S, Brown K A, Hamad-Schifferli K, Nano Lett, **2004** 4, 1925-1929.

[479] Parak W J, Pellegrino T, Micheel C M, Gerion D, Williams S C, Alivisatos A P, Nano Lett, **2002** 3, 33-36.

[480] Zhang Y, Zhou H, Ou-Yang Z C, Biophys J, **2001** 81, 1133-1143.

[481] Shlyakhtenko L S, Gall A A, Weimer J J, Hawn D D, Lyubchenko Y L, Biophys J, **1999** 77, 568-576.

[482] Klein J S, Bjorkman P J, PLoS Pathog, **2010** 6, e1000908.

[483] RCSB, Protein Data Bank, www.rcsb.org, August 4th 2014.

[484] Johnson L A, Welch G R, Theriogenology, **1999** 52, 1323-1341.

[485] Levinson G, Keyvanfar K, Wu J C, Fugger E F, Fields R A, Harton G L, Palmer F T, Sisson M E, Starr K M, Dennison-Lagos L, Hum Reprod, **1995** 10, 979-982.

[486] Parks J E, Lynch D V, Cryobiology, **1992** 29, 255-266.

[487] Gadella B M, Lopes-Cardozo M, van Golde L M, Colenbrander B, Gadella T W, Jr., J Cell Sci, **1995** 108 (Pt 3), 935-946.

[488] Leahy T, Gadella B M, Reproduction, **2011** 142, 759-778.

[489] Toshimori K, *J Electron Microsc*, **2011** *60*, S31-S42.

[490] Tsai P-S, Garcia-Gil N, van Haeften T, Gadella B M, *PLoS One*, **2010** *5*, e11204.

[491] Ehrenwald E, Parks J E, Foote R H, *Gamete Res*, **1988** *20*, 145-157.

[492] Travis A J, Kopf G S, *J Clin Invest*, **2002** *110*, 731-736.

[493] Shadan S, James P S, Howes E A, Jones R, *Biol Reprod*, **2004** *71*, 253-265.

[494] Pawar K, Kaul G, *Toxicol Ind Health*, **2012** *30*, 520-533.

[495] Welsher K, Yang H, *Nat Nano*, **2014** *9*, 198-203.

[496] Herce H D, Garcia A E, Litt J, Kane RS, Martin P, Enrique N, Rebolledo A, Milesi V, *Biophys J*, **2009** *97*, 1917-1925.

[497] Feugang J M, Youngblood R C, Greene J M, Fahad A S, Monroe W A, Willard S T, Ryan P L, *J Nanobiotechnology*, **2012** *10*, 45.

[498] Makhluf S B-D, Qasem R, Rubinstein S, Gedanken A, Breitbart H, *Langmuir*, **2006** *22*, 9480-9482.

[499] Makhluf S B, Abu-Mukh R, Rubinstein S, Breitbart H, Gedanken A, *Small*, **2008** *4*, 1453-1458.

[500] Park J-H, Gu L, von Maltzahn G, Ruoslahti E, Bhatia S N, Sailor M J, *Nat Mater*, **2009** *8*, 331-336.

[501] Medintz I L, Uyeda H T, Goldman E R, Mattoussi H, *Nat Mater*, **2005** *4*, 435-446.

[502] Kondoh M, Araragi S, Sato K, Higashimoto M, Takiguchi M, Sato M, *Toxicology*, **2002** *170*, 111-117.

[503] Derfus A M, Chan W C W, Bhatia S N, *Nano Lett*, **2003** *4*, 11-18.

[504] Erogbogbo F, Yong K, Roy I, Hu R, Law W-C, Zhao W, Ding H, Wu F, Kumar R, Swihart M, Prasad P N, *ACS Nano*, **2010** *5*, 413-423.

[505] Erogbogbo F, Tien C A, Chang CW, Yong KT, Law W C, Ding H, Roy I, Swihart M T, Prasad P, *Bioconjug Chem*, **2011** *22*, 1081-1088.

[506] Li Z F, Ruckenstein E, *Nano Lett*, **2004** *4*, 1463-1467.

[507] Rosso-Vasic M, Spruijt E, Popovic Z, Overgaag K, van Lagen B, Grandidier B, Vanmaekelbergh D, Dominguez-Gutierrez D, De Cola L, Zuilhof H, *J Mater Chem*, **2009** *19*, 5926-5933.

[508] Zhang X, Neiner D, Wang S, Louie A Y, Kauzlarich S M, *Nanotechnology*, **2007** *18*, 095601.

[509] Warner J H, Rubinsztein-Dunlop H, Tilley R D, *J Phys Chem B*, **2005** *109*, 19064-19067.

[510] Kůsová K, Cibulka O, Dohnalová K, Pelant I, Valenta J, Fučíková A, Žídek K, Lang J, Englich J, Matějka P, Štěpánek P, Bakardjieva S, *ACS Nano*, **2010** *4*, 4495-4504.

[511] Belomoin G, Therrien J, Smith A, Rao S, Twesten R, Chaieb S, Nayfeh MH, Wagner L, Mitas L, *Appl Phys Lett*, **2002** *80*, 841-843.

[512] Li X, He Y, Swihart M T, *Langmuir*, **2004** *20*, 4720-4727.

[513] Mangolini L, Thimsen E, Kortshagen U, *Nano Lett*, **2005** *5*, 655-659.

[514] Intartaglia R, Bagga K, Brandi F, Das G, Genovese A, Di Fabrizio E, Diaspro A, *J Phys Chem C*, **2011** *115*, 5102-5107.

[515] Abderrafi K, Garcia Calzada R l, Gongalsky M B, Suarez I, Abarques R, Chirvony V S, Timoshenko V Y, Ibanez R, Martinez-Pastor J P, *J Phys Chem C*, **2011** *115*, 5147-5151.

[516] Kuzmin P G, Shafeev G A, Bukin V V, Garnov S V, Farcau C, Carles R, Warot-Fontrose B, Guieu V, Viau G, *J Phys Chem C*, **2010** *114*, 15266-15273.

[517] Semaltianos N G, Logothetidis S, Perrie W, Romani S, Potter R J, Edwardson S P, French P, Sharp M, Dearden G, Watkins K G, *J Nanopart Res*, **2010** *12*, 573-580.

[518] Švrček V, Mariotti D, Nagai T, Shibata Y, Turkevych I, Kondo M, *J Phys Chem C*, **2011** *115*, 5084-5093.

[519] Choi J, Wang N S, Reipa V, *Bioconj Chem*, **2008** *19*, 680-685.

[520] Wang L, Reipa V, Blasic J, *Bioconj Chem*, **2004** *15*, 409-412.

[521] Seldon T A, Hughes K E, Munster D J, Chin D Y, Jones M L, *J Biomol Tech, 2011 22, 50-52*.

[522] Stutz H, *Electrophoresis*, **2009** *30*, 2032-2061.

[523] Coen M C, Lehmann R, Groning P, Bielmann M, Galli C, Schlapbach L, *J Colloid Interface Sci*, **2001** *233*, 180-189.

[524] Wang Z, Jin G, *J Biochem Biophys Met*, **2003** *57*, 203-211.

[525] Ikeda T, Hata Y, Ninomiya K, Ikura Y, Takeguchi K, Aoyagi S, Hirota R, Kuroda A, *Anal Biochem*, **2009** *385*, 132-137.

[526] Intartaglia R, Barchanski A, Bagga K, Genovese A, Das G, Wagener P, Di Fabrizio E, Diaspro A, Brandi F, Barcikowski S, *Nanoscale*, **2012** *4*, 1271-1274.

[527] Bagga K, Barchanski A, Intartaglia R, Dante S, Marotta R, Diaspro A, Sajti C L, Brandi F, *Laser Phys Lett*, **2013** *10*, 065603.

[528] Mariotto G, Das G, Quaranta A, Della Mea G, Corni F, Tonini R, *J Appl Phys*, **2005** *97 113502*.

[529] Daldosso N, Das G, Larcheri S, Mariotto G, Dalba G, Pavesi L, Irrera A, Priolo F, Iacona F, Rocca F, *J Appl Phys*, **2007** *101*, 113510.

[530] Ruiz-Chica A J, Medina M A, Sanchez-Jimenez F, Ramirez F J, *Nucl Acids Res*, **2004** *32*, 579-589.

[531] Das G, Mariotto G, Quaranta A, *J Electroch Soc*, **2006** *153*, F46-F51.

[532] Gu H-w, Xie P, Shen Z-r, Shen D-y, Zhang J-m,Zhang R-b, *Chinese J Polym Scir*, **2004** *22*, 463-468.

[533] Knopp D, Tang D, Niessner R, *Anal Chim Acta*, **2009** *647*, 14-30.

[534] Qin D, He X, Wang K, Zhao X J, Tan W, Chen J, *J Biomed Biotechnol*, **2007** *2007*, 89364.

[535] Cornell R M, Schwertmann U, *The Iron Oxides: Structure, Properties, Reactions, Occurrences and Uses*, Wiley, **2006**, Stuttgart.

[536] Binnewies M, Jäckel M, Willner H, Rayner-Canham G, *Allgemeine Und Anorganische Chemie*, Spektrum Akad. Verlag, **2010**, Heidelberg.

[537] Derfus A M, von Maltzahn G, Harris T J, Duza T, Vecchio K S, Ruoslahti E, Bhatia S N, *Adv Mater*, **2007** *19*, 3932-3936.

[538] Regmi R, Bhattarai S R, Sudakar C, Wani A S, Cunningham R, Vaishnava P P, Naik R, Oupicky D, Lawes G, *J Mater Chem*, **2010** *20*, 6158-6163.

[539] Xu H, Aguilar Z P, Yang L, Kuang M, Duan H, Xiong Y, Wei H, Wang A, *Biomaterials*, **2011** *32*, 9758-9765.

[540] Chen D, Ni S, Chen Z, *China Particuology*, **2007** *5*, 357-358.

[541] Zhang Z, Wei B Q, Ajayan P M, *Appl Phys Lett*, **2001** *79*, 4207-4209.

[542] Maity D, Agrawal D C, *J Magn Magn Mater*, **2007** *308*, 46-55.

[543] Rudzka K, Delgado Á V, Viota J L, *Mol Pharm*, **2012** *9*, 2017-2028.

[544] Kim D K, Zhang Y, Voit W, Rao K V, Muhammed M, *J Magn Magne Mater*, **2001** *225*, 30-36.

[545] Amendola V, Riello P, Meneghetti M, *J Phys Chem C*, **2011** *115*, 5140-5146.

[546] Amendola V, Meneghetti M, Granozzi G, Agnoli S, Polizzi S, Riello P, Boscaini A, Anselmi C, Fracasso G, Colombatti M, Innocenti C, Gatteschi D, Sangregorio C, *J Mater Chem*, **2011** *21*, 3803.

[547] Zhang C-L, Li S, Wu T-H, Peng S-Y, *Mater Chem Phys*, **1999** *58*, 139-145.

[548] Huang P, Kong Y, Li Z, Gao F, Cui D, *Nanoscale Res Lett*, **2010** *5*, 949-956.

[549] Merkle M, *Laser-Generierte und Bio-Konjugierte Eisenoxidnanopartikel zur Spezifischen Zellmarkierung*. Leibniz Univeristät Hannover, Thesis, **2013**, Hannover.

[550] Bartczak D, Sanchez-Elsner T, Louafi F, Millar T M, Kanaras A G, *Small*, **2011** *7*, 388-394.

[551] Bakshi M S, Sachar S, Kaur G, Bhandari P, Kaur G, Biesinger M C, Possmayer F, Petersen NO, *Cryst Growth Des*, **2008** *8*, 1713-1719.

[552] Pothukuchi S, Yi L, Wong C P, *Shape Controlled Synthesis of Nanoparticles and Their Incorporation into Polymers*, 54th Electronic Components and Technology Conference 2004 Proceedings. **2004**, pp. 1965-1967 Vol.1962.

[553] Pustovalov V K, *Chem Phys*, **2005** *308*, 103-108.

[554] Pustovalov V K, Babenko V A, *Laser Phys Lett*, **2004** *1*, 516-520.

[555] Yan Z, Bao R, Chrisey D B, *Phys Chem Chem Phys*, **2013** *15*, 3052-3056.

[556] Mazur A I, Marcsisin E J, Bird B, Miljkovic M, Diem M, *Anal Chem*, **2012** *84*, 1259-1266.

[557] López-Banet L, Santana M D, García G, Piernas M J, García L, Pérez J, Calderón-Casado A, Barandika G, *Eur J Inorg Chem*, **2013** *2013*, 4280-4290.

[558] Berg J M, Tymoczko J L, Stryer L, Häcker B, Held A, Lange C, Mahlke K, Maxam G, Seidler L, Zellerhoff N, *Stryer Biochemie*, Spektrum Akad. Verlag, **2009**, Heidelberg.

[559] Bagga K, *Functional Silicon Nanoparticles for Biomedical Applications: Laser Synthesis, Biofuntionalization and Productivity Studies*, Nanophysics. University of Genova, Thesis, **2013**, Genova.

[560] Albanese A, Chan W C W, *ACS Nano*, **2011** *5*, 5478-5489.

[561] Brandenberger C, Rothen-Rutishauser B, Muhlfeld C, Schmid O, Ferron G A, Maier K L, Gehr P, Lenz A G, *Toxicol Appl Pharmacol*, **2010** *242*, 56-65.

[562] Coulter J A, Jain S, Butterworth K T, Taggart L E, Dickson G R, McMahon S J, Hyland W B, Muir M F, Trainor C, Hounsell A R, O'Sullivan J M, Schettino G, Currell F J, Hirst D G, Prise K M, *Int J Nanomedicine*, **2012** *7*, 2673-2685.

[563] Fan J H, Hung W I, Li W T, Yeh J M, *Biocompatibility Study of Gold Nanoparticles to Human Cells*, in: 13th International Conference on Biomedical Engineering. 2009, Springer Berlin Heidelberg.

[564] Jan E, Byrne S J, Cuddihy M, Davies A M, Volkov Y, Gun'ko Y K, Kotov N A, *ACS Nano*, **2008** *2*, 928-938.

[565] Khan J A, Pillai B, Das T K, Singh Y, Maiti S, *Chembiochem*, **2007** *8*, 1237-1240.

[566] Lee J, Lilly G D, Doty R C, Podsiadlo P, Kotov N A, *Small*, **2009** *5*, 1213-1221.

[567] Li J J, Zou L, Hartono D, Ong C N, Bay B H, Lanry Yung L Y, *Adv Mater*, **2008** *20*, 138-142.

[568] Mironava T, Hadjiargyrou M, Simon M, Jurukovski V, Rafailovich M H, *Nanotoxicology*, **2010** *4*, 120-137.

[569] Murawala P, Phadnis S M, Bhonde R R, Prasad B L V, *Colloids Surf B*, **2009** *73*, 224-228.

[570] Pernodet N, Fang X, Sun Y, Bakhtina A, Ramakrishnan A, Sokolov J, Ulman A, Rafailovich M, *Small*, **2006** *2*, 766-773.

[571] Pfaller T, Puntes V, Casals E, Duschl A, Oostingh G J, *Nanotoxicology*, **2009** *3*, 46-59.

[572] Ponti J, Colognato R, Franchini F, Gioria S, Simonelli F, Abbas K, Uboldi C, James Kirkpatrick C, Holzwarth U, Rossi F, *Nanotoxicology*, **2009** *3*, 296-306.

[573] Qu Y, Lü X, *Biomedical Materials*, **2009** *4*, 025007.

[574] Shukla R, Bansal V, Chaudhary M, Basu A, Bhonde R R, Sastry M, *Langmuir*, **2005** *21*, 10644-10654.

[575] Simon-Deckers A, Brun E, Gouget B, Carrière M, Sicard-Roselli C, *Gold Bulletin*, **2008** *41*, 187-194.

[576] Singh S, D'Britto V, Prabhune A A, Ramana C V, Dhawan A, Prasad B L V, *New J Chem*, **2010** *34*, 294-301.

[577] Sun L, Liu D, Wang Z, *Langmuir*, **2008** *24*, 10293-10297.

[578] Tsoli M, Kuhn H, Brandau W, Esche H, Schmid G, *Small*, **2005** *1*, 841-844.

[579] Vijayakumar S, Ganesan S, *Toxicol Environ Chem*, **2013** *95*, 277-287.

[580] Villiers C, Freitas H, Couderc R, Villiers M-B, Marche P, *J Nanopart Res*, **2010** *12*, 55-60.

[581] Wei X-L, Mo Z-H, Li B, Wei J-M, *Colloids Surf B*, **2007** *59*, 100-104.

[582] Yen H-J, Hsu S-h, Tsai C-L, *Small*, **2009** *5*, 1553-1561.

[583] Cho W-S, Kim S, Han B S, Son W C, Jeong J, *Toxicology Lett*, **2009** *191*, 96-102.

[584] Alkilany A M, Nagaria P K, Hexel C R, Shaw T J, Murphy C J, Wyatt M D, *Small*, **2009** *5*, 701-708.

[585] Zhang W, Ji Y, Wu X, Xu H, *ACS Appl Mater Interfaces*, **2013** *5*, 9856-9865.

[586] Hillyer J F, Albrecht R M, *J Pharmaceut Sci*, **2001** *90*, 1927-1936.

[587] Sadauskas E, Jacobsen N R, Danscher G, Stoltenberg M, Vogel U, Larsen A, Kreyling W, Wallin H, *Chem Cent J*, **2009** *3*, 16.

[588] Souza G R, Molina J R, Raphael R M, Ozawa M G, Stark D J, Levin C S, Bronk L F, Ananta J S, Mandelin J, Georgescu M-M, Bankson J A, Gelovani J G, Killian T C, Arap W, Pasqualini R, *Nat Nano*, **2010** *5*, 291-296.

[589] Amendola V, Rizzi G A, Polizzi S, Meneghetti M, *J Phys Chem B*, **2005** *109*, 23125-23128.

[590] Barchanski A, Taylor U, Klein S, Petersen S, Rath D, Barcikowski S, *Reprod Domest Anim*, **2011** *46*, 42-52.

[591] Barcikowski S, Jakobi J, Petersen S, Hahn A, Baersch N, Chichkov N B, *Adding Functionality to Metal Nanoparticles During Femtosecond Laser Ablation in Liquids*, in: ICALEO 2007: 26th International Congress on Applications of Lasers & Electro-OpticsOctober 29 - November 1, 2007, Orlando, FL, USA. Laser Inst. of America, **2007**, Orlando, Florida.

[592] Bozon-Verduraz F, Brayner R, Voronov V V, Kirichenko N A, Simakin A V, Shafeev G A, *Quant Electron*, **2003** *33*, 714-720.

[593] Burzhuev S, Dâna A, Ortaç B, *Sensors Actuat A: Phys*, **2013** *203*, 131-136.

[594] Fong Y-Y, Gascooke J R, Metha G F, Buntine M A, *Aust J Chem*, **2012** *65*, 97.

[595] Giorgetti E, Giusti A, Laza S C, Marsili P, Giammanco F, *Phys Stat Sol(a)*, **2007** *204*, 1693-1698.

[596] Giorgetti E, Giusti A, Giammanco F, Marsili P, Laza S, *Molecules*, **2009** *14*, 3731-3753.

[597] Giorgetti E, Muniz-Miranda M, Marsili P, Scarpellini D, Giammanco F, *J Nanopart Res*, **2012** *14*, 1-13.

[598] Haustrup N, O'Connor G M, *Appl Phys Lett*, **2012** *101*, 263107.

[599] Kalyva M, Bertoni G, Milionis A, Cingolani R, Athanassiou A, *Microsc Res Tech*, **2010** *73*, 937-943.

[600] Maciulevičius M, Vinčiūnas A, Brikas M, Butsen A, Tarasenka N, Tarasenko N, Račiukaitis G, *Phys Procedia*, **2013** *41*, 531-538.

[601] Mafuné F, Kohno J-y, Takeda Y, Kondow T, *J Phys Chem B*, **2002** *106*, 7575-7577.

[602] Mafuné F, Kohno J-y, Takeda Y, Kondow T, *J Phys Chem B*, **2003** *107*, 12589-12596.

[603] Menéndez-Manjón A, Wagener P, Barcikowski S, *J Phys Chem C*, **2011** *115*, 5108-5114.

[604] Menendez-Manjon A, Baersch N, Barcikowski S, *Mobility of Nanoparticles Generated by Femtosecond Laser Ablation in Liquids and Its Application to Surface Patterning*, LPM 2008: 9th International Symposium onLaser Precision Microfabrication, June 16-20th 2008, Québec City, Québec, Canada.

[605] Nikov R G, Nikolov A S, Nedyalkov N N, Dimitrov I G, Atanasov P A, Alexandrov M T, *Appl Surf Sci*, **2012** *258*, 9318-9322.

[606] Petersen S, *In Situ Bioconjugation of Laser-Generated Gold Nanoparticles for the Rapid Design of Nanomarkers*. Gottfried Wilhelm Leibniz Universität Hannover, Thesis, **2009**, Hannover.

[607] Qian W, Murakami M, Ichikawa Y, Che Y, *J Phys Chem C*, **2011** *115*, 23293-23298.

[608] Saitow K-i, Yamamura T, Minami T, *J Phys Chem C*, **2008** *112*, 18340-18349.

[609] Saitow K-i, Okamoto Y, Yano Y F, *J Phys Chem C*, **2012** *116*, 17252-17258.

[610] Šišková K, Pfleger J, Procházka M, *Appl Surf Sci*, **2010** *256*, 2979-2987.

[611] Šmejkal P, Pfleger J, Vlčková B, *Appl Phys A*, **2010** *101*, 37-40.

[612] Van Overschelde O, Dervaux J, Yonge L, Thiry D, Snyders R, *Laser Phys*, **2013** *23*, 055901.

7. Annex

7.1. Supporting Information

Supporting chapter

Nanoparticle shape
Partly unpublished data
Petersen et al. 2012 [XIII], *Cooperation LZH-FLI*

Generally, each system aims to find an energetically favorable state. In terms of pure fluids this implies a reduction of surface area due to surface tension which results in a spheroid/droplet shape. However, if specific capping agents in a defined concentration are applied, the anisotropic growth of certain facets can be controlled.[551]
Thus, with controlled precipitation conditions, the shape of NPs can be modulated during CRM, yielding nanostructures such as nanocubes, nanorods or nanostars for the design of complex nanodevices.[552]

During ps-PLAL fabrication of ligand-free AuNPs in *Milli-Q* water, some deformed structures could also appear together with the spherical particles, especially if fluences > 0.5 J cm^{-2} are applied (**Figure 7.1**). The unshaped structures do not feature a consistent form apart from being elongated with a neck-like site and are well-distinguishable from the spheres.

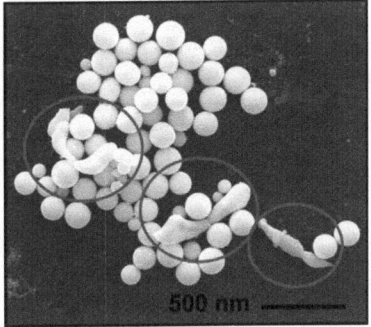

Figure 7.1. PLAL-fabricated AuNPs with anisotropic shape. Scanning electron micrographs of ligand-free gold nanoparticles fabricated with a fluence of 2 J cm^{-2}. Laser-sintered/welded structures are marked by red circles.

In the literature it is understood, that laser-ablated NPs are usually polycrystalline, which means that nuclei coalescence must take place when the interface energy is minimized by lattice rearrangements, e.g. during melting.[311]

In fact, during laser ablation with the TruMicro system, local lattice (T_{max}) beyond the critical point of water (647 K) are developed and the melting temperature (T_m, particle-size dependent, e.g. 1067 K for a 38 nm AuNP) can also be reached for fluences > 0.5 J cm^{-2} (**Table 7.1**), as determined by **eq 7.1**[553]:

$$T_{max} = T_\infty + \frac{3I_0\, t_p\, K_{abs}}{4\rho_0\, c_0\, r_0}$$

eq 7.1

T_{max} = maximal local lattice temperature, T_∞ = initial temperature equal to the surrounding medium temperature, I_0 = illumination power, t_p = pulse duration, K_{abs} = absorption efficiency factor of AuNPs approximated by Pustovalov et al.[554], ρ_0 = density of the particle material, c_0 = specific heat capacity of the particle material, r_0 = radius of the NP

Table 7.1. Calculated T_{max} of ligand-free AuNPs with respect to varied fluences. Molten particles are marked by grey coloration. Adapted with permission from Barchanski et al., copyright 2015 by the American Chemical Society.[428]

Target Position (mm) / Fluence (J cm^{-2})	T_{max}/K
-2 / 0.17	682
-1 / 0.24	937
0 / 0.5	1543
+1 / 1.51	3918
re-irradiation @ focal point in water	2.18x10^6

Thus, if the AuNPs are not quickly removed from the process zone with liquid agitation, the next laser pulse may provoke local heating and melting which leads to temporary deformation and size reduction of NPs and the formation of molten globules which in turn coalesce with gold ions and other fragments to form novel NPs (**Figure 7.2a**).

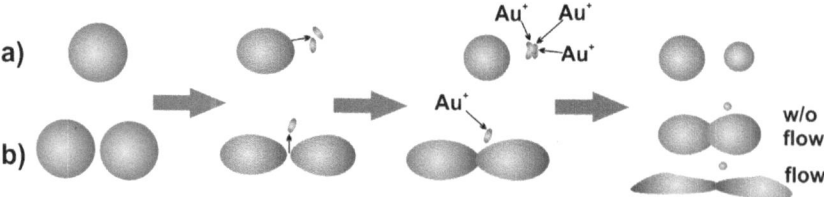

Figure 7.2. Schematic model of AuNP shape modulation. a) Shape modulation mechanism of a singular AuNP. **b)** Shape modulation mechanism of two close-distanced AuNPs. Both mechanisms are presented for laser-induced melting, resulting in size reduction and formation of new NPs **(a)** or particle welding/sintering **(b)**. w/o = without.

However, if two close-distanced AuNPs are molten, they may coalesce together by neck-like fusion yielding a flow-dependent, dumbbell or an unspecific elongated shape deformation (**Figure 7.2b**). In contrast to the cold, chemical welding described in **Chapter 4.1.2.** at which joining takes place without fusion at the interface, the laser-induced re-processing may also be termed *laser welding* or *sintering*[555] and is accomplished during the fabrication process.

In this context, Yan et al. reported on the generation of nanosheets after the ablation of a silver target in *Milli-Q* water with an UV excimer laser due to heating and surface sintering effects of the primary nanocrystals.[555] Because the solution was not agitated, the sheet formation was determined by the surface free energies of Ag crystals.

A similar shape modulation effect is observed during an *in situ* bioconjugation of AuNPs with high-concentrated penetratin ligands.

If *in situ* bioconjugation with ligands is performed in moderate concentration (0.1-1 µM, corresponding to a penetratin to an AuNP ratio of 6:1 and 66:1, respectively), a typical size quenching effect is registered (**Figure 7.3 & Figure 4.30**).

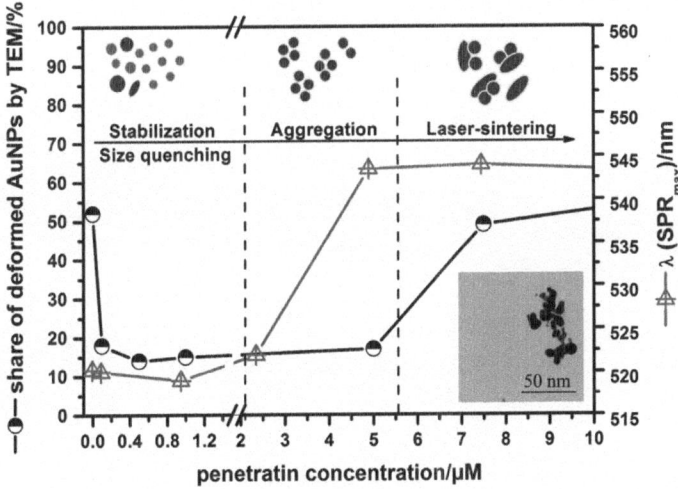

Figure 7.3. Share of deformed AuNPs, estimated by TEM, and shift in SPR$_{max}$ as function of the penetratin concentration. The share of deformed AuNPs, estimated by TEM micrographs, was defined as the share of AuNP with a form factor of < 0.9. TEM image of clusters generated in 20 µM penetratin is presented in the inset. Adapted with permission from Petersen et al., copyright 2011 by the American Chemical Society.[197]

Above a threshold concentration of 2 µM penetratin (penetratin to AuNP ratio of 130:1), the aggregation of nanoparticles is observed due to multilayer formation and charge shielding, which reduces the electrostatic repulsion between the particles while bringing them into close contact (**Chapter 4.2.2.3., Figure 4.30**). However, individual, spherical particles are still distinguishable (**Figure 7.3**). If the ligand concentration is further increased above a concentration of 5 μM (penetration to AuNP ratio of > 330:1), (**Figure 7.3**), the interparticle distance will most likely becomes so low that a ligand-induced laser sintering may appear, resulting in compact, deformed nano-networks (**Figure 7.3**, inset).

In any case, it is important to note, that the shape deformation found with laser- or ligand-induced sintering does not correlate with any consistent form and is not controllable in terms of a defined amount or size of the structures, thus this approach should be disregarded as potential method for controlled shape modulation. However, the appearance of anisotropic structures should be considered for high-fluence ablation and ablation with high-concentrated ligands because they might interfere with biological applications, making an additional purification step with mild centrifugation essential.

Supporting figures and tables

Table SI 1. Specific characteristics of the element gold.[48-50]

Atomic Number	79
Group, Period, Block	11, 6, d
Standard Atomic Weight	196.97 u
Phase	solid
Electron Configuration	[Xe] $4f^{14}5d^{10}6s^1$
Density at Room Temperature	19.32 g cm^{-3}
Melting Point	1337.33 K ($\sim$ 1064.18 °C)
Boiling Point	3109 K ($\sim$ 2836 °C)
Heat of Fusion	12.55 kJ mol^{-1}
Heat of Vaporization	324 kJ mol^{-1}
Molar Heat Capacity	25.42 J mol^{-1} K^{-1}
Specific Heat Capacity	0.129 J g^{-1} K^{-1}
Oxidation States	5, 4, **3**, 2, **1**, -1
Electronegativity	2.4 (Pauling scale)
Atomic Radius, Non-Bonded	214 pm
Covalent Radius	130 pm
van der Waals Radius	166 pm
Crystal Structure	lattice face centered cubic
Magnetic Ordering	dimagnetic
Thermal Conductivity at Room Temperature	318 W m^{-1} K^{-1}
Thermal Expansion at Room Temperature	14.2 μm m^{-1} K^{-1}
Vickers Hardness	216 MPa

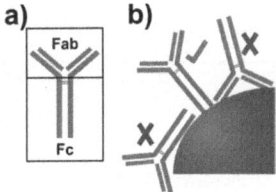

Figure SI 1. Schematic illustration of the antibody molecule structure and the binding orientation of antibodies to the gold surface. a) Illustration of the antibody molecule structure depicting the antigen-binding portion (Fab) and the non-targeting, constant portion (Fc). **b)** The random orientation of antibodies on the gold surface due to electrostatic attraction yields fictional (blue check) and non-functional (red crosses) molecules. Reprinted with permission from Barchanski et al., copyright 2015 by the American Chemical Society.[376]

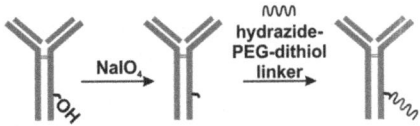

Figure SI 2. Schematic illustration of the simplified conjugation protocol for glycosylated antibodies with a heterobifunctional hydrazide-PEG-dithiol linker. Based on the established protocol of Kumar et al.[270] Reprinted with permission from Barchanski et al., copyright 2015 by the American Chemical Society.[376]

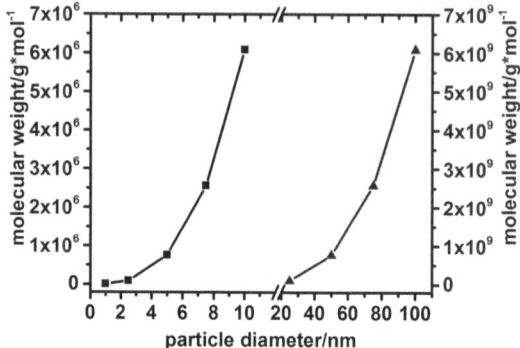

Figure SI 3. The exponential development of molecular weight with increasing nanoparticle diameter. Square data are corresponding to the left ordinate and triangle data are corresponding to the right ordinate.

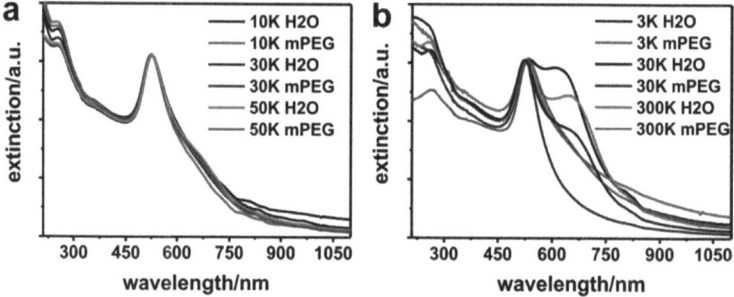

Figure SI 4. UV-vis spectra of filtrated AuNP samples. a) Normalized UV-vis spectra of the retentates after filtration with *Vivacon*® filtration tubes. **b)** Normalized UV-vis spectra of the retentates after filtration with *Nanosep*® filtration tubes.

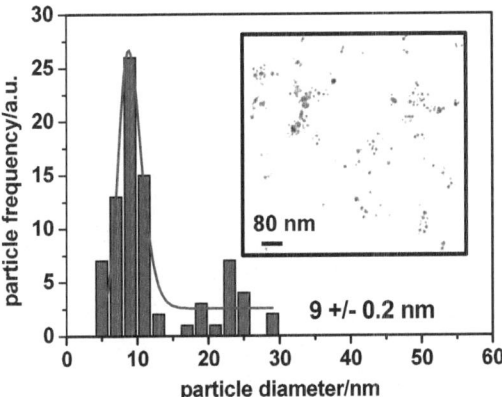

Figure SI 5. Particle size distribution of laser-generated AuNP-anti-IgG bioconjugates. Scanning electron micrograph is presented in the inset. Adapted with permission from Barchanski et al., copyright 2015 by the American Chemical Society.[428]

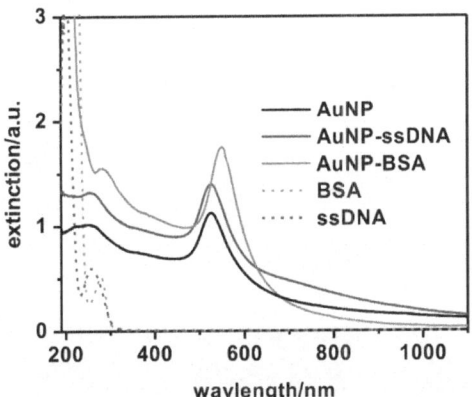

Figure SI 6. Verification of bioconjugation by UV-vis spectroscopy. UV-vis spectra of pure BSA (green dashed line) and ssDNA biomolecule (red dashed line) solutions, ligand-free AuNPs (black solid line) and purified AuNP-BSA (green solid line) and AuNP-ssDNA (red solid line) bioconjugate pellets. Adapted with permission from Barchanski et al., copyright 2015 by the American Chemical Society.[428]

Table SI 2. Overview of IR frequencies detected in the FT-IR spectra with the corresponding functional groups / motives.

Wavenumber/cm⁻¹	Functional Group / Motive
3000-3700	OH
1650, 1380-1420, 1050	C-O, C=O [431;432]
833, 1156, 1390, 1510, 1560, 1653	nucleosides (adenine, guanine, cytosine, thymine) [433]
1060	C-C sugar [434]
1234	phosphate PO₂⁻, B-form DNA marker [434]
1510	aromatic amine
1156	P-O, P=O [556]
1700-1600	amide I (C=O) [436;437]
1600-1500	amide II (N-H, C-N) [436;437]
1350-1200	amide III [436;437]
1650-1655, 1300	α-helix [438;439]
1663-1685	β-sheet, β-turn [438]
1644-1648	random chains [438]
1635-1639	extended chains [438]
1621-1632	extended chains plus beta sheet [438]
1235-1260	β-sheet [439]
1511	NO₂ [557]

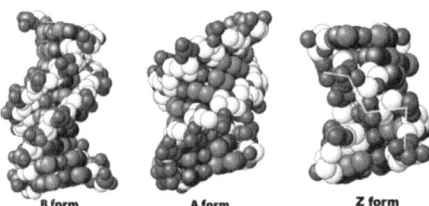

Figure SI 7. Schematic illustration of B-form, A-form and Z-form DNA. Reprinted from Berg et al., copyright 2009, Spektrum Akademischer Verlag.[558]

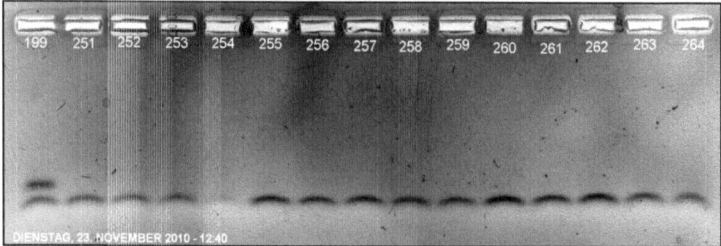

Figure SI 8. Photography of a gel after electrophoresis (exemplary data). Sample 199 represents the untreated ssDNA control, while the samples 251 – 264 are the ssDNAs that were *in situ* conjugated and subsequently separated from the particles by dithiotheriol (DTT). The untreated control is separated into two bands due to biomolecule dimerization by the thiol function, while the DTT-treated samples are all separated.

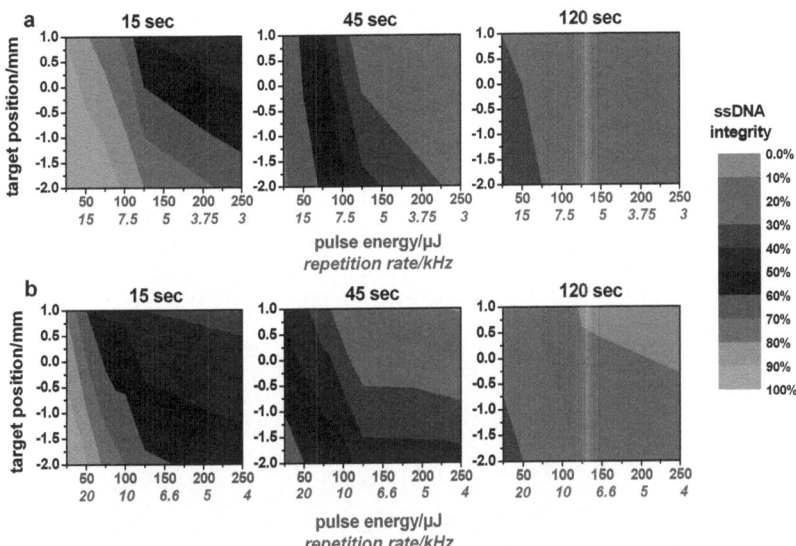

Figure SI 9. Integrity of ssDNA after 15 s, 45 s and 120 s for different target positions and pulse energy/repetition rate combinations for a laser power of 0.75 W (a) and 1 W (b). Target position 0 is defined as the position of the determined focal point in air, while target position 1 is defined as position 1 mm in front of position 0 and target positions -1 and -2 are defined as positions of 1 mm and 2 mm behind of position 0, respectively.[428]

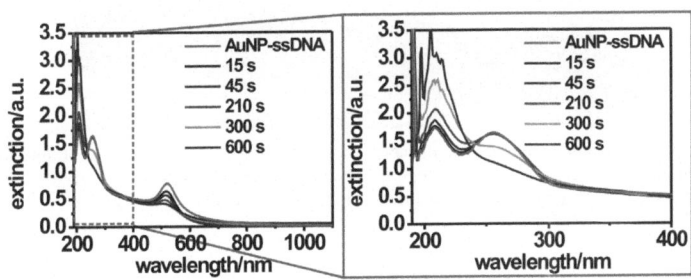

Figure SI 10. Extinction spectra of re-irradiated AuNP-ssDNA bioconjugates with magnification of the wavelength range indicating biomolecule decomposition. Reprinted with permission from Barchanski et al., copyright by the American Chemical Society.[428]

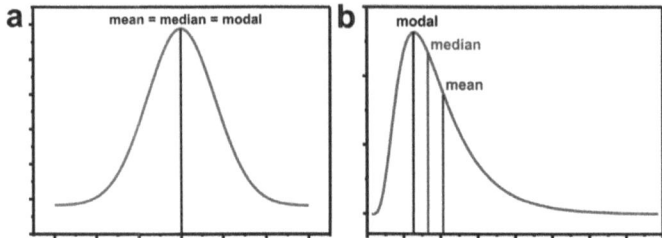

Figure SI 11. Definition of modal, median and mean particle sizes. a) Symmetric PSD where mean = median = mode. b) Asymmetric PSD where mean, median and mode are different.

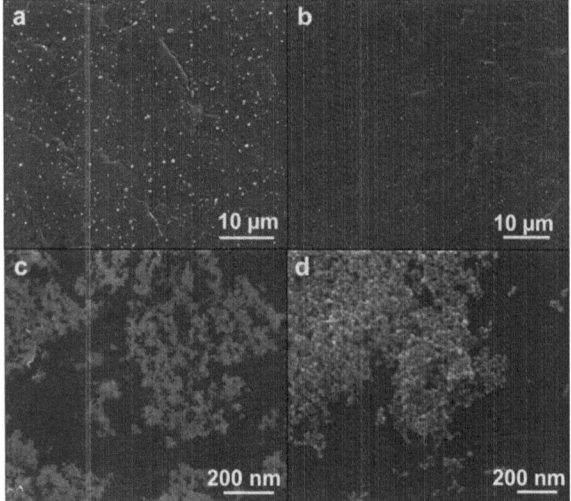

Figure SI 12. Scanning electron micrographs of AuNPs prior and after photofragmentation. a) Scanning electron micrograph of untreated AuNPs. b)-d) Scanning electron micrographs of photo-fragmented AuNPs after 180 s of irradiation (b), and after 600 s of irradiation (c, d).

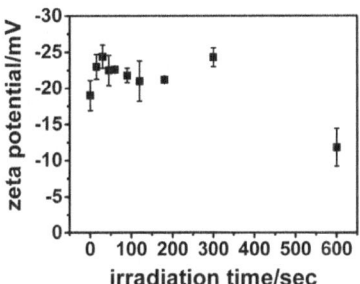

Figure SI 13. Zeta potential values of photofragmentation study.

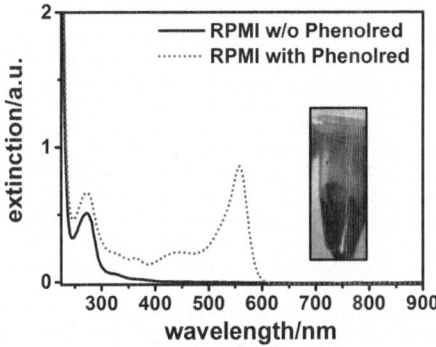

Figure SI 14. Extinction contribution of Phenol red. UV-vis spectra, comparing RPMI 1640 cell culture media with (standard, red dotted line) and without (special, black solid line) Phenol red.

Table SI 3. Chemical composition of RPMI 1640 cell culture media without Phenol red.

Substance	Conc./ mg L⁻¹	Substance	Conc./ mg L⁻¹	Substance	Conc./ mg L⁻¹
NaCl	6000	glycine	10	biotin	0.2
KCl	400	L-histidine	15	L-valine	20
Na₂HPO₄*7H₂O	1512	L-hydroxyproline	20	glutathione	1
MgSO₄*4H₂O	100	L-isoleucine	50	vitamine B12	0.005
Ca(NO₃)₂*4H₂O	100	L-leucine	50	D-Ca-pantothenate	0.25
D-Glucose	2000	L-lysine*HCl	40	cholinchloride	3
NaHCO₃	2000	L-methionine	15	folic acid	1
L-arginine	200	L-phenylalanine	15	myo-inositol	35
L-asparagine	50	L-proline	20	nicotinamide	1
L-aspartic acid	20	L-serine	30	p-aminobenzoic acid	1
L-cystine	50	L-threonine	20	pyridoxine*HCl	1
L-glutamine	300	L-tryptophan	5	riboflavin	0.2
L-glutamic acid	20	L-tyrosine	20	thiamine*HCl	1

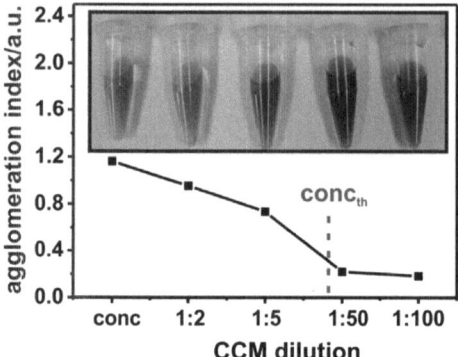

Figure SI 15. Agglomeration index of gold nanoparticle colloids fabricated in cell culture medium (CCM) as function of CCM concentration. Corresponding photographs of the colloids are presented in the inset. The threshold concentration ($conc_{th}$) for a maximum agglomeration index of 0.3 is marked by the red dashed line.

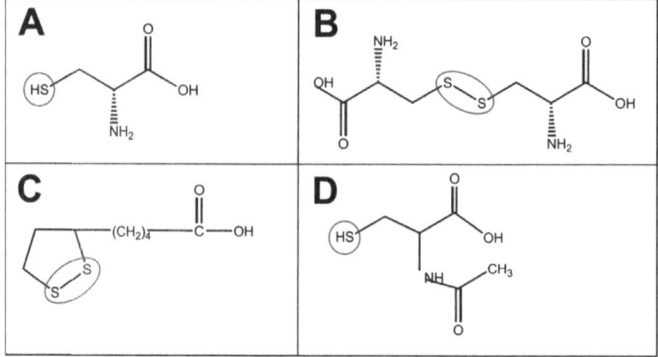

Figure SI 16. Schematic illustration of the molecular structures of four biomolecules. Molecular structures and functional groups (red circles) of biomolecules adopted for AuNP coordination experiments, with **A** = L-cysteine. **B** = L-cystine. **C** = DL-α-lipoic acid. **D** = N-acetyl-L-cysteine.

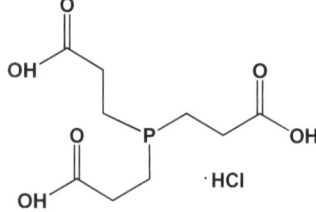

Figure SI 17. Chemical structure of TCEP (tris(2-carboxyethyl)phosphine hydrochloride).

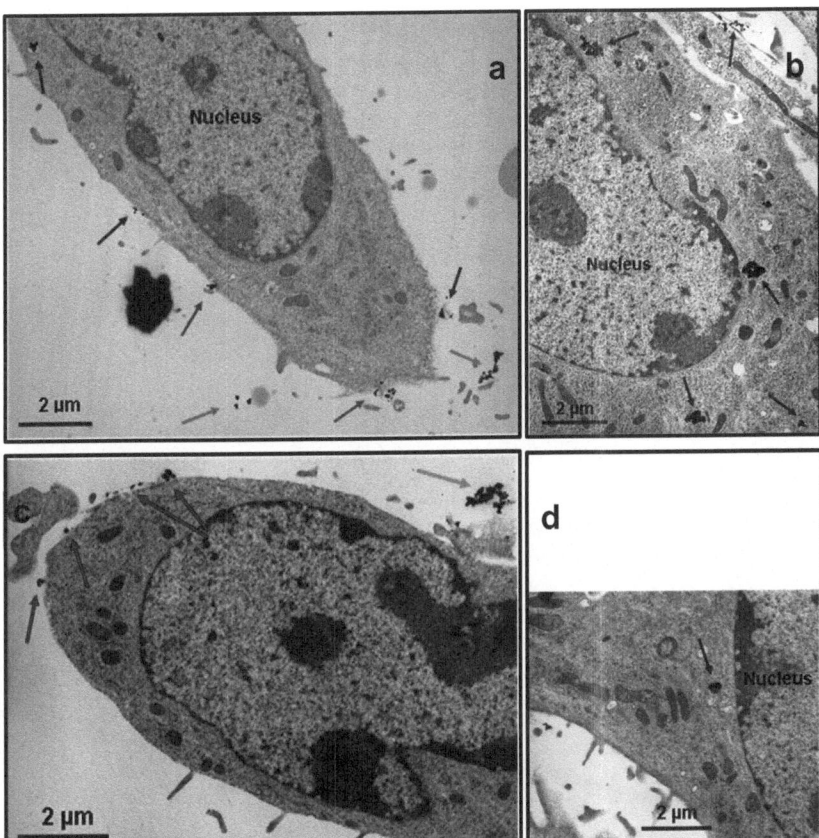

Figure SI 18. Transmission electron micrographs of M3E3/C3 cells after incubation with AuNP-penetratin bioconjugates. a) – b) Incubation with 1 µM primary nanobioconjugates. **c) – d)** Incubation with 5 µM nanobioconjugate clusters. Incubation times are 0.5 h for **a** & **c** and 4 h for **b** & **d**. Yellow arrows = agglomerates that are not associated with the cell membrane, blue arrows = agglomerates that are associated with the cell membrane, green arrows = agglomerates that are internalized into the cell, red arrows = internalized agglomerates in intracellular vesicles. Reprinted with permission from A. Barchanski, copyright 2009 by Annette Barchanski, Master's thesis.[474]

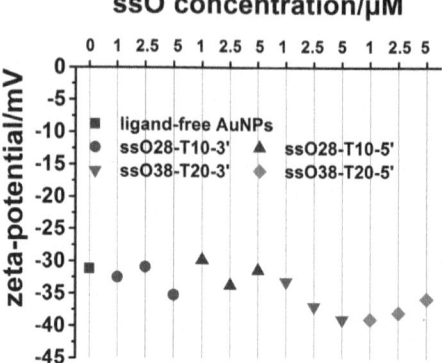

Figure SI 19. Zeta potential data of ligand-free and spacer-containing AuNP samples. Zeta potential values of ligand-free AuNPs (red square) and of AuNP-ssO bioconjugates (ssO28-T10-3' = green dots, ssO38-T20-3' = grey inverted pyramids, ssO28-T10-5' = purple pyramids, ssO38-T20-5' = orange diamonds) for ssO concentrations of 1 µM, 2.5 µM and 5 µM. Reprinted with permission from Barchanski et al., copyright 2012 by the American Chemical Society.[420]

Table SI 4. Average Feret diameter of ligand-free AuNPs and AuNP-ssO conjugates, determined by calculation of Haiss et al.[419] SD = standard deviation. Reprinted with permission from Barchanski et al., copyright 2012 by the American Chemical Society.[420]

Label	ssO28-T10-5'			ssO38-T20-5'			ssO28-T10-3'		
ssO conc./µM	1	2.5	5	1	2.5	5	1	2.5	5
Mean Size/nm	12	12	12	12	12	12	13	13	12
SD	0.7	0.9	0.6	1.3	1.0	0.6	1.0	0.9	1.2
Average Size/nm	12			12			13		
SD	0.1			0.3			0.6		

Label	ssO38-T10-3'			ssO18-3'			AuNP		
ssO conc/µM	1	2.5	5	1	2.5	5	0		
Mean Size/nm	13	13	15	13	13	11	15		
SD	1.9	0.7	0.3	0.2	0.1	0.6	0.3		
Average Size/nm	13			13			15		
SD	0.3			0.4			0.3		

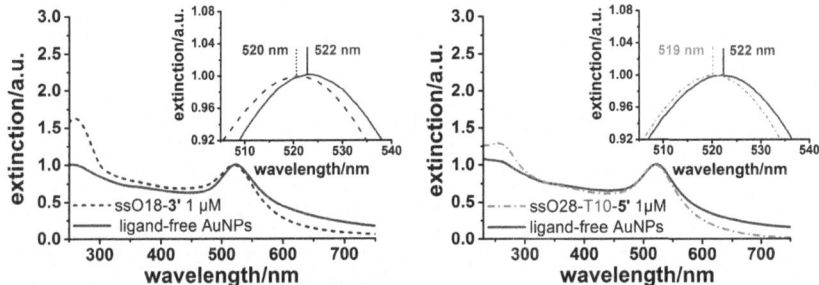

Figure SI 20. Verification of bioconjugation by UV-vis spectroscopy. Normalized extinction spectra of ligand-free AuNPs (solid red line), ssO18-3' conjugates (dotted blue line) and ssO28-T10-5' conjugates with 1 µM ssO concentration (dash-dotted orange line). Magnifications of SPR are presented in the insets. Modified with permission from Barchanski et al., copyright 2012 by the American Chemical Society.[420]

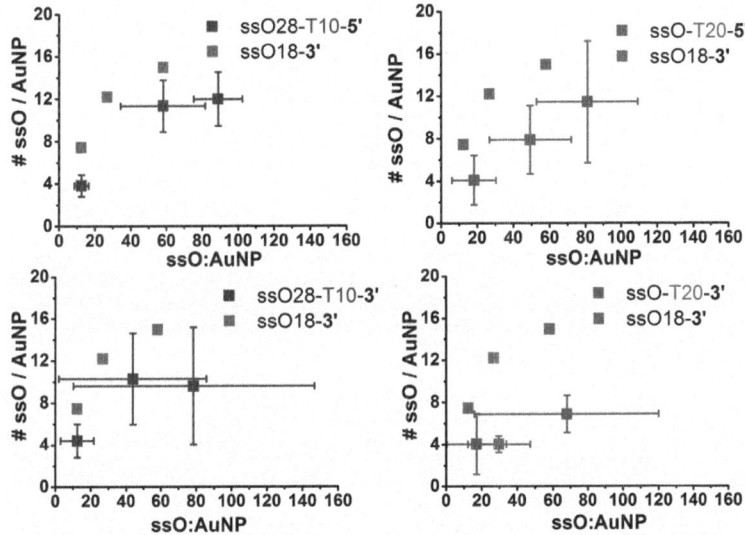

Figure SI 21. Calculated surface coverage values with standard deviations. Surface coverage values of ssO-T10 (red squares) and ssO-T20 (blue squares) spacer-containing conjugates in comparison to ssO18 spacer-less model conjugate (grey squares). Raw data including standard deviations are plotted against the ratio of ssO molecules to the number of AuNPs in solution. Reprinted with permission from Barchanski et al., copyright 2012 by the American Chemical Society.[420]

Table SI 5. Mean hydrodynamic diameter of ligand-free AuNPs and AuNP-ssO conjugates. SD = standard deviation. Reprinted with permission from Barchanski et al., copyright 2012 by the American Chemical Society.[420]

Label	ssO28-T10-5'			ssO38-T20-5'			ssO28-T10-3'		
ssO conc/µM	1	2.5	5	1	2.5	5	1	2.5	5
Mean Size/nm	79	80	96	86	87	98	74	71	98
SD	1.4	2.9	4.9	1.4	1.5	4.6	1.3	1.8	0.3

Label	ssO38-T10-3'			ssO18-3'		
ssO Conc/µM	1	2.5	5	1	2.5	5
Mean Size/nm	84	83	85	59	57	69
SD	2.1	3.9	4.3	0.9	1.1	0.7

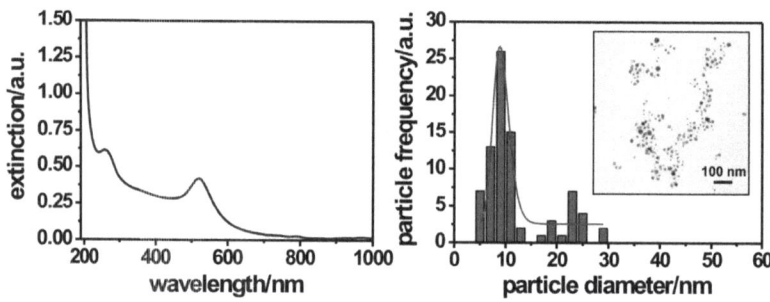

Figure SI 22. Verification of bioconjugation and particle size distribution. Extinction spectrum (left panel) and particle size distribution (right panel) of PLAL-generated AuNP-anti IgG bioconjugates with a number-weighted modal particle diameter of 9±0.2 nm. TEM micrograph is presented in the inset. Modified with permission from Barchanski et al., copyright 2015 by the American Chemical Society.[428]

Table SI 6. IDs, molecular masses and isoelectric points of the applied CPPs. Adapted with permission from Barchanski et al., copyright 2015 by the American Scientific Publishers.[198]

CPP	Sample ID	Molecular Mass/MW	pI	Zeta-potential/mV
Deca-Arginine	10R	1797.2	13.20	+6.2±0.5
Transactivator of Transcription	TAT	1614.0	12.90	+4.3±0.7
Simian-Virus 40 Large T Antigen Nuclear Localization Signal	NLS	1403.2	9.63	-10.1±1.7

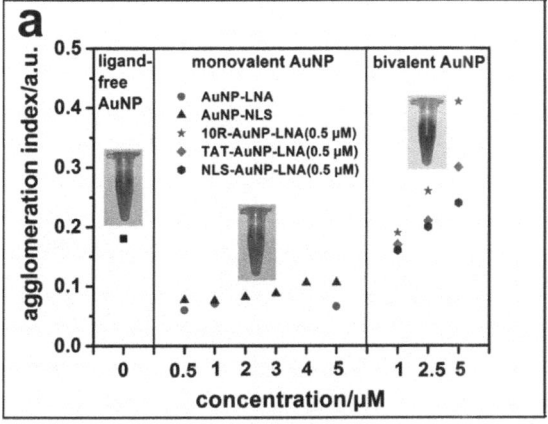

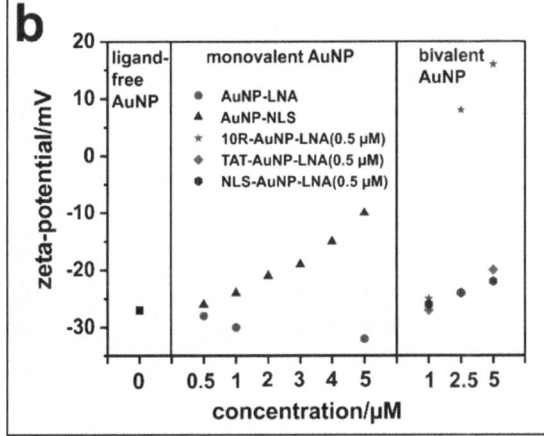

Figure SI 23. Agglomeration index and zeta potential values of AuNP probes for the spermatozoa penetration study. Agglomeration index and photographs **(a)** and zeta potential values **(b)** of ligand-free AuNPs (black square), monovalent AuNP-LNA bioconjugates (red dots) and bivalent CPP-AuNP-LNA bioconjugates with 1 μM, 2.5 μM and 5 μM concentrations. 10 R, TAT and NLS nanobioconjugates are presented as pink stars, green diamonds and purple hexagons, respectively. Reprinted with permission from Barchanski et al., copyright 2015 by the American Scientific Publishers.[198]

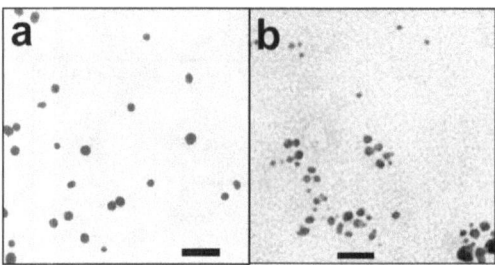

Figure SI 24. Verification of AuNP bioconjugate stability. Transmission electron micrographs of monovalent AuNP-LNA conjugates before **(a)** and after **(b)** transfer into salt-containing media. Scale bars = 50 nm. Modified with permission from Barchanski et al., copyright 2015 by the American Scientific Publishers.[198]

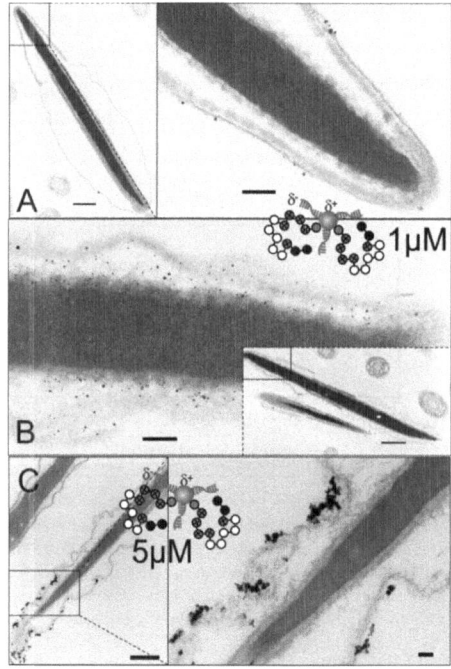

Figure SI 25. Transmission electron micrographs of bovine spermatozoa after co-incubation with bivalent CPP-AuNP-LNA bioconjugates, demonstrated on the example of 10R-AuNP-LNA. Nanobioconjugates with 1 µM CPP concentration are attached to the plasma membrane of acrosome-intact spermatozoa **(A)** and accumulated between the PAS and the NE of acrosome-reacted sperm cells **(B)**, while nanobioconjugates with 5 µM CPP concentrations are subject to ligand-induced agglomeration and detected as clusters on the micrographs **(C)**. Scale bars = 100 nm (overview images) and 50 nm (magnifications). Reprinted with permission from Barchanski et al., copyright 2015 by the American Scientific Publishers.[198]

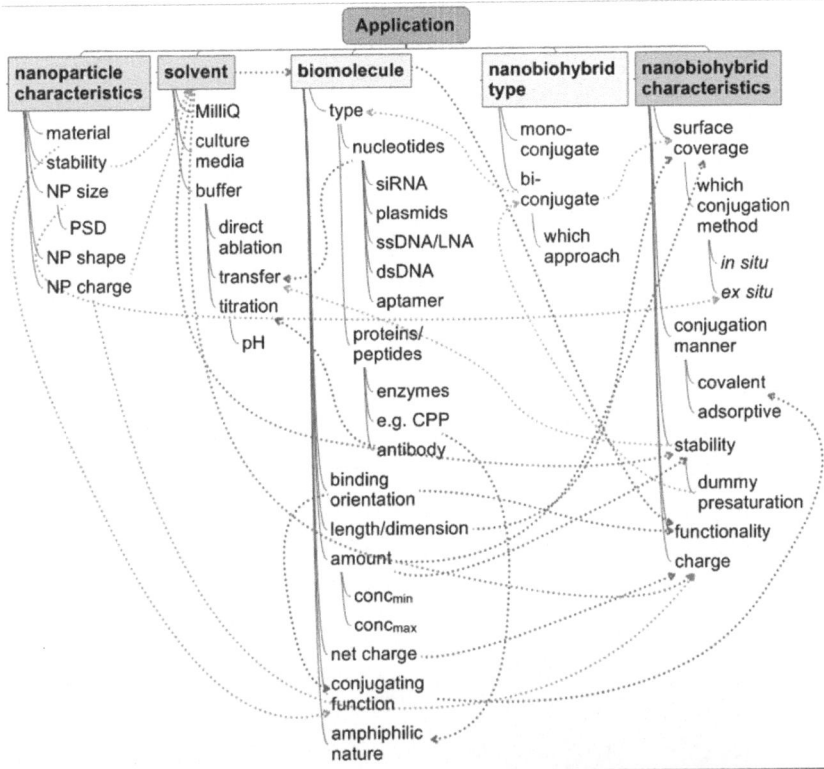

Figure SI 26. Mind map presenting the determined aspects and their impacts among each other that have to be considered prior fabrication of a customized nanobioconjugate for a specific application.

Table SI 7. Specific characteristics of the element silicon.[48-50]

Atomic Number	14
Group, Period, Block	14, 3, p
Standard Atomic Weight	28.09
Phase	solid
Electron Configuration	[Ne] $3s^2 3p^2$
Density at Room Temperature	2.33 g cm^{-3}
Melting Point	1687.15 K (1414 °C)
Boiling Point	3538.15 K (3265 °C)
Heat of Fusion	50.21 kJ mol^{-1}
Heat of Vaporization	359 kJ mol^{-1}
Molar Heat Capacity	19.99 J mol^{-1} K^{-1}
Specific Heat Capacity	0.71 J g^{-1} K^{-1}
Oxidation States	4, 3, 2, 1, -1, -2, -3, -4
Electronegativity	1.90 (Pauling scale)
Atomic Radius, Non-Bonded	210 pm
Covalent Radius	114 pm
van der Waals Radius	210 pm
Crystal Structure	diamond cubic (lattice spacing: 5.43 Å)
Magnetic Ordering	diamagnetic
Thermal Conductivity at Room Temperature	149 W m^{-1} K^{-1}
Thermal Expansion at Room Temperature	2.6 µm m^{-1} K^{-1}
Band Gap Energy at Room Temperature	1.12 eV

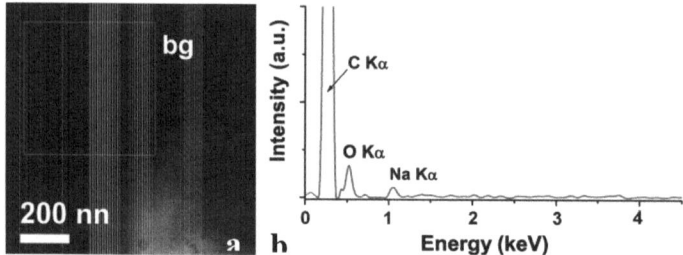

Figure SI 27. Negative control values. HAADF-STEM image of the measurement area (red-framed box) for STEM-EDX on the background region (bg) without SiNPs **(a)**, STEM-EDX elemental characterization results **(b)**, ICP-OES data of *Milli-Q* water, confirming the presence of potassium and sodium. Reprinted with permission from Intartaglia et al., copyright 2012 by the Royal Society of Chemistry.[526]

Table SI 8. ICP-OES measurement results of *Milli-Q* water used for SiNP fabrication by PLAL. Reprinted with permission from Intartaglia et al., copyright 2012 by the Royal Society of Chemistry.[526]

Element	Amount/ppm	Standard Deviation
Ca	0.0001	0.0002
K	0.0420	0.0009
Mg	0.0000	0.0001
Na	0.0270	0.0011

Figure SI 28. Confocal imaging of human fibroblasts incubated with anti-vinculin primary antibodies and ligand-free SiNPs as negative control. Blue color = Hoechst dye. Reprinted with permission from Bagga et al., copyright 2013 by IOP Publishing.[559]

Table SI 9. Specific characteristics of the element iron.[48-50]

Atomic Number	26
Group, Period, Block	8, 4, d
Standard Atomic Weight	55.85 u
Phase	solid
Electron Configuration	[Ar] $3d^6 4s^2$
Density at Room Temperature	7.87 g cm^{-3}
Melting Point	1811.15 K (~ 1538 °C)
Boiling Point	3134.15 K (~ 2861 °C)
Heat of Fusion	13.81 kJ mol^{-1}
Heat of Vaporization	340 kJ mol^{-1}
Molar Heat Capacity	25.1 J mol^{-1} K^{-1}
Specific Heat Capacity	0.45 J g^{-1} K^{-1}
Oxidation States	6, 5, 4, **3**, **2**, 1, -1, -2
Electronegativity	1.83 (Pauling scale)
Atomic Radius, Non-Bonded	204 pm
Covalent Radius	124 pm
van der Waals Radius	200 pm
Crystal Structure	body-centered cubic (a = 286.65 pm) = bcc (delta); face-centered cubic (between 1185 – 1667 K) = fcc (gamma)
Magnetic Ordering	ferromagnetic (@ 1043 K)
Thermal Conductivity at Room Temperature	80.4 W m^{-1} K^{-1}
Thermal Expansion at Room Temperature	11.8 μm m^{-1} K^{-1}
Band Gap Energy at Room Temperature	1.12 eV

Table SI 10. Oxides, oxide-hydroxides and hydroxides of the element iron.[535]

Oxide-Hydroxides and Hydroxides	Oxides
goethite α-FeOOH	hematite α-Fe$_2$O$_3$
lepidocrocite γ-FeOOH	magnetite Fe$_3$O$_4$ (Fe^{II}Fe$_2$^{III}O$_4$)
akaganéite β-FeOOH	B-Fe$_2$O$_3$
schwertmannite Fe$_{16}$O$_{16}$(OH)$_y$(SO$_4$)$_z$ * n H$_2$O	ε-Fe$_2$O$_3$
δ-FeOOH	wustite FeO
feroxyhyte δ'-FeOOH	
high pressure FeOOH	
ferrihydrite Fe$_5$HO$_8$ * 4 H$_2$O	
bernalite Fe(OH)$_3$	
Fe(OH)$_2$	
green rusts Fe$_x$^{III}Fe$_y$^{II}(OH)$_{3x+2y-z}$(A$^-$)$_z$:A=Cl$^-$:1/2 SO$_4$$^{2-}$	

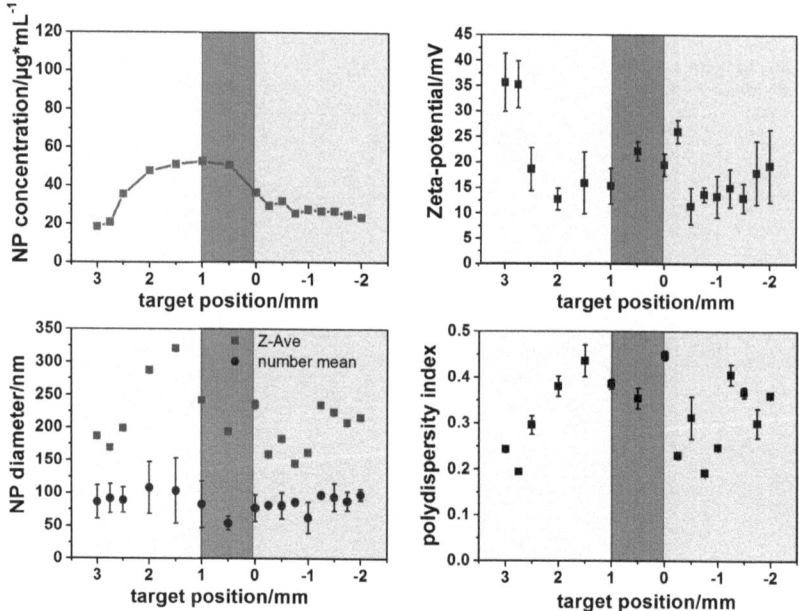

Figure SI 29. Results of parameter series for MNP fabrication. Laser parameters were fixed to 0.5 W, 100 μJ and 5 kHz and target position was varied from 3 to -2 mm. Target position 0 is defined as the position of the determined focal point in air, while positive and negative target positions are defined as positions in front of and behind the 0 position, respectively. NP concentration (upper left, red squares), zeta potential (upper right, blue squares), hydrodynamic diameter (lower left, Z-Ave = green squares, number mean = purple dots) and polydispersity index (upper right, black squares) are presented. Traffic-light coloration correlates to the parameter values that result in MNPs with optimal (green color), medium (yellow color) and poor (red color) quality. Modified with permission from M. Meißner, copyright 2013 Marita Meißner, Bachelor thesis.[415]

Table SI 11. Measurement data of Mössbauer analysis of IONP colloid. IS = isomeric shift, QS = quadrupol splitting, H = magnetic field. Reprinted with permission from M. Meißner, copyright 2013 Marita Meißner, Bachelor thesis.[415]

	IS /mm s⁻¹ (Rhodium)	IS /mm s⁻¹ (α-iron)	QS /mm s⁻¹	H /T
Doublet 1	0.890	0.999	0.712	0
Doublet 1	0.322	0.431	0.864	0
Sextet 1	-0.1094	0.0004	0	32.83

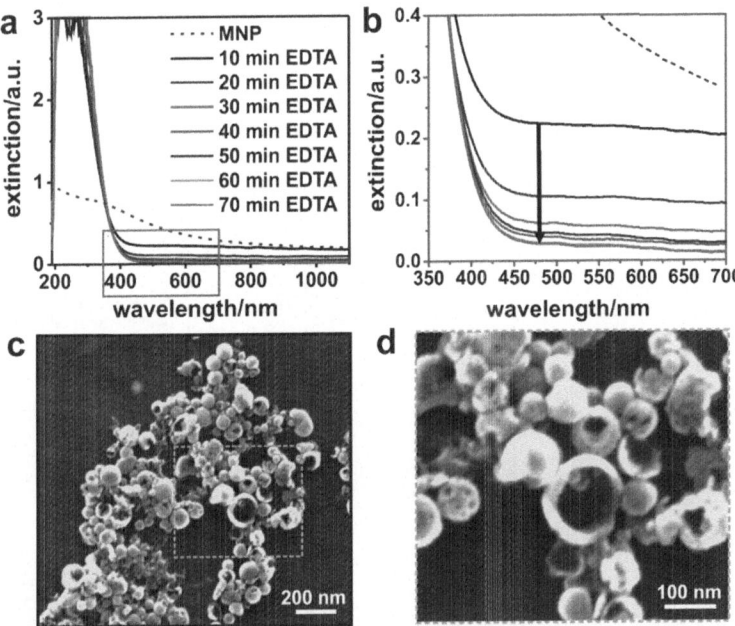

Figure SI 30. Impact of EDTA treatment on AuNP samples. Extinction spectra (**a & b**) and scanning electron micrographs (**c & d**) of EDTA-treated MNPs. Extinction spectra are demonstrating the reduction in NIR extinction of EDTA-treated samples (colored solid lines) in comparison to the untreated control (red dotted line), where (**b**) is a magnified view of the red box in chart (**a**). The aggressive particle hole formation of 60 min EDTA-treatment is presented on the SEM pictures where (**d**) is a magnified view of the blue dashed box in figure (**c**). Modified with permission from M. Meißner, copyright 2013 Marita Meißner, Bachelor thesis.[415]

Table SI 12. Overview of AuNP toxicity (w/o reviews). Ref. = reference, Surf. Mod. = surface modification, ROA = route of administration, Ex. Dur. = exposure duration, IV = intravenous, IP = intraperitoneal.

Ref.	Cell line	Surf. Mod.	ROA/NP Dose/Size	Ex. Dur.	Tests	Results
In vitro						
[560]	Human HeLa, A459, MDA-MB-435	None & Transferrin	0.2 nM ~ 1.2×10^{11} NP mL^{-1} 16, 32, 45, 21, 26, 49, 98 nm	2-24 h	XTT	No cytotoxic effect neither by monodisperse, nor by aggregated particles
[131]	Human primary leucocytes, HeLa	CTAB; Poly(ethylene oxide)	Rods: 1.45×10^7 L^{-1} 15 nm Spheres: 1.67×10^9 L^{-1} 50 nm Spheres 3.33×10^7 L^{-1} 15-50 nm	2 h	Cytokine detection, Gene expression	Surface chemistry has a strong effect on the activation state of macrophages after particle internalization. Carboxy groups on the particle surface induce expression of mRNAs encoding pro-inflammatory proteins, while amino groups on the particle surface induce mRNAs encoding anti-inflammatory proteins.
[561]	Human A549	Citrate	200-2000 µg 15 nm	4 & 24 h	Real-time PCR, ELISA	No adverse effects, no oxidative stress induction, no inflammatory cytokines
[151]	Human HeLa	Immunogenic peptides	0.1 µM - 0.4 mM 3-100 nm	N/A	MTT	No cytotoxicity up to 0.4 mM concentration
[120]	Human K562	Citrate, Biotin, L-cysteine, Glucose, CTAB	0-250 µM 4, 12, 18 nm	1 h-3 d	MTT	Toxic, only if modified with glucose and cysteine and nanoparticle-concentration > 25 µM
[562]	Human DU145, MDA-MB-231, L132	N/A	10 µg mL^{-1} - 2 mg mL^{-1} 1.9 nm	24 h	MTT	Cell-type-specific response. Reduced viability and apoptotic response found for MDA-MB-231 cells, ROS production in MDA-MB-231 and DU145 cells
[175]	Human hTERT-BJ1	Tiopronin, TAT peptide	0-15 µM 3 nm	24 h	MTT	No cytotoxic effects for 5 µM concentration, small effect for 10 µM
[127]	Human BGC 823	Chitosan	0.05-1 mg ml^{-1} N/A	44 h	MTT	Toxicity detected after exposure to 0.8 mg ml^{-1} AuNPs with a zeta potential of 40 mV, no toxicity with zeta potentials of 20 mV and 30 mV
[563]	Human hBMSCs, HuH-7	Citrate	0-75 µg mL^{-1} 5-30 nm	5 d	MTT, Apoptosis/ necrosis, ROS measurement	> 80 % cell survival for 15 & 30 nm AuNPs; decreased survival for 5 nm AuNPs; necrosis by ROS increased with AuNP concentration

Ref.	Cell line	Surf. Mod.	ROA/NP Dose/Size	Ex. Dur.	Tests	Results
[118]	Human MDA-MB-231	Coumarin-PEG-thiol	50-200 µg ml⁻¹ 10 nm	24 h	Cell Titer 96	No toxicity detected
[5]	Human Hep3B, Panc-1	None	1-67 µM 5 nm	4 h	MTT	No intrinsic cytotoxicity or anti-proliferative effects detected
[130]	Monkey COS-1, red blood cells Bacteria E.coli	Quaternary ammonium	0.38-3 µM 2 nm	1-24 h	MTT, hemolysis	Cationic particles – moderately toxic; Anionic particles – nontoxic
[176]	Human HeLa	PEG	0.08-100 µM 3.7 nm	6-72 h	MTT	NPs entered nucleus, no toxicity induced
[564]	Human HepG2	Cysteine	N/A 20 nm	1-6 h 1-4 d	Multiplexed cytotoxicity assay	AuNPs inhibited the proliferation and intracellular calcium release in cells
[565]	Human HeLa	Citrate	0.2-2 nM 18 nm	3-6 h	MTT	No changes in gene-expression patterns, no splicing of xbp1 mRNA
[566]	Human HepG2	Citrate, CTAB	98.5 µg mL⁻¹ 3.5 & 5.5 nm	24 h	Morphological examination, Trypanblue assay, LDH, MTT	Severe morphological changes and cell death in 2D & 3D culture (less) using CTAB AuNPs; Negligible effects using Citrate AuNPs
[567]	Human MRC-5	FBS	0.5-1 nM 20 nm	24-72 h	Oxidative stress PCR, Gene expression analysis	Oxidative damage induced upregulation of antioxidants, stress response genes and protein expression
[121]	Human HeLa	Citrate, BSA, ssDNA, dsDNA, dsRNA	10 nM 15 nm	24 h	Gene expression analysis, Cell-cycle analysis, Annexin assay	Citrate-stabilized NPs caused change in gene expression, disturbance of mitosis and ca. 20 % increase in apoptosis
[568]	Human CF-31	Citrate	0-189 µg mL⁻¹ 13 nm 0-26 µg mL⁻¹ 45 nm	2-17 d	Confocal microscopy, SEM, TEM	Cytoskeleton filament disruption; expression of ECM proteins was diminished; recovery of cells as function of AuNPs size, concentration and exposure time
[569]	Mouse NIH3T3	BSA	10⁻⁷-10⁻⁴ M 15 nm	3 h	MTT	No cytotoxic effect detected
[124;125]	Human HeLa; Sk-Mel-28 Mouse L929; J774A1	Triphenyl-phosphine derivates, Glutathione	1-10000 µM 1.1, 1.4, 8-15 nm	6-48 h	Microscopy, MTT, Annexin assay	Highest toxicity (three-fold higher than any other size) at 1.4 nm diameter by ROS production and upregulation of stress-related genes; no toxicity for Glutathione-capped AuNPs

Ref.	Cell line	Surf. Mod.	ROA/NP Dose/Size	Ex. Dur.	Tests	Results
[126]	Human A549; HepG2; Syrian hamster BHK21	Citrate	0-120 nM 33 nm (d_{hyd})	36-72 h	Microscopy, PI, MTT, Cleavage of poly(ADP-ribose) polymerase	Toxicity detected in A549 cells only after exposure to 10 nM gold
[570]	Human CF-31	Citrate	0-0.8 mg mL^{-1} 14 nm	2-6 d	Confocal microscopy, TEM, migration assay	Abnormal actin filaments and extracellular matrix constructs; decreasing cell proliferation, adhesion and motility
[571]	Human A549, Jurkat, THP-1, NHBE	Citrate	2 nM 4-13 nm	24-48 h	CellTiter Blue®, Toxilight®	No cytotoxicity detected
[572]	Human HepG2 Dog MDCK	Polyvinyl-pyrrolidone	10-1000 nM 8 nm	24 h	Neutral red uptake, colony forming efficiency test	No cytotoxicity detected
[573]	Human HDF-f	Citrate	10-300 µM 10-50 nm	72 h	MTT, histology	20 nm-sized NPs were not toxic, even at 300 µM concentration
[117]	Human MCF7	Poly-N-isopropyl-acryl-amide-co-acrylamide co-polymer	0-0.74 nM 18 nm	48 h	Sulfo-rhodamine staining	No cytotoxic effects up to 0.74 mM concentration
[116]	Human MDA-MB-231	Coumarin-PEG- thiol	50-200 µg/ml 10 nm	24 h	Cell Titer 96, MTT	No toxicity detected
[574]	Mouse RAW264.7 macrophage	Lysine, poly-L-Lysine	10-100 µM 3.5 nm	24-72 h	MTT	Au(0)NP – non-cytotoxic, non-immunogenic, antioxidant effect
[575]	Human HepG2 Rat NRK-52E	Citrate	0-200 µg mL^{-1} 40 nm	24 h	MTT	No cytotoxic effects determined
[576]	Human HepG2	Glycolipid	100 µM 10 nm	3 h	MTT, comet assay	No cytotoxicity up to 100 µM concentration
[577]	Human HeLa	CALNN peptide	0.02-0.32 nM 13-60 nm	24 h	Trypan blue staining	No cytotoxicity up to 0.04 nM concentration; 95 % cytotoxicity for 0.32 nM concentration
[122]	Bovine GM7373	none	0-50 µM 15 nm	96 h	Microscopy, PI, TUNEL-assay, XTT	Toxicity detected only in XTT-assay after exposure to 50 µM gold (22 % loss in sperm motility)

Ref.	Cell line	Surf. Mod.	ROA/NP Dose/Size	Ex. Dur.	Tests	Results
[119]	Monkey COS-7	PEI2	N/A N/A	6 h, 42 h	MTT	20-30 % loss of viability
[123]	Human HeLa, HepG2 Mouse 3T3/ NIH	BSA, 4 targeting peptides	~ 150 pM 20-25 nm	3 h	LDH	5 % loss of viability
[578]	Mouse MC3T3-E1 Human U-2OS, SK-ES-1, MOR/P, CCD-919Sk, BLM, MV3, SMel-28, HeLa, Hek-12, MOR/CPR	Triphenly-phosphine mono-sulfonate	0-0.4 µM 1.4 & 18 nm	72 h	MTT	MV3 and BLM showed most significant sensitivity to Au55. 100 % cell death at 0.4 µM concentration; 18 nm AuNPs are much less toxic than 1.4 nm Au55 clusters
[579]	Human PC-3, MCF-7 Hamster CHO22	Citrate	10-130µg mL^{-1} 3-45 nm	24 h	LDH, neutral red cellular uptake assay, reduction assay	3, 8 and 30 nm-sized NPs were more sensitive to the cell lines and caused gradual cell death within 24 h at higher concentrations, other sizes were non-toxic up to 3-4 fold concentrations and longer exposure
[580]	Dendritic cells from C57BL/6 mouse	Polyvinyl-pyrrolidone	0.5 mM 10 nm	4-48 h	Flow cytometry	No cytotoxicity even at high concentrations; no activation of DCs; modified secretion of cytokines
[581]	Human HepG2	Paclitaxel	1 nmol L^{-1} - 1.2 mol L^{-1} 25 nm	24 h	MTT, QCM, flow cytometry	Low cytotoxicity by enhancement of apoptosis
[582]	Mouse J774 A1	None	1-10 ppm 2.8, 5.5, 38 nm	24-72 h	Cell number count, gene expression analysis	Cytotoxic effect for concentration of 10 ppm, small particles upregulate expression of IL-1, IL-6 and TNF-alpha
In vivo						
[159]	Human (n = 10)	Citrate	Oral 30 mg day^{-1} > 20 nm (Aurasol®)	1-4 weeks 1-5 months	Blood test	IL6, TNF-alpha, IgG, IgM and rheumatic factor were significantly suppressed
[133]	Male Wistar rats (n = 30)	Citrate	IV Up to 0.015 mg kg^{-1} 20 nm	1-7 d 1-2 months	RNA micro-array analysis, Organ index	Up- and down-regulation of genes

Ref.	Cell line	Surf. Mod.	ROA/NP Dose/Size	Ex. Dur.	Tests	Results
[151]	Male mice BALB/c (n = 6)	Citrate	IP 8 mg kg^{-1} 3-100 nm	>1200 h	Physical and behavioral examination	Lethality induction found to be NP size dependent 3-5 nm did not induce sickness, while other sizes did
[152;583]	Mice BALB/c (n = N/A)	PEG-SH	IV 170 µg kg^{-1} - 4.26 mg kg^{-1} = 1×10^{10}-1.5×10^{14} particles 4-100 nm & 13 nm	30 min 1-7 d	Histology, TEM, gene analysis, Immuno-histochemis-try, TUNEL assay	13 nm AuNPs induced inflammation and apoptosis in the liver tissue 30 min after injection; 4 nm and 100 nm sized particles showed similar biological effects (gene expression)
[154]	Juvenile swine (n = 9)	Gum Arabic, maltose	IV 2 mg kg^{-1} 6-10 nm 15-20 nm	1-24 h, 7-32 d	Tissue distri-bution, Se-rum analysis	No abnormalities in serum analysis
[155;205]	Mice BALB/c (n = N/A)	Citrate	IV 7-700 mg kg^{-1} 1.35-2.7 g kg^{-1} 1.9 nm	5 min 24 h 2 weeks	Radiography, Blood test	No toxicity detected
[156]	Pigs (n = 3)	Arabic gum	IV 0.8-1.9 mg kg^{-1} 15-20 nm	0.5-24 h	Histology, AAS, NAA, X-ray, CT contrast measure-ments	No hematological or renal side effects
[157]	Mice C57BL/6 (n = N/A)	Citrate	IV 1000 µg kg^{-1} 20 & 100 nm	24 h	Histology, TEM, TUNEL assay, MTT, Immuno-cytochemis-try, Western blotting	No cytotoxic effect detected
[158]	Male mice C57/BL6 (n = N/A)	Citrate	IP 40-400 mg kg^{-1} day^{-1} 12.5 nm	8 d	Histology, serum bio-chemical / hematologi-cal analysis	No evident toxicity
[153]	Mytilus edulis	Citrate	1 mM 1-13 nm	24 h	Oxidative stress, cata-lase activity, neutral red retention, 2DE gels	AuNPs induced oxidative stress in bivalves, especially in digestive gland
[96]	White Rat (n = 30) Rabbit (n = 10)	PEG	IV 0.3 mg kg^{-1} 15, 50 nm	24 h 72 h	Histology, AAS, TEM	Expressed changes of the inner organs observed

Ref.	Cell line	Surf. Mod.	ROA/NP Dose/Size	Ex. Dur.	Tests	Results
[160]	Male mice ICR (n = 36)	Citrate	Oral, IP, IV 137.5 µg kg^{-1} - 2.2 mg kg^{-1} 13.5 nm	14-28 d	Organ index	Low concentrations do not show any effect, high concentrations induce decreases in body weight, red blood cells & hematocrit; oral & IP administration showed highest toxicity, IV the lowest
			Reproduction-relevant			
[145]	Zebrafish embryos (n = 33)	Citrate	0.25-250 µM Suspends in egg-water 3-100 nm	120 h	INAA, Porbit method analysis	No appreciable toxicity
[146]	Zebrafish embryos (n = 60-64)	Citrate	0.025-1.2 nM Suspends in egg-water 11.6 ± 0.9 nm	24-120h	Histology, TEM	Slight increase in deformities
[138]	Human sperm (n = 10)	PVP	30-500 µM In medium 50 nm	60-120 min	Eosin Y test	Dose-dependent effect on motility and viability, increased membrane impairment
[150]	Transgenic zebrafish embryo	Triphenyl-phosphine (TPPMS), L-Gluta-thione, Aurovist®	50 µM-2 mM Diluted in medium 1.2-1.4 nm	24 h	LD$_{50}$, ICP-MS, induction of heat shock protein and HSP promotor	Toxicity was dependent on size and ligands with covalent ligands being less toxic than labile ligands. Transgenic zebrafish responded similar to wildtype zebrafish in terms of teratogenicity but were 20-fold more sensitive in reporting hepatotoxicity of AuNP (1.4 nm)
[148]	Chicken egg (n = 200)	N/A	50 ppm Injection in ovo (0.3 mL) < 100 nm	18 d	Gene analysis	No toxic effects
[149]	Murine embryos (n = 107)	Ligand-free	~ 1000 NP/embryo Microinjection into blasto-mere of two-cell-stage embryo 11 nm	N/A	.LSCM, Assessment of embryo development, real-time PCR	No abnormal development of blastocysts, no influence on gene expression
[137]	Bovine sperm (n = 21)	Ligand-free, ssDNA	140-14,000 NP/sperm In medium 7 – 11 nm	2h	TEM, IVOS sperm analysis, phase contrast microscopy, FACS, ROS/RNS	Sperm morphology and viability remained unimpaired; decrease in sperm motility and fertilizing ability using ligand-free AuNPs

Ref.	Cell line	Surf. Mod.	ROA/NP Dose/Size	Ex. Dur.	Tests	Results
[136]	Porcine oocytes (n = 350) Boar sperm (n = 7)	BSA	Oocytes: 10 µg mL^{-1} & 30 µg mL^{-1} In medium 6 nm & 8 nm Spermatozoa: 10 µg mL^{-1} In medium 20 nm	46 h oocytes 2 h sperm	CLSM	No toxic effects
[135]	Human sperm (n = 1)	N/A	N/A In medium 9 nm	15 min	Microscopy, motility & morphology analysis	20 % loss in sperm motility & morphological defects of spermatozoa
[139]	Mouse epididymal sperm (n = N/A)	N/A	0.5 & 1 x 10^{15} NP/mL In medium 2.5 nm	2 h	Microscopy	Disruption of nuclear chromatin decondensation
[147]	Chicken egg (n = 120)	N/A	N/A Injection in ovo N/A	0-20 d	Morphological evaluation, Serum analysis, LDH Assay	No toxic effects

Table SI 13. Overview of cellular uptake and intracellular fate of AuNPs (w/o reviews).
Ref. = reference, Surf. Mod. = surface modification, ROA = route of administration, Uptake Char. = uptake characteristics, IV = intravenous, IP = intraperitoneal.

Ref.	NP Shape	NP Diameter	Surf. Mod.	Cell Line Animal	ROA/NP Dose/ Uptake Char.	Intracellular Fate
				Size		
[560]	Spheres	15, 30, 45 nm monodisperse; 26, 49, 98 nm aggregates	Transferrin, BSA	Human HeLa, MDA-MB-435, A549	HeLa & A549: Transferrin receptor-mediated endocytosis; AuNP aggregation reduces uptake. MDA-MB-435: unknown mechanism, receptor-independent	Endosomes; no AuNPs in cytosol
[82]	Spheres Rods	Spheres: 14, 30, 50, 74, 100 nm Rods: 40x14 nm, 74x14 nm	Citric acid, CTAB	Human HeLa	Most efficient uptake of 50 nm particles; Lower uptake of rods	Endosomes
[168]	Spheres Rods	Spheres: 14, 50 nm Rods: 20x30 nm, 14x50 nm, 7x42 nm	Transferrin	Human HeLa, SNB19 Mouse Sto	Uptake efficiency depending on cell line; Most efficient uptake of 50 nm spheres	Endosomes
[169]	Spheres	2-100 nm	Citric acid, Herceptin (HER)	Human SK-BR-3	Size-dependent receptor-mediated endocytosis and cell regulation; HER-AuNPs are internalized more efficient than unmodified AuNPs; most efficient uptake for 25-50 nm particles	Endosomes, multivesicular bodies
[173]	Spheres	10, 25, 50 nm	Citric acid	Rat NRK	Larger particles were more readily internalized than smaller ones	N/A
[174]	Spheres	13, 45, 70, 110 nm	Thiolized ssDNA	Human CL1-0, HeLa	13 & 45-nm-AuNPs entered cells through endocytosis and accumulated in endocytic vesicles; cellular uptake decreased with the increase of particle size; 70 & 110-nm-AuNPs moved to the top of cells	Endosomes
				Shape		
[584]	Rods	Aspect ratio 1.0, 2.1, 2.6, 2.9, 3.4, 4.1	CTAB, PAA, PAH	Human HT-29	Receptor-mediated uptake, varying extent of uptake: PAH > PAA > CTAB	N/A

Ref.	NP Shape	NP Diameter	Surf. Mod.	Cell Line Animal	ROA/NP Dose/ Uptake Char.	Intracellular Fate
[82]	Spheres Rods	Spheres: 14, 30, 50, 74, 100 nm Rods: 40x14 nm, 74x14 nm	Citric acid, CTAB	Human HeLa	Most efficient uptake of 50 nm particles; Lower uptake of rods	Endosomes
[168]	Spheres Rods	Spheres: 14, 50 nm Rods: 20x30 nm, 14x50 nm, 7x42 nm	Transferrin	Human HeLa, SNB19 Mouse Sto	Uptake efficiency depending on cell line; Most efficient uptake of 50 nm spheres	Endosomes
[85]	Rods	< 100 nm	CTAB, CTAB-poly(styrene-sulfonate) + IgG	Human Gingival epithelioid cells, Oral cancer cells	Endocytotic uptake of conjugates after serum protein adsorption	N/A
[585]	Rods	13.5 × 57 nm	poly diallyl-dimethyl ammonium-chloride	Human MDA-MB-231	Endocytotic uptake into vesicles after 15 min incubation by receptor-mediated endocytosis; with prolonged incubation, nanorods were found in all classic lysosome maturation states; after 6 hours nanorods appear in residual bodies	Occasionally escape of nanorods from lysosomes observed but without accumulation of rods in the cytoplasm; escaped nanorods are recycled back into the lysosomal system; with time the rods were exocytosed and re-endocytosed
Surface charge						
[171]	Spheres	2 nm (core size) ~ 10 nm (+ ligands)	⁰AuNP, ⁺AuNP, ⁻AuNP, ⁺⁻AuNP	Human CP70, A2780, BECs, ASM	Uptake of ⁺AuNP was found highest; uptake dependent on plasma membrane potential	N/A
[550]	Spheres	17 nm	BPPP dehydrate dipotassium salt, 3 small peptides	Human Endothelial cells	Particles of similar size and charge but different peptide functionalization have different receptor-uptake mechanisms	Endosomes
[131]	Spheres Rods	15 – 50 nm	CTAB, Poly(ethylene oxide)	Human Primary leucocyte, HeLa	Faster uptake of CTAB-AuNPs by macrophages and monocytes compared to HeLa; uptake of rods more efficient than for spheres; PEO hindered the uptake; no difference between positively/negatively charged particles; uptake mechanism: macropinocytosis	N/A

Ref.	NP Shape	NP Diameter	Surf. Mod.	Cell Line Animal	ROA/NP Dose/ Uptake Char.	Intracellular Fate
[180]	Spheres	18 nm	Citrate acid, PVA, PAA	Human SK-BR-3	5 – 10 times higher up-take of positively charged AuNPs, neutral AuNPs showed lowest uptake level	N/A
[127]	Spheres	140 nm	Chitosan	Human BGC 823	The higher the zeta po-tential (+20 - +40 mV), the higher the cellular uptake; endocytotic up-take mechanism	Endosomes
[181]	Spheres	18, 35, 65 nm	EDA, glu-cose-amine, HPA, tau-rine, PEG	Human HDMEC	Positively charged AuNPs were internalized to a greater extent than negative or neutral AuNPs	N/A
[182]	Spheres	5, 10, 20 nm	PtBA, PDMDOM AA, PBAEAM, PNIPAM	Human Caco-2	Positively charged AuNPs were internalized to a greater extent than negative and those to a greater extent than neu-tral AuNPs	N/A
[122]	Spheres	5 – 65 nm distribution	Ligand-free, positively charged	Cattle GM7373	No inhibition of particle uptake at 4 °C, suggest-ing diffusion as entrance mechanism	Particles were locat-ed in the cytosol, surrounded by lyso-somal-like structures
[183]	Spheres	2 nm (core) 6 – 10 nm (+ ligands)	0AuNP, +AuNP, -AuNP	Plant seedlings	Positively charged AuNPs were found to be most readily taken up by plant roots while nega-tively charged AuNPs are most efficiently translocated into plant shoots from the roots. Efficiency of NP uptake was plant-dependent.	N/A
Functionalization						
[550]	Spheres	17 nm	BPPP de-hydrate di-potassium salt, 3 small peptides	Human Endothe-lial cells	Particles of similar size and charge but different peptide functionalization have different receptor-uptake mechanisms	Endosomes; fate not analyzed
[131]	Spheres Rods	15-50 nm	CTAB, Poly(ethyl-ene oxide) = PEO	Human Primary leuco-cytes, HeLa	Faster uptake of CTAB-AuNPs by leucocytes compared to HeLa; up-take of rods more effi-cient than for spheres; PEO hindered the up-take; no difference be-tween positive-ly/negatively charged particles; uptake mecha-nism: macropinocytosis	Fate not analyzed

Ref.	NP Shape	NP Diameter	Surf. Mod.	Cell Line Animal	ROA/NP Dose/ Uptake Char.	Intracellular Fate
[200]	Spheres	5 nm 35 nm	Tat, PEG, PEG-Tat	Human hTERT-BJ1	Translocation across cell membrane; Efficient uptake of all conjugate species	5 nm particles found in nucleus; 35 nm particles imaged in cytoplasm
[190]	Spheres	13 nm	Pen, TAT, LSP L1, LSP L2	Hamster CHO	Selective and efficient delivery of conjugates into the lysosomes; Pen-conjugates more efficient than TAT-conjugates	AuNPs with Pen/TAT: cytosol delivery, AuNPs with L1/L2: lysosomal delivery
[175]	Spheres	2.8 nm	Tiopronin, Tiopronin-Tat	Human hTERT-BJ1	Endocytotic uptake	Cytoplasm after endosomal escape; Nucleus
[176]	Spheres	3.7 nm	PEG, NH2-PEG-NH2	Human HeLa	Cellular uptake was found to be receptor- and time-dependent; nuclear penetration after 24 h detected	Nucleus
[177]	Spheres	5.1 nm	Cationic lipids, siRNA	Human MDA-MB-435	Uptake efficiency was dependent on the lipid layer	siRNA delivery into the nucleus was demonstrated by gene silencing
[201]	Spheres	14 nm	PEG, TAT, Pntn, NLS	Human HeLa	Efficient cellular uptake	NP found in cytosol, nucleus and mito-chondria, only few in vesicles; within 10-24 h particles moved into vesicles; after 24 h particles were exo-cytosed from the cell
[196]	Spheres	5 nm	CPP, NLS, CPP-NLS	Human HOS TE85	All conjugates entered the cells compared to ligand-free AuNPs; Up-take efficiency of CPP-NLS conjugates was highest.	most conjugates were found inside endosomes; No nuclear translocation observed
[83]	Spheres	16 nm	Citrate-BSA, CALNN-BSA, Tat, Penetratin, NLS, Liposomes (30-400nm)	Human HeLa	Endocytotic uptake of conjugates	Most conjugates found trapped in endosomes; Tat & Penetratin conjugates escaped from endosomes; NLS-conjugated AuNPs were imaged within the nucleus
[178]	Spheres	2.4, 5.5, 8.2, 16, 38, 89 nm	PEG, CPP	Monkey COS-1	Uptake is CPP-, and particle size-dependent; no entry was observed for 16 nm and larger AuNPs	Intracellular destina-tion is NP size-dependent; 2.4 nm AuNPs localized in the nucleus, inter-mediate 5.5.-8.2 nm AuNPs were deliv-ered into cytoplasm

Ref.	NP Shape	NP Diameter	Surf. Mod.	Cell Line Animal	ROA/NP Dose/ Uptake Char.	Intracellular Fate
[199]	Spheres	12 nm	Citrate, SAP	Human HeLa	No uptake of citrate-stabilized AuNPs, vesicular uptake of SAP-AuNPs	Particles found in multivesicular bodies
[179]	Spheres	5, 15 nm	SV40 large T antigen, BSA	Human HeLa	Receptor-dependent uptake; enhanced internalization with increased amount of ligands and with increased NP size	Nuclear localization
[574]	Spheres	3.5 nm	Lysine, poly-L-Lysine	Mouse RAW 264.7	AFM: suggests pinocytosis CFLSM/TEM: indicate internalization in lysosomal bodies	Perinuclearly arranged lysosomal bodies
[123]	Spheres	22 nm	BSA+NLS, BSA+Tat, BSA+AFP, BSA+IBDP	Human HeLa, HepG2 Mouse 3T3/NIH	Endocytotic uptake due to enlarged size of nanoparticle bioconjugates	NLS & Tat conjugates found in cytoplasm only; AFP conjugates found in cytoplasm in 3T3/NIH cells & in nucleus of HeLa; IBDP conjugates found in nuclei of two cell lines
[189]	Spheres	4.3-4.9 nm (core)	MUS, 2:1 MUS:OT, 1:2 MUS:OT, 2:1 MUS:brOT	Mouse DC2.4	AuNPs with striations of altering anionic and hydrophobic ligands penetrate the plasma membrane w/o bilayer disprution; AuNPs with random distribution of ligands were trapped in endosomes	Cytosol / Endosomes
Barrier crossing / Tissue distribution						
[133]	Spheres	20 nm	Citrate	Male Wistar rats	IV 0-0.15 mg kg⁻¹	Particle accumulation in liver and spleen, in lung (after 1 d), in kidney & testis (after 1 month) Crossed the blood-testis barrier
[202]	Spheres	10-250 nm	Citrate	Male Wistar rats	IV 77-120 µg	Particle accumulation in spleen and liver (all sizes) and in brain, heart, kidney, testis and thymus (10 nm)
[154]	Spheres	6-10 nm 15-20 nm	Gum Arabic, maltose	Juvenile swine	IV 2 mg kg⁻¹	AuNP accumulation in macrophages of the liver (Gum Arabic-coated) and lung (maltose-coated)

Ref.	NP Shape	NP Diameter	Surf. Mod.	Cell Line Animal	ROA/NP Dose/ Uptake Char.	Intracellular Fate
[155;205]	Spheres	1.9 nm	Citrate	BALB/c mice	IV 700 mg kg^{-1} 1.35-2.7 g kg^{-1}	AuNPs found in all organs; high levels delivered to tumors; after 5 h detected in urine
[586]	Spheres	13 nm 4-58 nm	Citrate	BALB/c mice	IP 20 µg g^{-1} Oral 200 mg kg^{-1}	Gastrointestinal uptake by persorption. More readily for smaller particles
[156]	Spheres	15-20 nm	Arabic gum	Pigs	IV 0.8-1.88 mg kg^{-1}	Nanoparticles accumulated in lung and liver
[157]	Spheres	20 nm 100 nm	Citrate	C57BL/6 mice	IV 1 mg kg^{-1}	AuNPs crossing blood-retinal barrier (20 nm) and distributed in all retinal layers
[158]	Spheres	12.5 nm	Citrate	Male C57/BL6 mice	IP 40-400 mg kg^{-1} day^{-1}	Particle accumulation in spleen, kidney, liver and neural tissue after crossing the brain barrier
[203]	Spheres	5 nm	PEG-SH	Rats	IV, intratracheal 570-870 µg kg^{-1}	Particle accumulation in liver and spleen
[144]	Spheres	10, 15, 30 nm	PEG-SH	Human placenta (in vitro)	Perfusions 2 x 10^9-7.9 x 10^{11} NP	No crossing of placental barrier
[587;143]	Spheres	2, 40, 100 nm	Citrate	Pregnant & non-pregnant C57BL/6 mice	IV, IP, intratracheal 12-60 µg mL^{-1} = 1.4-1.6 µg kg^{-1}	Particles found in macrophages after 1h, at moderate exposure primarily in Kupffer cells (2-40 nm); no crossing of placental barrier
[141]	Spheres	1.4 nm Au55 cluster, 18 nm	^{198}Au	WKY rats	IV, intratracheal 3rd trimester 54-530 µg kg-1	IT: 1.4 nm AuNPs crossed air/blood barrier (liver, kidney, blood, urine, skin, carcass) while 18 nm AuNPs were trapped in the lung; IV: NPs found in all organs; crossing of placental barrier
[204]	Spheres	15-200 nm	Citrate	Male ddY mice	IV 1 g kg^{-1}	Particle accumulation in liver, spleen, lung (all sizes); heart, stomach, kidney, brain (15 and 50 nm); crossing of blood-brain barrier

Ref.	NP Shape	NP Diameter	Surf. Mod.	Cell Line Animal	ROA/NP Dose/ Uptake Char.	Intracellular Fate
[588]	Spheres	15 nm	HAS, Poly-allylamine hydrochloride, Polystyrene-4-sulfonate	Male CD1 mice	IV 150-200 µL	Particle accumulation in the hippocampus, thalamus, hypothalamus and cerebral cortex
[140]	Spheres	5 nm 30 nm	^{198}AuNP	Pregnant Wistar rats	IV 0.02 mg on GD 19	Crossed placental barrier (5 nm more than 30 nm); transferred rate to fetus was small
[96]	Spheres	15 nm 50 nm	PEG	White rat Rabbit	IV 0.3 mg kg^{-1}	15 nm AuNPs detected in all organs with smooth distribution over liver, spleen and blood in both animal models; small AuNPs circulated longer in the organism; expressed changes of the inner organs observed; crossing of blood-brain barrier
[160]	Spheres	20, 40, 80 nm	PEG-TA, PEG-SH	Female BALB/c mice	Oral, IP, IV 137.5-2200 µg kg^{-1}	20 nm AuNPs showed best blood pool activity and tumor uptake and extravasation, while 40-80 nm AuNPs were cleared readily by uptake in liver and spleen
[206]	Spheres	2.1 nm 8.2 nm primary size, 5-30 nm 40-80 nm aggregates	BSA, GSH	Female mice	IP 7550 µg kg^{-1}	BSA-AuNP aggregates accumulated in liver and spleen, GSH-AuNPs in low concentration in all organs; GSH-AuNP aggregates were more efficiently excreted by urine
[207]	Spheres	2 nm	GSH, BPPP, cysteine	Balb/c mice	IV 100 µL = 9 mg mL^{-1}	Low concentration of GSH-AuNPs found in liver (3%), most in urine (50%) while bis-PP-AuNPs and cysteine-AuNPs were hardly excreted into urine

Table SI 14. Overview of AuNP synthesis by PLAL (w/o reviews). Ref. = reference.

Ref.	Solvent	Laser	Details
[589]	Toluene	9 ns at 10 Hz of a Nd:YAG laser (Quantel YG980E) @ 1064 nm	Core-shell, carbon/graphite matrix
[321]	DMSO, AN, THF	9 ns at 10 Hz of a Nd:YAG laser (Quantel YG980E) @ 1064/532 nm	Stable colloids
[340]	ddH$_2$O, BSA	9 ns at 10 Hz of a Nd:YAG laser (Quantel YG980E) @ 1064 nm	BSA bioconjugation Size reduction by fragmentation
[590]	ddH$_2$O, Aptamer, Penetratin, TAT	120 fs at 5 kHz of a Spitfire Pro laser @ 800 nm (100 µJ)	Bi-conjugation with 2 different biomolecules, bioconjugation by fast ex situ technique
[591]	n-hexane, ethanol	120 fs at 5 kHz of a Spitfire Pro laser @ 800 nm (100 µJ) & 6 ps TruMicro @ 1030 nm	Conjugation with 1-dodecanthiole & PVP, silica-shell with TEOS
[360]	ddH$_2$O	120 fs at 5 kHz of a Spitfire Pro laser @ 800 nm (100 µJ) & 6 ps TruMicro @ 1030 nm	Ablation in liquid flow
[292]	ddH$_2$O	120 fs at 5 kHz of a Spitfire Pro laser @ 800 nm (100 µJ)	Stable colloids, Au-Ag alloys
[69]	ddH$_2$O	140 fs at 1 kHz of a Ti:Saphire laser (Hurricane) @ 800 nm (0.5 J)	Two-step method, fabrication and fragmentation
[330]	ddH$_2$O	140 fs at 1 kHz of a Ti:Saphire laser (Hurricane) @ 800 nm (0.5 J)	Bioconjugation with Chitosan, alpha-omega-dithiol, PNIPAM, PEG, Dextran
[592]	ddH$_2$O, ethanol	20 ns at 15 kHz of a copper vapor laser @ 510,6 nm	Stable colloids
[593]	ddH$_2$O	100 ns at 1 kHz of a ND:YLF laser @ 527 nm (16 mJ)	Stable colloids
[320]	Liquid alkanes n-pentane to decane	5 ns at 10 Hz of a Nd:YAG laser @ 532 nm	Fluence-& chain-length-dependent shape modification
[329]	Liquid alkanes, dodecanethiol	5 ns at 10 Hz of a Nd:YAG laser @ 532 nm	Stable colloids, conjugation with dodecanthiol
[308]	ddH2O, ethanol, C$_2$H$_4$Cl$_2$	20 ns at 15 kHz of a Cu vapor laser @ 510,5 nm	Size-dependence on solvent and fluence
[331;594]	ddH$_2$O, SDS	7 ns at 10 Hz of a Nd:YAG laser (Quantel) @ 1064 & 532 nm	Kinetic models, Size quenching by surfactant
[463]	ddH$_2$O, CTAB, CTAC	7 ns at 10 Hz of a Nd:YAG laser (Quantel) @ 1064 & 532 nm	Oxidation state dependence on surfactant nature (cationic/anionic)
[595]	ddH$_2$O, PAMAM G5 dendrimer	25 ps at 10 Hz of a Nd:YAG laser (EKSPLA) @ 1063 & 532 nm	Wavelength dependence, stable colloids with stabilizer
[596]	ddH$_2$O, PAMAM G5 dendrimer	25 ps at 10 Hz of a Nd:YAG laser (EKSPLA) @ 1063 & 532 nm	Stabilized colloids using PAMAM
[597]	Acetone,	25 ps at 10 Hz of a Nd:YAG laser (EKSPLA) @ 1063 & 532 nm	Stable colloids

Ref.	Solvent	Laser	Details
[598]	ddH$_2$O	500 fs @ 1030 & 515 & 343 nm	Dependence on laser wavelength
[294]	ddH$_2$O, propanol, hexane	Ruby laser @ 694 nm (100 shots)	Stable colloids
[317]	ddH$_2$O	6 ns at 200 Hz on a Nd:YAG laser (Edgewave) @ 1064 nm	Nanoparticle formation mechanism
[332;325]	Alpha-, beta-, gamma-Cyclodextrin	110 fs at 1 kHz of a Ti:Saphire laser (Hurricane) @ 800 nm	Size quenching by stabilizer
[298]	ddH$_2$O	110 fs at 1 kHz of a Ti:Saphire laser (Hurricane) @ 800 nm	Stable colloids, fluence-dependent size
[599]	ddH$_2$O	100 fs at 10 kHz of a Ti:Saphire laser (Coherent) @ 800 nm 7 ns at 10 Hz of a Nd:YAG laser (Quanta-Ray) @ 1064 & 532 nm	Size dependence on laser parameters
[346]	ddH$_2$O	7 nm at 10 Hz of a Nd^{3+}-YAG laser (Continuum) @ 1064 nm 5-7 ns at 10 Hz of a Nd^{3+}-YAG laser (Quantel) @ 266, 355, 532, 1064 nm	Size dependence on wavelength and re-irradiation; long-term stability of colloids
[345;600]	ddH$_2$O, glucose	10 ns at 10 kHz of a Nd:YVO4 laser (Baltic HP) @1064 nm	Stable colloids, size fragmentation, real-time absorption measurement
[316;601]	SDS	At 10 Hz of a Nd:YAG laser (Quanta-ray) @ 1064 nm	Size dependence on fluence and laser shots, size fragmentation
[319;601]	SDS	At 10 Hz of a Nd:YAG laser (Quanta-ray) @ 1064 & 532nm	Size quenching by surfactant
[353]	SDS	At 10 Hz of a Nd:YAG laser (Quanta-ray) @ 1064 & 532nm	Growth process of AuNPs by gold cluster aggregation
[602]	ddH$_2$O, SDS	At 10 Hz of a Nd:YAG laser (Quanta-ray) @ 1064 & 532nm	Formation of gold nanonetworks
[603]	ddH$_2$O, acetone, ethanol	120 fs at 5 kHz of a Ti:Saphire laser (Spitfire) @ 800 nm	Optimizing ablation conditions in terms of liquid layer, focal length and lens position
[350]	ddH$_2$O	120 fs at 5 kHz of a Ti:Saphire laser (Spitfire) @ 800 nm	Size-dependence on the water temperature
[351]	ddH$_2$O	120 fs at 5 kHz of a Ti:Saphire laser (Spitfire) @ 800 nm	Size-dependence on the laser intensity profile, ablation in liquid flow
[344]	ddH$_2$O	120 fs at 5 kHz of a Ti:Saphire laser (Spitfire) @ 800 nm	Size-dependence on the laser repetition rate
[604]	ddH$_2$O, acetone	120 fs at 5 kHz of a Ti:Saphire laser (Spitfire) @ 800 nm	Stable colloids, electrophoretic mobility
[295]	ddH$_2$O, NaBr, NaCl, NaF, NaS, LiCl, KCl, CaCl$_2$	8-10 ns at 100 Hz of a Nd:YAG laser (Innolas) @ 1064 nm	Ion-effects on AuNP stabilization
[605]	ddH$_2$O	15 ns at 10 Hz on a Nd:YAG laser @ 1064 nm	Alteration of solution on storage

Ref.	Solvent	Laser	Details
[606;359;39]	ddH$_2$O, oligonucleotides	120 fs at 5 kHz of a Ti:Saphire laser (Spitfire) @ 800 nm	In situ bioconjugation, optimal parameters, size quenching
[197]	Penetratin	120 fs at 5 kHz of a Ti:Saphire laser (Spitfire) @ 800 nm	In situ bioconjugation, pH-dependence
[607]	PEG	700 fs at 100 kHz of a Ytterbium-doped fiber laser (FCPA, IMRA) @ 1045 nm	Controlled PEGylation
[318]	NaCl, NaBr, NaF, NaP, Hepes, EDTA, BSA, Androhep	8-10 ns at 100 Hz of a Nd:YAG laser (Innolas) @ 1064 nm	Particle size control by low ionic strength, delayed bioconjugation by fast ex situ technique
[347]	NaCi	40 fs – 200 ps at 100 Hz of a Ti:Saphire laser @ 800 nm	Size dependence on fluence and pulse duration
[608]	Supercritical fluid – CO$_2$	9 ns at 20 Hz of a Nd:YAG laser @ 532 nm	Size and shape dependence on supercritical CO2 density
[609]	Supercritical fluid – CHF$_3$	8 ns at 20 Hz of a Nd:YAG laser @ 532 nm	Gold nanonetworks, fractal structure, dependence on fluid density
[354]	ddH$_2$O, oligonucleotides	120 fs at 5 kHz of a Ti:Saphire laser (Spitfire) @ 800 nm	Bioconjugation by fast ex situ technique
[361]	TRIS, TAT	40 ns at 3 kHz @ 1064 nm	Bioconjugation by fast ex situ technique, size-dependence on delay time of bioaddition
[610]	TPyP chloroform	6 ns at 10 Hz of a Nd:YAG laser @ 1064 nm	Photodecomposition of chloroform, Stabilization of colloid by TPyP
[611]	Liquid ammonia (233 K)	6 ns at 10 Hz of a Nd:YAG laser @ 1064 & 532 & 355 nm	Stable colloids
[348;349]	ddH$_2$O	100 fs at 1 kHz of a Ti:Saphire laser (Hurricane) @ 800 nm	Size-dependence on pulse energy, pulse repetition frequency and ablation spot size
[324]	NaCl, KCl, NaNO$_3$, HCl, NaOH, n-propylamine	120 fs at 1 kHz of a Ti:Saphire laser (Hurricane) @ 800 nm	Partial oxidation and charging of AuNP surface, size control by salts, surface functionalization
[309]	ddH$_2$O	120 fs at 1 kHz of a Ti:Saphire laser (Hurricane) @ 800 nm	Size-dependence on plasma formation
[612]	ddH$_2$O, ethanol, 1-octanethiol	6 ns at 1-300 Hz of an excimer laser (ATLEX) @ 248 nm	Size-dependence on pulse number, fluence and solvent
[269]	ddH$_2$O, Aptamer	120 fs at 5 kHz of a Ti:Saphire laser (Spitfire) @ 800 nm	In situ bioconjugation

7.2. List of Abbreviations

$(dd)H_2O$	(double distilled) water
10R	Deca-Arginine
A	adenine
Ab	antibody
ASTM	American Society for Testing and Materials
AuNP	gold nanoparticle
BE	binding energy
Bi-con	bi-conjugate
Bio_{bound}	bound biomolecules
Bio_{input}	applied biomolecules
$Bio_{unbound}$	unbound biomolecules
BSA	bovine serum albumin
BSE	backscattered electrons
C	cytosine
CA	consideration area
CCM	cell culture media
CD	cluster of differentiation
CE	conjugation efficiency
CLSM	confocal laser scanning microscopy
CMC	critical micelle concentration
CML	chronic myeloid leukemia
conc	concentration
$conc_{max}$	maximum concentration
$conc_{min}$	minimum concentration
$conc_{th}$	threshold concentration
CPP	cell-penetrating peptides
CRM	chemical reduction method
CT	computer tomography
CTAB	cetyl trimethylammonium bromide
CTAC	cetyltrimethylammonium chloride
Cy	cyanine
DAPI	4',6-Diamidin-2-phenylindol
DCD	drop coating deposition

DIC	differential interference contrast
DL-alpha	DL-α-lipoic acid
DLS	dynamic light scattering
DMEM	Dulbecco's modified Eagle's medium
DNA	deoxyribonucleic acid
dsDNA	double-stranded DNA
DTT	Dithiotheritol
DVLO	Derjaguin-Verwey-Landau-Overbeek
EDC	1-Ethyl-3-(3-dimethylaminopropyl)carbodiimide
EDTA	ethylenediaminetetraacetic acid
EDXS	energy-dispersive X-ray spectroscopy
EGFR	epidermal growth factor receptor
ELISA	enzyme-linked immunosorbent assay
EPR	enhanced permeation and retention
ESCA	electron spectroscopy for chemical analysis
ET	Everhart-Thornley
EtBr	Ethidiumbromide
Fab	antigen-binding fragment
FACS	fluorescence-activated cell sorting
Fc	crystallizable fragment
Fcc	face-centered cubic
FCS	fetal calf serum
FCS	forward scatter
FISH	fluorescence-*in-situ*-hybridization
FITC	Fluorescein isothiocyanate
FLI	Friedrich-Loeffler Institute
FluM	fluorescence microscopy
FRET	fluorescence resonance energy transfer
fs	femtosecond
FT-IR	Fourier-Transform infrared
G	guanine
GMO	genetically modified organism
HAADF	high-angle annular dark-field imaging
HEPES	4-(2-hydroxyethyl)-1-piperazineethane-sulfonic acid
HIV-1	human immunodeficiency virus type-1
HOMO	highest occupied molecular orbital

HOPG	highly-oriented pyrolytic graphite substrate
hrGFP	humanized renilla green fluorescent protein
HTP	high-temperature-and-pressure
IAS	inner acrosomal membrane
ICP-OES	inductively coupled plasma atomic emission spectroscopy
IgG	immunoglobulin
IIT	Italian Institute of Technology
IONP	iron-oxide nanoparticle
IR	infrared
IS	isomeric shift
IVF	*in vitro* fertilization
LAL	laser ablation in liquids
LDV	laser Doppler velocimetry
LNA	locked nucleic acid
LSP	localized surface plasmons
LSPR	localized surface plasmon resonances
LUH	Leibniz University Hannover
LUMO	lowest unoccupied molecular orbital
MHH	Medizinische Hochschule Hannover
MNP	magnetic nanoparticles
mPEG	methoxyl (poly)ethylene glycol
MRI	magnetic resonance imaging
MWCO	molecular weight cut off
NA	numerical aperture
NaH_2PO_4	sodium dihydrogen phosphate
Na_2HPO_4	disodium hydrogen phosphate
NALC	N-acetyl-L-cysteine
NE	nuclear envelope
NH	amide
NHS	N-hydroxysuccinimide
NIR	near infrared
NIST	National Institute of Standards and Technology
NLS	nuclear localization signal
NP	nanoparticle
ns	nanosecond
OAS	outer acrosomal membrane

OCT	optical coherence tomography
OD	optical density
OES	monocarboxy-(1-mercaptoundec-11-yl)hexaethylene glycol
OPPS	orthopyridyldisulfide
PAS	post-acrosomal sheath
PBS	phosphate buffered saline
PCS	photon correlation spectroscopy
PDB	protein data base
PDI	polydispersity index
Pen	penetratin
pI	isoelectric point
PI	propidium iodide
PLAL	pulsed laser ablation in liquids
PM	plasma membrane
PMIDA	N-phosphonomethyl iminodiacetic acid hydrate
PMT	photomultiplier
PPTT	plasmonic photodynamic therapy
ps	picosecond
PSD	particle size distribution
PTD	protein transduction domain
PVDF	polyvinylidene fluoride
QD	quantum dot
QELS	quasi-elastic light scattering
QS	quadrupole splitting
RA	rheumatoid arthritis
RES	reticulo-endothelial system
ROS	reactive oxygen species
RPMI	Roswell Park Memorial Institute medium
SA:V	surface area to volume
SAM	self-assembled monolayers
SC	surface coverage
SDS	sodium dodecyl sulphate
SE	secondary electrons
SEM	scanning electron microscopy
SERS	surface-enhanced Raman spectroscopy
SHG	second-harmonic generation

SiNP	silicon nanoparticle
siRNA	short-interfering RNA
SMGT	sperm-mediated gene transfer
SSC	side scatter
ssDNA	single-stranded DNA
ssO	single-stranded oligonucleotides
STEM	scanning transmission electron microscopy
SV40	Simian Virus 40
T	thymine
TAT	transactivator of transcription
TBE	Tris/Borate/EDTA
TCEP	Tris(2-carboxyethyl)phosphine hydrochloride
TE	Tris-EDTA
TEM	transmission electron microscopy
vdW	van der Waals
w	with
w/o	without
XPS	X-ray photoelectron spectroscopy
XRD	X-ray diffraction

7.3.   List of Formula Symbols

$2n\sin\beta$	numerical aperture
A_{450}	absorbance at 450 nm wavelength
A_{spr}	absorbance at the SPR peak
B_1, B_2	fitting parameters (3.00, 2.20)
c	speed of light in vacuum
c_0	specific heat capacity of the particle material
C_{abs}	absorption cross section
C_{ext}	extinction cross section
C_{sca}	scattering cross section
D	diffusion coefficient
d	lattice spacing in a crystalline sample
d	nanoparticle diameter
deg	deflection angle
d_{hyd}	hydrodynamic diameter
D_x	degree of degraded biomolecules
E	electric field strength
$E_{binding}$	BE of the electron
$E_{kinetic}$	measured kinetic energy of electrons
E_{photon} (hv)	energy of the X-ray photons being used
I	intensity of scattered light
I_0	illumination power
I_0	intensity of reference
I_x	intensity of acquired gel bands
K	biomolecule footprint
k	Boltzmann constant
K_{abs}	absorption efficiency factor of AuNPs approximated by Pustovalov
m_e	effective mass of an electron
N	density of free electrons in the nanoparticle
n	refraction index
N_r	average number of biomolecules per nanoparticle for given radius
q	an integer number
R	radius of footprint approximation on the nanoparticle surface
$r = r_0$	hydrodynamic nanoparticle radius
RL	resolution limit
T	absolute temperature
T_{max}	maximal local lattice temperature
T_∞	initial temperature equal to the surrounding medium temperature
t_p	pulse duration

v_w	electrophoretic mobility
w_p	bulk plasma frequency
β	half opening angle of the objective
ε	dielectric constant of the solvent
ζ	zeta potential
η	dynamic viscosity (of the solvent)
θ	diffraction angle
λ	(laser) wavelength
λ_p	bulk plasmon resonance wavelength of gold
$\lambda_{SPR\,max}$	wavelength of the SPR peak of gold nanoparticles
ϱ_0	density of the particle material
Φ	adjustable instrumental correction factor
ϵ_0	permittivity in vacuum
ϵ_∞	high-frequency dielectric constant of gold
ϵ_m	dielectric constant of the surrounding medium

7.4.   List of Figures

7.5. List of Tables

7.6. Equipment

Table 7.2. Lab equipment used for experiments.

Device	Company
fs-pulsed laser system Spitfire Pro	Newport Spectra Physics 64291 Darmstadt, Germany
Positioning axis system micro FS 150-2	3D-Micromac AG 09126 Chemnitz, Germany
ps-pulsed laser system TruMicro 5050	Trumpf GmbH + Co.KG 71254 Ditzingen, Germany
Scanner head HurrySCAN II-14	Scanlab AG 82178 Puchheim, Germany
F-theta lens S4LFT0055/126	Sill Optics GmbH & Co. KG 90530 Wendelstein, Germany
fs-pulsed laser system Legend Elite	Coherent Italia 20154 Milano, Italy
Rotation system T-cube DC Servo motor controller	Thorlabs Inc NJ 07860 USA
Spectrophotometer UV 1650	Shimadzu Europe GmbH 47269 Duisburg, Germany
Spectrophotometer Cary 6000	Agilent Technologies Inc. CA 95051 USA
Fluorescence spectrometer Fluoromax-4	Horiba Jobin Yvon Srl 20090 Opera Milano, Italy
Fluorescence spectrometer Fluoroskan Ascent	Thermo Fisher Scientific GmbH, 58239 Schwerte, Germany
Fluorescence microplate reader Infinite M200 PRO	Tecan Deutschland GmbH 74564 Crailsheim, Germany
Raman microscope inVia	Renishaw Spa. 10044 Torino, Italy
Fourier transform infrared spectrometer Spectrum 100	PerkinElmer LAS GmbH 63110 Rodgau, Germany
Freeze dryer Epsilon 2-4 LSC	Christ Gefriertrocknungsanlagen GmbH 37520 Osterrode am Harz, Germany
KBr table press	S.T. Japan-Europe GmbH 50226 Frechen, Germany
Zetasizer Nano ZS	Malvern Instruments GmbH 71083 Herrenberg, Germany
Lab centrifuge 1 MiniSpin with F-45-12-11 constant angle rotor	Eppendorf AG 22339 Hamburg, Germany
Lab centrifuge 2 Universal 320	Hettich Holding GmbH & Co. oHG 32278 Kirchlengern, Germany
Ultracentrifuge	Thermo Fisher Scientific GmbH

Device	Company
Sorvall MTX-150 with S120-AT3 constant angle rotor	58239 Schwerte, Germany
Ultracentrifuge Optima Max	Beckmann Coulter GmbH 47807 Krefeld, Germany
Clean bench 2F120-II GC used for human fibroblast cultivation	Integra Biosciences GmbH 35463 Fernwald, Germany
Incubator HERAcell 150 used for human fibroblast cultivation	DJB Labcare Ltd Buckinghamshire, MK16 9QS, England
Clean bench Hera Safe used for M3E3/C3 cultivation	Thermo Fisher Scientific GmbH 58239 Schwerte, Germany
Incubator BD 150 used for M3E3/C3 cultivation	Binder GmbH 78532 Tuttlingen, Germany
Fluorescence microscope Axio Imager A1/Z1 & Axio Imager M1	Carl Zeiss AG 73447 Oberkochen, Germany
Confocal microscope Axioplan 200 & LSM510	Carl Zeiss Microscopy GmbH 37081 Göttingen, Germany
Confocal microscope A1	Nikon Instruments Europe B.V. 1076 ER Amsterdam, The Netherlands
Scanning electron microscope Quanta 400F	FEI Company 5651 GG Eindhoven, The Netherlands
Transmission electron microscope EM 10 C	Carl Zeiss AG 73447 Oberkochen, Germany
Transmission electron microscope 400T	Philips Research 5656 AE Eindhoven, The Netherlands
Transmission electron microscope Jem 1011	JEOL USA Inc. MA 01960 USA
Transmission electron microscope Jem 2200FS	JEOL USA Inc. MA 01960 USA
Rotation microtome UltraCut E	Leica Microsystems Vertrieb GmbH 35578 Wetzlar, Germany
Diamond knife for rotation microtome	Diatome US PA 19440 USA
EDS X-ray analyzer JED-2300 Si	JEOL USA Inc MA 01960 USA
X-ray diffractometer D8 Advance	Burker AXS 58165 Mannheim, Germany
Mössbauer spectrometer MIMOS-II	University of Mainz 55122 Mainz, Germany
Flow cytometer FACScan	BD Biosciences 69126 Heidelberg, Germany
Flow cytometer	BD Biosciences

Device	Company
FACSCalibur	69126 Heidelberg, Germany
HF converter	Himmelwerk Hoch- und Mittelfrequenzanla-
HU 2000	gen GmbH
	72072 Tübingen, Germany
Lab balance	Sartorius AG
R160P	37075 Göttingen, Germany
Lab balance 2	Denver Instrument GmbH
Si-234	37075 Göttingen, Germany
Microgram balance	Sartorius AG
CPA2P	37075 Göttingen, Germany
Shaker	Heidolph Instruments GmbH & Co. KG
Duomax 1030	91126 Schwabach, Germany
Vortex shaker	IKA® Werke GmbH & Co. KG
lab dancer	79219 Staufen, Germany
Tumbling mixer	VWR International GmbH
	64295 Darmstadt, Germany
Magnetic stirrer	Heidolph Instruments GmbH & Co. KG
MRHei-Standard	91126 Schwabach, Germany
with EKTHei-Con temperature control	
Ultrasonic bath	Bandelin electronic GmbH & Co. KG
Sonorex Super RK-510	12207 Berlin, Germany
Milli-Q water device	Merck Chemicals GmbH
Direct Q®	65824 Schwalbach, Germany

7.7. Materials

Table 7.3. Materials that were used for experiments.

Materials	Company
Gold foil	Goodfellow GmbH 61213 Bad Nauheim, Germany
Iron foil	Advent Research Materials OX29 4JA Oxford, England
Silicon cylinder	Alfa Aesar GmbH & Co KG 76057 Karlsruhe, Germany
Ultrafiltration tube *Vivacon®* 500 (10, 30, 50 kDa)	Sigma Aldrich Chemie GmbH 89555 Steinheim, Germany
Ultrafiltration tube *Nanosep®* (3, 30, 300 kDa)	Sigma Aldrich Chemie GmbH 89555 Steinheim, Germany
Pierce PVDF membrane	Thermo Fisher Scientific GmbH 58239 Schwerte, Germany
Commercial gold nanobioconjugates Dressed Gold®	Bioassay Works LLC, MD 21754 USA
Low volume quartz cuvette	Hellma GmbH & Co. KG 79379 Müllheim, Germany
Formvar-covered 300/200-mesh copper grid & carbon-coated sample disc for TEM preparation	Plano GmbH 35578 Wetzlar, Germany
Dip cell for zeta potential measurement	Malvern Instruments GmbH 71083 Herrenberg, Germany
Polyimide foil Kapton®	Chemplex Industries Inc. SW 2829, USA
HOPG substrate ZYA quality	NT-MDT Co. Moscow 124482, Russia
24-/48-well plates	Sarstedt AG & Co. 51582 Nürnbrecht, Germany
6-well dishes	TPP Techno Plastic Products AG, 8219 Trasading-en, Switzerland
Polysine glass slides	Gerhard Menzel GmbH 38116 Braunschweig, Germany
Counting chamber Neubauer improved	Paul Marienfeld GmbH & Co. KG 97922 Lauda-Königshofen, Germany
Cell culture flasks T75	Sarstedt AG & Co. 51582 Nürnbrecht, Germany

7.8.   Chemicals

All chemicals used in the experiments were purchased from Fluka Chemie AG (82024 Taufkirchen, Germany) and Sigma-Aldrich Chemie GmbH (89555 Steinheim, Germany), except:

Table 7.4. Chemicals that were used for experiments.

Chemicals	Company
uranyl acetate	Serva Electrophoresis GmbH
	69115 Heidelberg, Germany
lead citrate	Leica Microsystems CMS GmbH
	68165 Mannheim, Germany
Orange Dye	Thermo Fisher Scientific GmbH
	58239 Schwerte, Germany
Dulbecco's modified Eagle's medium	PAA Laboratories GmbH
(DMEM)	35091 Cölbe, Germany
Roswell Park Memorial Institute medium	PAA Laboratories GmbH
(RPMI-1640)	35091 Cölbe, Germany
Dulbecco's PBS	PAA Laboratories GmbH
	35091 Cölbe, Germany
heat-inactivated fetal calf serum	Invitrogen GmbH
(FCS)	76131 Karlsruhe, Germany &
	PAA Laboratories GmbH
	35091 Cölbe, Germany
penicillin/streptomycin	PAA Laboratories GmbH
	35091 Cölbe, Germany &
	Biochrom AG
	12247 Berlin, Germany
complete medium 199	Invitrogen GmbH
	76131 Karlsruhe, Germany
DAPI II Counterstain	Vysis Inc.
	IL 60515 USA
mounting medium	Vector Laboratories Inc.
Vectashield	CA 94010 USA
trypan blue	Serva Feinbiochemica GmbH & Co.
(in 0.9 % NaCl)	69115 Heidelberg, Germany
trypsin-EDTA	Invitrogen GmbH
	76131 Karlsruhe, Germany
tween 20	Merck Chemicals GmbH
	65824 Schwalbach, Germany

7.9.   Buffer Composition

Sodium phosphate buffer (NaH$_2$PO$_4$/Na$_2$HPO4, 100 mM, pH 8)
93.2 mL Na$_2$HPO$_4$ (1M), 6.6 mL NaH$_2$PO$_4$ (1M) + 900.2 mL *Milli-Q* water.

Phosphate buffered saline (PBS, 10 mM)
137 mM NaCl, 2.7 mM KCl, 10 mM Na$_2$HPO$_4$, 1.8 mM KH$_2$PO$_4$, pH 7.4.

Tris/Borate/EDTA buffer (TBE, pH 8)
89 mM TRIS base, 89 mM boric acid, 2 mM EDTA-Na$_2$.

Buffer for bull semen extension
200 mM Tris(hydroxymethly)aminomethane, 65 mM citric acid monohydrate, 96 mM D-fructose, pH 6.9

7.10. Biomolecules

Table 7.5. Biomolecules that were used for experiments.

Name/ID	Classification/Sequence	Func-tional Groups	Dye	MW/ g mol⁻¹	Company
Penetratin **Pen**	peptide CRQIKI-WFQNMRRKWKK(Ac)	cysteine	/	2,391.8	PANATecs GmbH 72070 Tübingen
Deca-Arginine **10R**	peptide C-Ahx-RRRRRRRRRR	cysteine	/	1,796.1	PSL Laboratories GmbH 69120 Heidelberg
Transactivator of transcription **TAT**	peptide C-AhxGRKKRRQRRRC	cysteine	/	1,613	PSL Laboratories GmbH 69120 Heidelberg
Simian Virus Large T-antigen Nuclear Localization Signal **SV-40 NLS**	peptide C-Gly-Gly-Gly-Pro-Lys-Lys-Lys-Arg-Lys-Val-Glu-Asp	cysteine	/	1,403.2	AnaSpec, EGT Corporate Headquarters CA 94555 USA
single-stranded DNA **ssDNA**	oligonucleotide GGC GAC TGT GCA AGC AGA	thiol	/	5,770.8	Purimex 34393 Grebenstein
ssDNA-Cy5	oligonucleotide GGC GAC TGT GCA AGC AGA(Cy5)	thiol	Cy5		Purimex 34393 Grebenstein
ssDNA-5'-T10	oligonucleotide ssDNA-10xT	disulfide 5'	/	8,810	Eurofins MWG GmbH 85560 Ebersberg
ssDNA-5'-T20	oligonucleotide ssDNA-20xT	disulfide 5'	/	11,852	Eurofins MWG GmbH 85560 Ebersberg
ssDNA-3'-T10	oligonucleotide ssDNA-10xT	disulfide 3'	/	8,768	Eurofins MWG GmbH 85560 Ebersberg
ssDNA-3'-T20	oligonucleotide ssDNA-20xT	disulfide 3'	/	11,810	Eurofins MWG GmbH 85560 Ebersberg
aptamer directed against streptavidin **miniStrep**	Aptamer TCT GTG AGA CGA CGC ACC GGT CGC AGG TTT TGT CTC ACA G-10xT-$(CH_2)_3$-S-S-$(CH_2)_6$OH	disulfide 3'	/	15,589	Biospring GmbH 60386 Frankfurt am Main
LNA	oligonucleotide OOG YCG XCT YTG ZAA YCA YA(C_6NH_2); O = HEGL,Y = LNA G; X = LNA A; Z = LNA C	thiol 5'	/	9,348; TT 8,812.68	Purimex 34393 Grebenstein
mPEG-SH	polymer $CH_3O(CH_2CH_2O)$ nCH_2CH_2SH	thiol	/	~ 5,000	Laysan Bio Inc AL 35016 USA

Name/ID	Classification/Sequence	Func- tional Groups	Dye	MW/ g mol^{-1}	Company
Protein A from *Staphylococcus aureus*	antibody-specific protein	Fc-specific region	/	~ 42,000	Sigma Aldrich Chemie GmbH 89555 Steinheim
IgG from rabbit serum	immunoglobulin	/		~ 150,000	Sigma Aldrich Chemie GmbH 89555 Steinheim
goat anti-rabbit IgG **anti-IgG-FITC**	antibody	Fab re-gion	FITC	~ 150,000	Sigma Aldrich Chemie GmbH 89555 Steinheim
rabbit anti-human Vinculin IgG **anti-Vinculin**	antibody	Fab re-gion	/	~ 116,000	Sigma Aldrich Chemie GmbH 89555 Steinheim
OPPS-PEG-NHS	linker	OPPS, NHS	/	2,000	Creative PEGWorks NC 27113 USA
albumin from bovine serum-Alexa594 **BSA-Alexa594**	serum protein	cysteine	Alexa594	~ 66,000	Life Technologies GmbH 64293 Darmstadt
albumin from bovine serum **BSA**	serum protein	cysteine	/	~ 66,000	Sigma Aldrich Chemie GmbH 89555 Steinheim

7.11.   List of own Publications

Regular scientific publications in peer-reviewed journals

[I]   A. Barchanski,   D. Funk,   O. Wittich,   C. Tegenkamp,   B. N. Chichkov,
C. L. Sajti: "Picosecond Laser Fabrication of Functional Gold-Antibody Nanoconjugates for
Biomedical Applications". *J. Phys. Chem. C*, **2015**, Accepted Manuscript, ahead of print.
Doi:10.1021/jp511162n.

[II]   A. Barchanski*,   U. Taylor*,   C. L. Sajti,   L. Gamrad,   W. A. Kues,   D. Rath,
S. Barcikowski: "Bioconjugated gold nanoparticles penetrate into spermatozoa depending on
plasma membrane status". *J. Biomed. Nanotech.*, 2015, 11(1-11). Accepted Manuscript, ahead of
print. Doi:10.1166/jbn.2015.2094.

[III]   Z. Pikramenou,   C. McCallion,   S. Carreira,   P. Dobson,   K. Brown,   Y. A. Diaz Fernandez,
M. Abdollah,   D. Zhou,   D. Sun,   S. Moise,   L. Litti,   L. L. Yung,   S. Borsley,   N. Dragneva,
A. Barchanski,   M. El-Sayed,   A. Heuer-Jungemann,   R. M. Pallares,   E. Tsang,   N. Barry,
S. Mitchell, N. T. K. Thanh, M. Thanou, I. Parkin, P. Ray, R. Jones: "Other Nanoparticles: general discussion". *Faraday Discuss.*, **2014**, 175: 289-303.
Doi:10.1039/C4FD90077D.

[IV]   U. Taylor,   W. Garrels,   **A. Barchanski**,   S. Petersen,   C. L. Sajti,   U. Baulain,   S. Klein,
W. A. Kues, S. Barcikowski, D. Rath: "Injection of Ligand-Free Gold and Silver Nanoparticles
into Murine Embryos Does Not Impact Preimplantation Development". *Beilstein J. Nanotech.*,
2014, 5: 677-688. Doi:10.3762/bjnano.5.80.

[V]   U. Taylor,   **A. Barchanski**,   S. Petersen,   W. A. Kues,   U. Baulain,   L. Gamrad,   C. L. Sajti,   S.
Barcikowski, D. Rath: "Gold nanoparticles interfere with sperm functionality by membrane adsorption without penetration". *Nanotoxicology*, **2014**, Suppl.I: 118-127.
Doi: 10.3109/17435390.2013.859321.

[VI]   K. Bagga*,   **A. Barchanski**,   R. Intartaglia,   S. Dante,   R. Marotta,   A. Diaspro,   C. L. Sajti,   F.
Brandi: "Laser-assisted synthesis of Staphylococcus aureus protein-capped silicon quantum dots
as bio-functional nanoprobes". *Laser Phys. Lett.*, **2013**, 10(6): 065603(8pp).
Doi:10.1088/1612-2011/10/6/065603.

[VII] S. Willenbrock,   M. C. Durán,   **A. Barchanski**,   S. Barcikowski,   K. Feige,   I. Nolte,   H. M.
Escobar: "Evaluation of Pulsed Laser Ablation in Liquids Generated Gold Nanoparticles as
Novel Transfection Tools – Efficiency and Cytotoxicity". *P SPIE, Frontiers in Ultrafast Optics: Biomedical,   Scientific,   and   Industrial   Applications   XIV*,   **2014**,   8972:   89720D-1-89720D-9.
Doi:10.1117/12.2038453.

[6] A. Barchanski, C. L. Sajti, C. Sehring, S. Petersen, S. Barcikowski: *"Design of Bioconjugated Gold Nanoparticles by Femtosecond Laser Ablation"*. LPM, Stuttgart, Germany, **2010**.

Scientific posters

[1] A. Barchanski, C. L. Sajti, B. N. Chichkov: *"Multifunctional and Theranostic Nanoparticles by Laser-Based, One-Step Fabrication and their Physicochemical Functionalization Aspects"*. Faraday Discussion 175: Physical Chemistry of Functionalized Biomedical Nanoparticles, Bristol, UK, **2014**.

[2] A. Barchanski, U. Taylor, S. Klein, A. Mittag, D. Rath, S. Barcikowski: *"Sperm Vitality and Blastocyst Development after Incubation with Ligand-Free and Bioconjugated Gold Nanoparticles"*. Nanomed, Berlin, **2010**.
Winner of 1st poster prize

[3] N. Hashimoto*, A. Barchanski*, S. Petersen, S. Barcikowski: *"Impact of Spacer Length on Laser-Generated Gold Nanomarker Design"*. ANGEL, Engelberg, Schweiz, **2010**.
Winner of 2nd poster prize

[4] A. Barchanski, S. Petersen, N. von Neuhoff, B. Schlegelberger, S. Barcikowski: *"Design, Biofunctionalization and Cell Penetration of Gold Nanomarker Made by Pulsed Laser Ablation"*. Bioplasmonics, Lugano, Italy, **2010**.

Danksagung

Ich möchte mich bei den zahlreichen Personen bedanken, die zum Gelingen dieser Arbeit beigetragen haben.

Mein besonderer und größter Dank für die Ermöglichung und Betreuung dieser Arbeit geht an meinen Doktorvater Prof. Dr.-Ing. Stephan Barcikowski, der an mich glaubte, mich förderte, mich geduldig motivierte, aber auch kritisch hinterfragte und mich im Ganzen sehr viel lehrte.

Des Weiteren bedanke ich mich ganz herzlich bei Prof. Dr. Detlef Rath für die freundliche Übernahme des Koreferates und bei Prof. Dr. Sebastian Schlücker für die Annahme des Prüfungsvorsitzes.

Ebenfalls bedanke ich mich bei meinen Vorgesetzten Prof. Dr. Boris Chichkov und Dr. Csaba László Sajti vom Laser Zentrum Hannover e.V. für die Unterstützung meiner Karriere und für das Verständnis sowie die Ermöglichung der Versuchsdurchführung am Rande meiner WiMi Tätigkeit.

Bester Dank geht auch im Allgemeinen an das Laser Zentrum Hannover e.V. für die wissenschaftliche Förderung sowie an den Exzellenzcluster REBIRTH der Deutschen Forschungsgemeinschaft, an die NBank und an die Masterrind GmbH für die finanzielle Förderung meiner Forschung.

So manche Analyse wäre ohne entsprechende Techniken und Geräte nicht möglich gewesen. Ganz besonders bedanke ich mich daher bei Kerstin Rohn vom Institut für Pathologie (TiHo Hannover) für die Präparation der TEM-Schnitte, bei Dr. Olga Wittich (LUH) für die XRD-Messungen, bei Dr. Sabine Klein (FLI, Neustadt), Dr. Rudolph Bauerfeind und Wolfgang Poschelt (beide MHH) für die Unterstützung bei der Konfokalmikroskopie sowie bei Prof. Christoph Tegenkamp (LUH) für die Anfertigung der XPS-Aufnahmen.

Ein Auslandsaufenthalt am Italian Institut of Technology ermöglichte es mir spannende und wertvolle Einblicke in das internationale Forschungsgeschehen zu erhalten und andere Arbeitsweisen kennenzulernen. Für die Ermöglichung dieser Gelegenheit bedanke ich mich herzlich bei Prof. Dr.-Ing. Stephan Barcikowski und Prof. Dr. Fernando Brandi.

Ferner danke ich:

- meinen brillanten Kolleginnen und Kollegen: Svea Petersen, Jurij Jakobi, Anne Hahn, Ana Menendez-Manjon, Andreas Schwenke, Tatjana Melnyk, Olga Kufelt, Camilla Sehring, Rupert Rosenfeld, Philipp Wagener, Christoph Rehbock, Niko Bärsch, Daniel Bartke, Naomi Hashimoto, Sabrina Schlie-Walter, Lothar Koch, Heiko Meyer, Anastasia Koroleva, Dag Heinemann
- den besten Kooperationspartnern überhaupt: Ulrike Taylor, Sabine Klein, Detlef Rath und Wilfried Kues (FLI, Neustadt); Nils von Neuhoff, Brigitte Schlegelberger, Makito Emura, Michael Bock, Nora Fekete, Michael Heuser, Nidhi Jyotsana (MHH); Maria Carolina Duran, Hugo Murura Escobar (TiHo Hannover); Romuald Intartaglia, Komal Bagga, Fernando Brandi (IIT), Gerald Draeger (LUH); Björn Meermann (BfG)
- den unbezahlbaren technischen Angestellten: Andrea Deiwick (Zellkultur), Rainer Gebauer (SEM), Andre Mittag (Gel Elektrophorese)
- und meinen fleißigen Studenten: Sebastian Werneburg, Camilla Sehring, Jovana Mihajlovic, Michelle Svetkoff, Justina Tam, Irene Sevilla de la Llave, Marita Meißner, Dominik Funk, Gaurav Das, Hauke Dieken, Matthias Merkle, Alexander Gußahn, Malte Worzischek, Magdalena Robacha, Matthias Pallus

für die grandiose und unkomplizierte Zusammenarbeit, die angenehme Arbeitsatmosphäre und all die wertvollen Gespräche.

Ein großer Dank geht an meine Korrekturleser Rebekka von Fintel, Andreas Schwenke, Marita Meißner, Niko Bärsch, Olga Kufelt, Ana Menéndez-Manjón, Christian Hennigs und Christoph Rehbock.

Ganz herzlich bedanke ich mich bei meinen Freunden, die mich alle auf ihre Art stets motivierten, aber auch schöne Ablenkung brachten wenn mir die Decke auf den Kopf zu fallen drohte.

Und danke Marcel für deine liebevolle Unterstützung auf der Zielgeraden und dass du immer für mich da bist. Du hast mir stets die notwendige Erdung verliehen und mir immer wieder vor Augen geführt, dass es neben Nanopartikeln auch noch andere wichtige Dinge im Leben gibt.

Letztlich widme diese Arbeit meinen wunderbaren Eltern Emilie und Bernhard als Dank für ihre grenzenlose, unvergleichliche Unterstützung und ihr inniges Vertrauen, das weit über den Rahmen dieser Arbeit hinausgeht.

Vielen Dank !